The Close Linkage between Nutrition and Environment through Biodiversity and Sustainability

The Close Linkage between Nutrition and Environment through Biodiversity and Sustainability

Local Foods, Traditional Recipes and Sustainable Diets

Special Issue Editor

Alessandra Durazzo

MDPI • Basel • Beijing • Wuhan • Barcelona • Belgrade

MDPI

Special Issue Editor
Alessandra Durazzo
CREA-Research Centre for Food and
Nutrition
Italy

Editorial Office
MDPI
St. Alban-Anlage 66
4052 Basel, Switzerland

This is a reprint of articles from the Special Issue published online in the open access journal *Sustainability* (ISSN 2071-1050) from 2017 to 2019 (available at: https://www.mdpi.com/journal/sustainability/special_issues/Nutrition_food_sustainability)

For citation purposes, cite each article independently as indicated on the article page online and as indicated below:

LastName, A.A.; LastName, B.B.; LastName, C.C. Article Title. *Journal Name* **Year**, *Article Number*, Page Range.

ISBN 978-3-03921-383-2 (Pbk)
ISBN 978-3-03921-384-9 (PDF)

ⓒ 2019 by the authors. Articles in this book are Open Access and distributed under the Creative Commons Attribution (CC BY) license, which allows users to download, copy and build upon published articles, as long as the author and publisher are properly credited, which ensures maximum dissemination and a wider impact of our publications.

The book as a whole is distributed by MDPI under the terms and conditions of the Creative Commons license CC BY-NC-ND.

Contents

About the Special Issue Editor

Alessandra Durazzo was awarded a master's degree in chemistry and pharmaceutical technology (cum laude) in 2003, and she received her PhD in horticulture in 2010. Since 2005, she has been a researcher at the CREA-Research Centre for Food and Nutrition. The core of her research is the study of chemical, nutritional, and bioactive components of food, particularly regarding the wide spectrum of substance classes and their nutraceutical features. For several years she was involved in national and international research projects evaluating several factors (agronomic practices, processing, etc.) that affect food quality, the levels of bioactive molecules and total antioxidant properties, as well as on their possible impact on the biological role played by bioactive components in human physiology. Her research activities are also addressed towards the development, management, and updating of the Food Composition Database as well as Bioactive Compounds and Food Supplements databases. Particular attention has been given towards the harmonization of analytical procedures as well as the classification and codification of food preparation and food supplements.

sustainability

MDPI

Editorial

The Close Linkage between Nutrition and Environment through Biodiversity and Sustainability: Local Foods, Traditional Recipes, and Sustainable Diets

Alessandra Durazzo

CREA Research Centre for Food and Nutrition, Via Ardeatina 546, 00178 Rome, Italy;
alessandra.durazzo@crea.gov.it; Tel.: +39-065-149-4430

Received: 14 May 2019; Accepted: 18 May 2019; Published: 21 May 2019

Abstract: This special issue, "The Close Linkage between Nutrition and Environment through Biodiversity and Sustainability: Local Foods, Traditional Recipes, and Sustainable Diets" is focused on the close correlation between the potential benefits and "functional role" of a food and the territory, including papers on the characterization of local foods and traditional recipes, on the promotion of traditional dietary patterns and sustainable diets.

Keywords: nutritional composition; bioactive components; typical/local foods; environmental and socio-demographic factors; traditional recipes; sustainable diets; traditional dietary patterns; Food Composition Databases

Introduction

There is a close correlation between the potential benefits and the "functional role" of a food and the territory. Nutrition science should support sustainable ecosystems, ecological resources, and healthy environments. Nutrition and environmental sustainability are strictly linked throughout the food system.

Appropriate farming based land use, protection of animal health and welfare, environmental conservation as linked to climate knowledge, soil quality, and landscaping, lead to the improvement of product quality.

The valorization of the typical/local/traditional products by identifying and evaluating food nutritional quality and safety characteristics represents an important goal for the preservation of local agro-biodiversity. Thus, supporting sustainable ecosystems and productive system. In this context, it is becoming important to address consumers towards a sustainable diet and environment-friendly foods and recipes. The determination of the nutritional composition of composite dishes has a key role for defining daily nutrient intakes at the population level and their association with health effects [1]. Specific databases created to promote and preserve the nutritional characteristics of some national traditional foodstuffs and traditional recipes are needed [2]. Studies on the nutritional profile and bioactive components of typical foods, of a territory, as well as traditional recipes are presented in this special issue.

The evaluation of the influence of agricultural practices, wild species, intra-species biodiversity, and environmental factors as well as the elucidation and clarification of the relationships between the environment, the food quality, and health benefits were discussed throughout this special issue.

As instance Miranda-Granados et al. [3] describes the alternative use as antimicrobial of extracts of Chipilín Leaves (*Crotalaria longirostrata* Hook. & Arn), a wild plant that grows in the state of Chiapas (Mexico) and is traditionally used as food.

Lisciani et al. [4] focused their work on the carbohydrates component of some Italian local landraces of garlic, whereas El Riachy et al. [5] studied the chemical and sensorial characteristics of olive oil produced from the Lebanese olive variety 'Baladi'.

An example of food of animal origin was given by Manzo et al. [6], that investigated the influence of ripening on chemical characteristics of a traditional Italian cheese, Provolone del Monaco.

The interesting review of Pacifico [7] showed how upland potatoes satisfied consumer demand for high quality foods linked to traditional areas of origin, new specialties, and niche products endowed with added nutritional value. It is commonly thought that the environmental synergy of the crop improves the potential beneficial properties of the tuber, giving it a special taste and a renowned quality.

It is worth mentioning the work by Blanco-Salas et al. [8] on classification and cataloguing of wild plants potentially used in human food in the protected area Sierra Grande de Hornachos of Extremadura in Spain, that gives an example of how botany knowledge can effectively promote development, enhancement of territories, and biodiversity preservation, by guiding tourism and consumers towards an haute cuisine, new gastronomy, and environment-friendly recipes.

The overall benefits can be described in terms of increasing the nutritional value of local/traditional foods in the various European countries, ensuring their quality and education of consumers regarding the use of those foods, fitting into a recommended dietary food pattern. The envisaged promotion of traditional products throughout environmentally sustainable techniques further contributes to environmental protection. This strategy should represent a valid tool for promotion of socio-economic development, enhancement of territories, and biodiversity preservation.

In this context, as marked in the work of Garanti and Berberoglu [9], the main challenge of a changing marketplace is to ensure that young generations continue consuming traditional products. Indeed in the area of social sustainability of traditional foods, the authors Garanti and Berberoglu [9] studied the perceptions of post-millennials towards traditional cheese products from a cultural perspective, concluding that loyalty towards traditional food products amongst post-millennials is built based on: -the memories that surrounded the food, -the rituals that preparing and eating a food involved, and -the identity that it builds, therefore allowing people to feel a sense of belonging to their ethnic group.

Mabhaudhi et al. [10], by reviewing the potential of indigenous crops in bringing about such transformative change to South Africa's food system, concluded how indigenous crop, being nutrient dense food and adapted to marginal conditions, is an example of sustainable agriculture and can support the existing food system for local farmers. In the perspective of evaluation of contribution to sustainable diets, livelihood needs, and environmental sustainability of the species throughout the geographical range, Maroyi et al. [11] reviewed the use, cultivation, and management of *Schinziophyton rautanenii*, a multipurpose plant species in Southern Africa: *S. rautanenii* offers enormous potential for contributing to the fulfillment of the 2030 Agenda for sustainable development goals adopted by the United Nations General Assembly, resulting in improved food security, household nutrition and health, income, livelihoods, ecological balance, sustainable diets, and food systems.

I would like to acknowledge the efforts of the authors of the publications in this special issue.

Conflicts of Interest: The authors declare no conflict of interest.

References

1. Durazzo, A.; Lisciani, S.; Camilli, E.; Gabrielli, P.; Marconi, S.; Gambelli, L.; Aguzzi, A.; Lucarini, M.; Maiani, G.; Casale, G.; et al. Nutritional composition and antioxidant properties of traditional Italian dishes. *Food Chem.* **2017**, *218*, 70–77. [CrossRef] [PubMed]
2. Marconi, S.; Durazzo, A.; Camilli, E.; Lisciani, S.; Gabrielli, P.; Aguzzi, A.; Gambelli, L.; Lucarini, M.; Marletta, L. Food composition databases: considerations about complex food matrices. *Foods* **2018**, *7*, 2. [CrossRef]

3. Miranda-Granados, J.; Chacón, C.; Ruiz-Lau, N.; Vargas-Díaz, M.E.; Zepeda, L.G.; Alvarez-Gutiérrez, P.; Meza-Gordillo, R.; Lagunas-Rivera, S. Alternative Use of Extracts of Chipilín Leaves (*Crotalaria longirostrata* Hook. & Arn) as Antimicrobial. *Sustainability* **2018**, *10*, 883.

4. Lisciani, S.; Gambelli, L.; Durazzo, A.; Marconi, S.; Camilli, E.; Rossetti, C.; Gabrielli, P.; Aguzzi, A.; Temperini, O.; Marletta, L. Carbohydrates components of some Italian local landraces: garlic (*Allium sativum* L.). *Sustainability* **2017**, *9*, 1922. [CrossRef]

5. El Riachy, M.; Bou-Mitri, C.; Youssef, A.; Andary, R.; Skaff, W. Chemical and sensorial characteristics of olive oil produced from the Lebanese olive variety 'Baladi'. *Sustainability* **2018**, *10*, 4630. [CrossRef]

6. Manzo, N.; Santini, A.; Pizzolongo, F.; Aiello, A.; Marrazzo, A.; Meca, G.; Durazzo, A.; Lucarini, M.; Romano, R. Influence of ripening on chemical characteristics of a traditional Italian cheese: Provolone del Monaco. *Sustainability* **2019**, *11*, 2520. [CrossRef]

7. Pacifico, D. Upland Italian Potato Quality—A Perspective. *Sustainability* **2018**, *10*, 3939. [CrossRef]

8. Blanco-Salas, J.; Gutiérrez-García, L.; Labrador Moreno, J.; Ruiz-Téllez, T. Wild plants potentially used in human food in the Protected Area "Sierra Grande de Hornachos" of Extremadura (Spain). *Sustainability* **2019**, *11*, 456. [CrossRef]

9. Garanti, Z.; Berberoglu, A. Cultural perspective of traditional cheese consumption practices and its sustainability among post-millennial consumers. *Sustainability* **2018**, *10*, 3183. [CrossRef]

10. Mabhaudhi, T.; Chibarabada, T.P.; Petrova Chimonyo, V.G.; Murugani, V.G.; Pereira, L.M.; Sobratee, N.; Govender, L.; Slotow, R.; Modi, A.T. Mainstreaming indigenous crops into food systems: A South African perspective. *Sustainability* **2019**, *11*, 172. [CrossRef]

11. Maroyi, A. Contribution of *Schinziophyton rautanenii* to sustainable diets, livelihood needs and environmental sustainability in Southern Africa. *Sustainability* **2018**, *10*, 581. [CrossRef]

sustainability

MDPI

Article

Mass Spectrometry-Based Metabolomics of Agave Sap (*Agave salmiana*) after Its Inoculation with Microorganisms Isolated from Agave Sap Concentrate Selected to Enhance Anticancer Activity

Luis M. Figueroa [1], Liliana Santos-Zea [1], Adelfo Escalante [2] and Janet A. Gutiérrez-Uribe [1,*]

[1] Tecnologico de Monterrey, Escuela de Ingeniería y Ciencias, Ave. Eugenio Garza Sada 2501, Col. Tecnológico, 64849 Monterrey, N.L., Mexico; luisfigueroa8605@gmail.com (L.M.F.); lilianasantos@itesm.mx (L.S.-Z.)
[2] Departamento de Ingeniería Celular y Biocatálisis, Instituto de Biotecnología, Universidad Nacional Autónoma de México (UNAM), Av. Universidad 2001, Col. Chamilpa, 62210 Cuernavaca, Mor., Mexico; adelfo@ibt.unam.mx
*   Correspondence: jagu@itesm.mx; Tel.: +52-81-8358-2000 (ext. 1802)

Received: 25 September 2017; Accepted: 4 November 2017; Published: 16 November 2017

Abstract: Saponins have been correlated with the reduction of cancer cell growth and the apoptotic effect of agave sap concentrate. Empirical observations of this artisanal Mexican food have shown that fermentation occurs after agave sap is concentrated, but little is known about the microorganisms that survive after cooking, or their effects on saponins and other metabolites. The aim of this study was to evaluate the changes in metabolites found in agave (*A. salmiana*) sap after its fermentation with microorganisms isolated from agave sap concentrate, and demonstrate its potential use to enhance anticancer activity. Microorganisms were isolated by dilution plating and identified by 16S rRNA analysis. Isolates were used to ferment agave sap, and their corresponding butanolic extracts were compared with those that enhanced the cytotoxic activity on colon (Caco-2) and liver (Hep-G2) cancer cells. Metabolite changes were investigated by mass spectrometry-based metabolomics. Among 69 isolated microorganisms, the actinomycetes *Arthrobacter globiformis* and *Gordonia* sp. were used to analyze the metabolites, along with bioactivity changes. From the 939 ions that were mainly responsible for variation among fermented samples at 48 h, 96 h, and 192 h, four were correlated to anticancer activity. It was shown that magueyoside B, a kammogenin glycoside, was found at higher intensities in the samples fermented with *Gordonia* sp. that reduced Hep-G2 viability better than controls. These findings showed that microorganisms from agave sap concentrate change agave sap metabolites such as saponins. Butanolic extracts obtained after agave sap fermentation with *Arthrobacter globiformis* or *Gordonia* sp. increased the cancer cell growth inhibitory effect on colon or liver cancer cells, respectively.

Keywords: agave sap; actinomycetes; anticancer activity; saponins; metabolomics

---

1. Introduction

Agave sap concentrate is a food produced by thermal treatment of the fresh sap ("aguamiel"). Agave sap is composed of water, sugars (glucose, fructose, and sucrose), proteins, gums, and mineral salts [1]. It is obtained from the mature plant, which is between eight and 10 years old. *Agave americana*, *A. salmiana*, and *A. atrovirens* are agave species known as "maguey pulquero", and used for production of "aguamiel" [2]. Regarding the concentrate, the high concentration of soluble solids (about 70%) confers stability to environmental conditions and resistance to microbial attack [3].

Saponins are the most commonly bioactive compounds found in *Agave* species, and they have shown biological activity against many types of cancer [4–6]. Saponins have been isolated mainly from the leaves of the plant [4], but recently they were reported in agave sap concentrate [5,6], and fresh agave sap [7]. These molecules were identified as glycosides of kammogenin, manogenin, gentrogenin, and hecogenin, and are the main source of variability among agave sap concentrates [6]. Particularly, a kammogenin glycoside identified as magueyoside B has been correlated with reduction in viability, as well as the induction of apoptosis in colon cancer cell lines [5,6].

Mass spectrometry (MS)-based metabolomics have been used as a platform for determining biochemical changes in food during processing, including fermentation [8–10]. In addition, the data obtained can be statistically processed using principal component analysis (PCA) to identify the most important biomarkers or informative metabolites in samples [9].

The high soluble solids content in the agave sap concentrate creates an environment that prevents the development of most microbial species. However, empirical observations of this artisanal Mexican food have shown that fermentation occurs after the agave sap is concentrated. The microbiota of agave sap and its fermented product, known as pulque, has been studied previously. Bacterial diversity in these products is dominated mainly by lactic acid bacteria and aerobic mesophiles belonging to α- and γ-proteobacteria [11]. However, little is known about the microorganisms that survive after cooking, or their effects on saponins and other metabolites. Additionally, environmental factors may affect the composition of the microbial community present in the soil of a particular site of cultivation, which could in turn affect the native microbial community found in the sap and its concentrate [12]. Therefore, the aim of this study was to evaluate the changes in metabolites found in agave sap (*A. salmiana*) after its fermentation with microorganisms isolated from agave sap concentrate. These microorganisms were selected based on a screening of viability reduction using agave sap butanolic extracts on colon (Caco-2) and liver cancer cells (Hep-G2).

2. Materials and Methods

2.1. Biological Material

Agave sap concentrate was provided by the local producer AGMEL S.A de C.V (Monterrey, Mexico), and stored at 4 °C until use for the isolation of microorganisms. Agave sap was obtained from *A. salmiana* plants grown in the state of Coahuila, Mexico. It was transported in dry ice and stored at −20 °C until use. Before use, the agave sap was autoclaved at 121 °C for 15 min.

2.2. Microorganism Isolation from Agave Sap Concentrate

Yeast Mold (YM, Difco^TM, Sparks, MD, USA), Potato Dextrose Agar (PDA, Difco^TM, Sparks, MD, USA), De Man, Rogosa and Sharpe (MRS, Difco^TM, Sparks, MD, USA), Reasoner's 2A (R2A, Sigma-Aldrich, St. Louis, MO, USA) and NZ amine A media were used to isolate microorganisms from agave sap concentrate. NZ amine medium was prepared with NZ amine A (2 g/L) (Sigma-Aldrich, St. Louis, MO, USA) and Noble agar (15 g/L) (Difco^TM, Sparks, MD, USA). The rest of the media were prepared according to the instructions of the manufacturers. A culture broth was used for the preservation of isolates and to prepare the inoculum used in fermentations. This culture broth was composed of casein peptone (0.5 g/L), yeast extract (0.5 g/L), glucose (0.5 g/L), soluble starch (0.5 g/L), magnesium sulfate (0.024 g/L), and dibasic potassium phosphate (0.3 g/L).

For isolation, agave sap concentrate (10 g) was diluted with 90 mL of peptone (0.1%) and stirred for 30 min at 200 rpm. Dilutions from 10^{-1} to 10^{-6} were prepared, plated in Petri dishes, and incubated at 30 °C for seven days. MRS medium cultures were incubated at 37 °C. Culture media without inoculum were used as controls. For preservation, each colony was placed in culture broth supplemented with glycerol (20% *v/v*) and stored at −80 °C.

2.3. Experimental Strategy for the Selection of Microorganisms that Enhanced Agave Sap Anticancer Activity

For the first step, inoculums of each isolate were prepared and used to ferment sterile agave sap in a final volume of 3 mL. Each isolate was grown in 10 mL of culture broth, and incubated at 30 °C and 200 rpm for 96 h. These cultures were used to inoculate sterile agave sap, and were incubated individually at 30 °C and 200 rpm for 48 h, 96 h, or 192 h. Sterile agave sap without inoculum was incubated in the same conditions and used as control. Fermentations were performed by triplicate.

Butanolic extracts of fermentations were prepared and fractioned by solid phase extraction (SPE) to obtain saponin-rich extracts. These saponin-rich extracts were tested on liver (Hep-G2, ATCC® HB-8065™) and colon (Caco-2, ATCC® HTB-37™) cancer cell lines. According to the results of the biological assays, five isolates were selected and identified by 16S rRNA gene sequence analysis. Two of them were used to scale up to a final volume of 40 mL under the same fermentation conditions. Saponin-rich extracts were obtained and tested on liver and colon cancer cell lines. The effect on cancer cell lines was contrasted to that observed in murine fibroblasts (NIH-3T3, ATCC® CRL-1658™), which were used as the non-cancer control. Detailed procedures are described below.

2.4. Saponin-Rich Extracts Preparation

After fermentation, saponin-rich extracts were obtained using butanol (1:1), as described by Leal-Diaz and colleagues [7]. Butanol was removed under vacuum at 40 °C in a Rocket evaporator (Genevac, Ltd., Ipswich, UK). Then, dry samples (120 mg) were dissolved in 20 mL of high-performance liquid chromatography (HPLC) grade water (BDH, Poole, UK), filtered through a 0.45 μm polytetrafluoroethylene (PTFE) syringe filter (Agilent Technologies, Santa Clara, CA, USA) and fractionated by SPE to obtain a saponin-rich fraction. Strata C18 (8B-S001-LEG, Phenomenex, Torrance, CA, USA) cartridges were washed twice with 20 mL of HPLC grade water, and then twice with 20 mL of an HPLC grade methanol:water (60:40 v/v) solution. Saponins were eluted with two volumes of 20 mL HPLC grade methanol (BDH, Poole, UK) and dried under vacuum. Dry samples were dissolved in 50% methanol in water (HPLC grade) and stored at −20 °C until biological assays and metabolomics analysis.

2.5. Biological Assays

Saponin-rich extracts were tested on Hep-G2 and Caco-2 cancer cell lines, as well as in murine fibroblasts (NIH-3T3) non-cancer cell line. Bioassays were performed as described by Antunes-Ricardo et al. [13], using a final extract concentration of 50 μg/mL in all assays. Dulbecco's Modified Eagle Medium (DMEM, Thermo Fisher Scientific, Waltham, MA, USA) supplemented with 5% of fetal bovine serum was used as negative control. Assays were performed by triplicate.

2.6. Microorganism Identification

The bacterial chromosomal DNA of selected isolates was extracted with the UltraClean® extraction kit (Mo Bio Laboratories, Inc., Carlsbad, CA, USA), and DNA amplification was conducted as described by Escalante et al. [11], with slight modifications. Briefly, each sample of DNA extracted was used as template for 16S rRNA gene fragment amplification with the primers targeted to Eubacteria, Eu530F, and Eu1449R [11,14]. The protocol used for most of the isolates consisted of an initial denaturation step (95 °C for 5 min), followed by 30 cycles of denaturation at 95 °C for 1 min, primer annealing at 42 °C for 30 s, and elongation at 72 °C for 1 min, plus an additional 5 min cycle. For isolate 31, the temperature of primer annealing was 55 °C. PCR reaction was performed in a Gradient 96 Robocycler (Stratagene, La Jolla, CA, USA). Samples of chromosomal DNA and PCR products were analyzed by 1% agarose gel electrophoresis in TBE 1× buffer. The gel was stained with ethidium bromide, and the bands were visualized under UV illumination.

PCR products were purified using the GeneJET Purification Kit (Thermo Fisher Scientific, Waltham, MA, USA), and quantified in a 2000c NanoDrop (Thermo Fisher Scientific, Waltham, MA, USA). After quantitation, PCR products were sequenced in an automated DNA sequencer of 16 capillaries (Applied Biosystems, model 3130xl, Foster City, CA, USA) at the Sequencing Unit from the Institute of Biotechnology-Universidad Nacional Autonoma de Mexico (UNAM). Forward and reverse sequences were analyzed with Chromatogram Explorer Lite and assembled using the

DNA Baser program (http://www.dnabaser.com/). The obtained sequences were submitted to the non-redundant nucleotide database at GeneBank using the Basic Local Alignment Search Tool (BLAST) program (www.ncbi.nlm.nih.gov) to identify the isolates. A multiple alignment of 16S rRNA gene sequences from the microorganisms isolated with those retrieved from the GeneBank database was performed using the ClustalW function contained in the MEGA7 program [15]. A distance matrix calculation of nucleotide substitution rates and a phylogenetic tree were constructed with the Kimura (two parameters) algorithm and the neighbor-joining method, respectively, which are also included in the program MEGA7 [15]. The sequences of microorganisms *Sulfolobus acidocaldarius* strain ATCC 33909 (NR_043400) and *Escherichia coli* strain EcSD4 (KC504012) were included as outgroups.

2.7. Metabolomics and Saponins Analysis by Mass Spectrometry

Only samples from scale up fermentations, performed by triplicate, were used for metabolomics analysis. Methanol (100%) fractions obtained from SPE were analyzed according to the chromatographic conditions described by Leal-Diaz et al. [7]. Samples were analyzed by a high-performance liquid chromatographic system coupled to a time-of-flight (TOF) mass spectrometer equipped with an electrospray source (ESI) (HPLC/ESI-MS/TOF). Separation was performed at 25 °C in a Zorbax Eclipse XDB-C18 column 4.6 × 150 mm with a guard column of the same material 12.5 mm × 4.6 mm (Agilent Technologies, Santa Clara, CA, USA). Samples were injected randomly. Blanks were included between samples to reduce noise and eliminate potential contaminants from the column.

Saponin quantitation was performed using a 1200 Series HPLC coupled to an evaporative light scattering detector (HPLC/ELSD) (Agilent Technologies, Santa Clara, CA, USA), as previously reported [7]. Saponin concentrations were obtained as protodioscin equivalents (PE) using a standard curve from 10 to 500 ppm. Concentrations were reported as μg PE/mg extract. For saponin characterization, extracted ion chromatograms were analyzed considering the exact masses reported for saponins and sapogenins, as well as their fragmentation patterns (±0.05 units) using the Analyst QS 1.1 software (Applied Biosystems) [6,7,16].

For metabolomic analysis, raw mass spectrometry data were processed using Mass Hunter A.02.00 software (Agilent Technologies, Santa Clara, CA, USA). Information about retention time, exact mass, m/z, and the abundance of all ions from each sample were analyzed. Data acquired from Mass Hunter were processed using Excel. First, all of the ions from blanks were summed using an Excel table. Ions of all the samples were considered to exclude ions of contaminants. Protodioscin standard and the characteristic peak of kammogenin found in all the samples were used as references to determine the reliability of the mass, and the retention times of ions between different blanks or samples. A safety of 20 s for the retention time and 40 ppm for the exact mass was considered, and sample ions with a signal/noise ratio above five were picked. Ions with the same mass and retention time than those found in blanks were subtracted from the samples and ions were aligned manually. A single list with all of the ions detected in the samples with their corresponding exact masses, m/z, and abundances was obtained. Finally, a two-dimensional data matrix (mass versus abundance) was generated for all of the samples. Data analysis was conducted using PCA to select the ions that generated more variability among samples. Metabolites that were related to the decrease in viability were characterized by their retention times and mass spectra and compared with previous literature reports [6,7,16], and the Metlin database [17].

2.8. Data Analysis

Results of bioassays were analyzed using a three-factor ANOVA (fermentation time, strain, and cell line). Saponin quantitation was analyzed using a two-factor ANOVA (fermentation time and strain). Comparisons were made using Tukey's honest significant difference (HSD) tests, values of $p < 0.05$ were accepted as statistically significant. All statistical analyses were conducted using the JMP Version 12 software (SAS Institute Inc., Cary, NC, USA).

3. Results and Discussion

3.1. Microorganisms Isolation, Screening, and Identification

Agave sap microorganisms were mainly grown in R2A and NZ amine A media with 38 and 29 isolates, respectively. No growth was detected in the MRS medium. Only one isolate was detected in YM and PDA media, which was indicative that the presence of yeasts in the concentrate is limited. Fermentations conducted in a final volume of 3 mL showed that 15 of 69 isolates produced changes in agave sap that reduced cancer cell viability when their corresponding extracts were tested in vitro. Particularly, extracts obtained from 48 h agave sap fermentation using isolates 16, 28, 31, and 44 had a significant reduction in the viability of Caco-2 cells (Supplementary Materials Figure S1). In the case of Hep-G2 cells, isolates that produced extracts with the highest bioactivity were 4, 16, 28, and 38 (Supplementary Materials Figure S2).

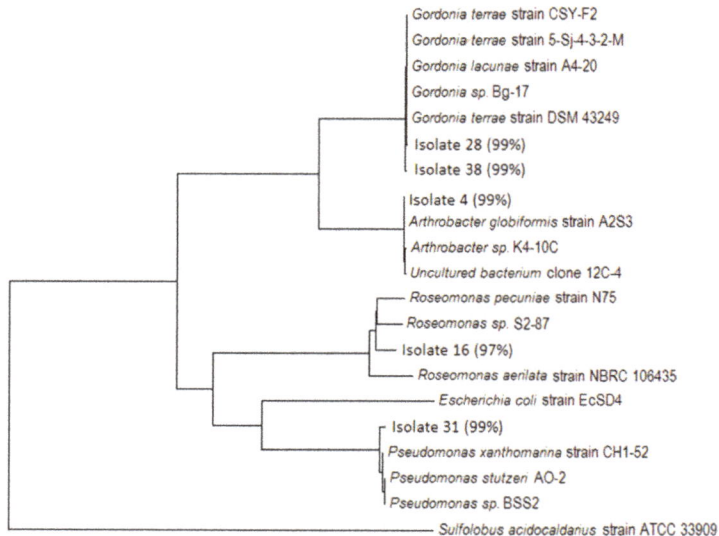

Figure 1. Phylogenetic tree of 16S rDNA sequences from isolates and their closest neighbors from bacteria or environmental clones available in the National Center for Biotechnology Information (NCBI) database. The 16S rRNA gene sequence of *Sulfolobus acidocaldarius* served as outgroup. Identity percentage with closest reference to 16S rRNA gene clones in the database is indicated in parenthesis.

The phylogenetic analysis of the 16S rRNA gene sequences revealed that isolates 16 and 31 formed a branch from the genus *Pseudomonas* and *Roseomonas*, respectively (Figure 1). Isolate 16 showed a closer identity with *Roseomonas aerilata* and *Roseomonas pecuniae*, whereas isolate 31 was more similar to *Pseudomonas stutzeri* and *Pseudomonas xanthomarina*. *R. aerilata* had been isolated from the air [18], whereas *R. pecuniae* had been isolated from the surface of a copper-alloy coin [19]. *P. stutzeri* had been isolated from diverse environments [20,21]. *P. xanthomarina* had been isolated from samples of sea squirts, and it was able to grow in 8% NaCl [22]. Isolate 4 was within the same branch as *Arthrobacter globiformis* (Figure 1). The genus *Arthrobacter* includes coryneform non-pathogenic bacteria isolated from environmental sources such as soil, water, and plant material. *A. globiformis* possesses the enzyme choline oxidase, which catalyzes the oxidation of choline to glycine betaine, a compound that prevents dehydration and plasmolysis in hyperosmotic environments [23]. *A. globiformis* has been included in the European Food and Feed Cultures Association (EFFCA) and the International Dairy Federation (IDF) inventory of microorganisms with documented history of use in human *globiformis* could be included in the European Food and Feed Cultures Association (EFFCA) and the International Dairy Federation (IDF) inventory of microorganisms with documented history of use in human food [24]. Isolates 28 and 38 were placed in the same branch as *Gordonia terrae* and *G. lacunae* (Figure 1). The first strains of the genus *Gordonia* were detected as opportunistic pathogens from human patients, but other strains play a role in the bioremediation and biodegradation of xenobiotics [25]. Isolate 4 (*A. globiformis*), along with isolates 28 and 38 (*Gordonia* sp.), belong to the group of actinomycetes. Strains of *A. globiformis* have shown the ability to produce a compound with high antioxidant capacity [26]. The actinomycete *G. lacunae* exhibited antibiotic activity [25], whereas *G. terrae* was reported as a carotenoids producer [27]. Interestingly, none of the genera identified in this study had previously been reported in agave sap or pulque (fermented agave sap).

3.2. Metabolomics

Based on the screening results, the actinomycetes *A. globiformis* and *Gordonia* sp. (isolates 4 and 28, respectively) were selected for scaled up fermentations. Mass spectrometry data analysis showed the presence of 2913 ions in all of the samples. After analyzing the abundance of all of the ions, 939 ions were mainly responsible for the variability observed among the saponin-rich extracts obtained from 48 h, 96 h, or 192 h fermentations. According to the principal component 1, extracts from agave sap fermented with *Gordonia* sp. (isolate 28) were separated from those of control agave sap and fermented with *A. globiformis* (isolate 4) (Figure 2).

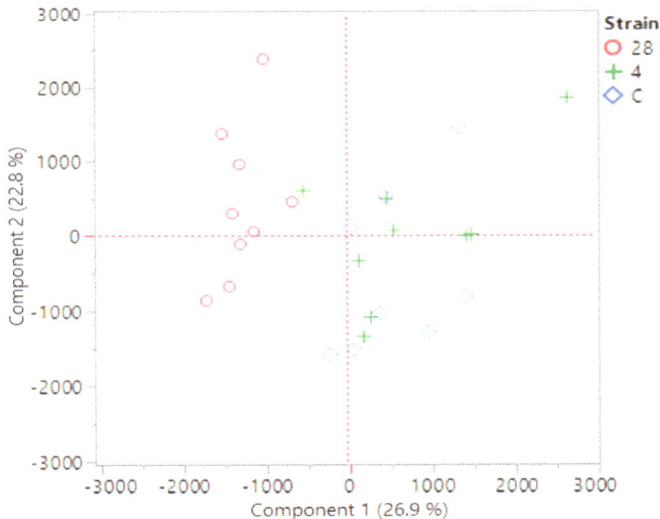

Figure 2. Principal Component Analysis (PCA) score plot of the 939 ions that were among those mainly responsible for the variability observed in the saponin-rich extracts obtained from 48 h, 96 h, or 192 h fermentations with *A. globiformis* (isolate 4), *Gordonia* sp. (isolate 28), and in the absence of agave sap (C).

The 939 ions responsible for the variability between samples were numbered based on the ion abundance detected by MS. Ions 1–4 were correlated with the inhibitory activity on the cancer cell lines, and were further investigated for identification based on their fragmentation patterns and previous reports in the literature (Table 1). Ion 1 was tentatively identified as trihydroxy-phytosphingosine, a sphingolipid found as a structural component of the membrane in plants and yeasts [28]. Ions 2 and 4, along with other ions detected at the same retention time, were tentatively identified as kammogenin glycosides [6,7]. On the other hand, ion 3 corresponded to the kammogenin aglycone, which was previously found in agave sap [7].

9

Table 1. Suggested identification of compounds found in agave sap fermented with *Gordonia* sp.

Compound	Suggested Compounds	Accurate Mass	m/z
1	Dehydro-phytosphingosine	315.27	316.28 (M + H)$^+$, 338.25 (M + Na)$^+$
2	Magueyoside A	1194.48	445.29 (M-3Hex-2Pen + H)$^+$, 607.34 (M-2Hex-2Pen + H)$^+$, 769.35 (M-1Hex-2Pen + H)$^+$, 1217.51 (M + Na)$^+$
	Magueyoside B	1062.44	445.29 (M-3Hex-1Pen + H)$^+$, 607.34 (M-2Hex-1Pen + H)$^+$, 769.36 (M-1Hex-1Pen + H)$^+$, 1085.48 (M + Na)$^+$
3	Kammogenin	444.28	445.29 (M + H)$^+$, 467.25 (M + Na)$^+$, 483.23 (M + K)$^+$, 911.52 (2M + Na)$^+$
4	Kammogenin glycoside 1 (KG1)	1092.44	445.29 (M-4Hex + H)$^+$, 607.34 (M-3Hex + H)$^+$, 769.36 (M-2Hex + H)$^+$, 931.40 (M-1Hex + H)$^+$, 1115.51 (M + Na)$^+$
	Kammogenin glycoside 2 (KG2)	1224.48	445.29 (M-4Hex-1Pen + H)$^+$, 607.34 (M-3Hex-1Pen + H)$^+$, 769.36 (M-2Hex-1Pen + H)$^+$, 1247.47 (M + Na)$^+$

Mass spectra analysis of compound 2 showed that two kammogenin glycosides were co-eluting. These saponins exhibited characteristic ions at m/z 1217.51 ([M + Na]$^+$) and 1085.48 ([M + Na]$^+$), and both were isolated previously from flowers of *A. offoyana* and identified as magueyoside A and magueyoside B, respectively [16]. Similarly, at the retention time of compound 4, two kammogenin glycosides were found. The saponin that showed the ion at m/z 1247.51 ([M + Na]$^+$) was also previously detected in agave sap [7], and the ion at m/z 1115.51 ([M + Na]$^+$) was similar but without the pentose residue. According to mass spectra and the Metlin database [17], compound 1 tentatively corresponded to dehydro-phytosphingosine. Dehydro-phytosphingosine belongs to the sphingolipids, a class of lipids found in all living organisms as essential structural membrane constituents. Some sphingolipids from plants have shown anti-inflammatory properties and reduced the activity of several signaling pathways [28]. Particularly, dehydro-phytosphingosine from maize seeds has showed to induce apoptosis in Caco-2 cells [29].

3.3. Saponins Analysis and Effects on Cancer Cell Viability

The extract from agave sap fermented for 48 h with *A. globiformis* (4) reduced the viability of mouse fibroblasts (NIH-3T3) in comparison with control, but the change on the viability of cancer cells was not significant (Figure 3a). The effect on Caco-2 cells viability for the sample obtained from the 48 h fermentation with *Gordonia* sp. (28) was similar to that obtained for control, but it reduced the viability of Hep-G2 cells from 78.9 % to 60.6%. At 48 h fermentation, *A. globiformis* (4) did not enhance the anticancer activity of agave sap butanolic extracts.

Extracts obtained after 96 h fermentation with *A. globiformis* (4) reduced the viability of Caco-2 cells from 70.2% to 56.0% (Figure 3b). NIH-3T3 cell viability was not negatively affected after the treatment with the butanolic extracts obtained after 96 h fermentation; only the sample obtained after 192 h fermentation with *A. globiformis* (4) reduced the viability of non-cancer cells (Figure 3c). This extract also reduced the hepatic cancer cells (Hep-G2) viability from 81.4% to 66.6%, and did not decrease the viability of Caco-2 cells in comparison with control (Figure 3c).

(a)

(b)

(c)

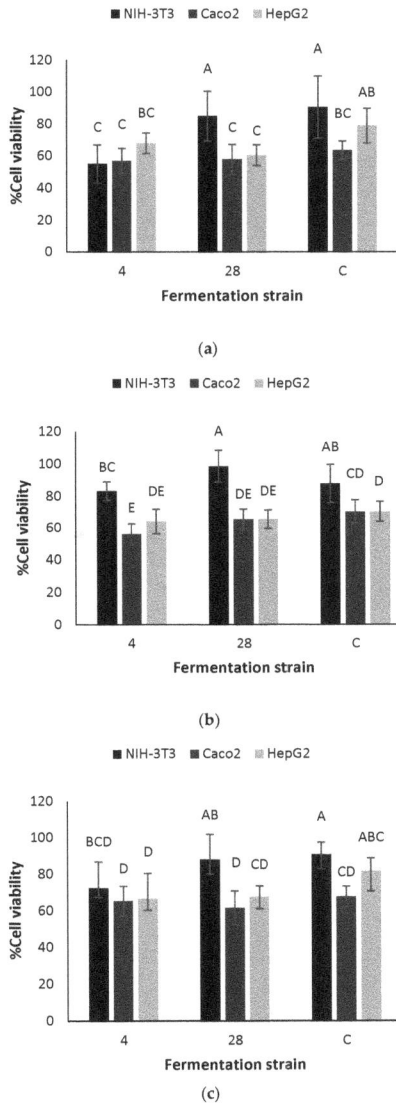

Figure 3. Effect of extract from agave sap fermented with *A. globiformis* (4) and *Gordonia* sp. (28) on three different cell lines at (**a**) 48 h; (**b**) 96 h; and (**c**) 192 h fermentation. Non-fermented controls (C) were included. All assays were performed at a final concentration of 50 μg/mL. Bars not connected with the same letter were significantly different.

Therefore, considering cytotoxicity on NIH-3T3, *Gordonia* sp. (28) enhanced the anticancer activity of agave sap in Hep-G2 cells after 48 h fermentation. At 96 h, *A. globiformis* (4) enhanced the anticancer effect on Caco-2 cells; but 192 h fermentation with *A. globiformis* (4) would not be recommended to enhance the bioactivity of butanolic extracts, since it similarly affected Hep-G2 cells and NIH-3T3 cells.

Saponins derived from kammogenin, manogenin, gentrogenin, and hecogenin were detected in all of the extracts. Despite the correlation of ion abundance with inhibitory activity, no differences were found in the concentration of the most abundant saponins found in butanolic extracts (Table 2). Magueyosides A and B co-eluted, and were the most abundant saponins in all of the samples, which agrees with previous reports [5–7]. Based on the relative intensity of the extracted ion chromatograms for these compounds, there was an increase of magueyoside B concentration and a slight reduction of magueyoside A in the extracts obtained from *Gordonia* sp. fermentation (Figure 4A,B). These findings, in conjunction with the results of the bioassays, indicate that the inhibitory activity of the extracts depends not only on the total concentration of saponins, but also on the saponins' profile; a method with higher resolution would be useful to quantitate them separately. Additionally, Magueyoside B has been correlated with a significant reduction in the viability of colon cancer cells (Caco-2) [6], and it has been shown to induce apoptosis in colon cancer cell line HT-29 [5].

Table 2. Concentration of saponins in fermented and unfermented agave sap *.

Strain	Fermentation Time (h)	MagA + MagB [§]	KG1 + KG2
		(µg PE/mg Extract)	
A. globiformis	48	2.99 ± 0.68 [a]	2.05 ± 0.30 [a,b]
	96	2.73 ± 0.58 [a]	1.81 ± 0.32 [a,b]
	192	2.33 ± 0.71 [a]	1.71 ± 0.57 [a,b]
Gordonia sp.	48	2.71 ± 0.29 [a]	2.04 ± 0.21 [a,b]
	96	3.09 ± 0.65 [a]	2.43 ± 0.67 [a,b]
	192	2.70 ± 1.01 [a]	2.80 ± 0.71 [a]
Control	48	3.56 ± 0.05 [a]	2.51 ± 0.11 [a,b]
	96	2.57 ± 0.39 [a]	1.84 ± 0.20 [a,b]
	192	1.70 ± 0.59 [a]	1.27 ± 0.41 [b]

[§] Magueyoside A and magueyoside B; Kammogenin glycosides with m/z of 1115.51 $(M + Na)^+$ and 1247.51 $(M + Na)^+$; **PE**: protodioscin equivalents. * Mean values ± SD of replicate samples analyzed in triplicate. Mean values within the same column followed by different letters a to b are significantly different ($p < 0.05$).

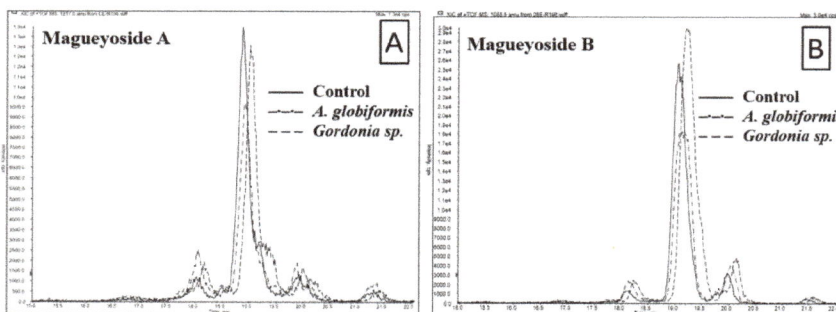

Figure 4. (**A**) Comparison of extracted ion chromatograms for magueyoside A after 96 h of agave sap fermentation with *A. globiformis* or *Gordonia* sp; (**B**) Comparison of extracted ion chromatograms for magueyoside B after 96 h of agave sap fermentation with *A. globiformis* or *Gordonia* sp.

4. Conclusions

Microorganisms that survive the thermal treatment of agave sap have the potential to modify agave sap metabolites such as saponins. Then, their anticancer enhancement effect depended on the fermentation time and the cancer cell line tested. *Gordonia* sp. enhanced the effect on Hep-G2 cells after 48 h fermentation. Agave sap fermentation after 96 h fermentation with *A. globiformis* increased the Caco-2 cancer cell growth inhibitory effect of the corresponding saponin-rich extract. It was shown that magueyoside B, a kammogenin glycoside, was found at higher intensities in the samples fermented with *Gordonia* sp. that reduced Hep-G2 viability better than controls. However, the fact that the concentration of these saponins was not affected after fermentation indicates that the

Sustainability **2017**, *9*, 2095

anti-cancer activity depends not only on the concentration of total saponins, but also on the profile of these compounds within each sample. The changes in the saponins profile observed in this experiment demonstrate that it is important to characterize the microorganisms that survive thermal treatment of foods and affect the phytochemicals composition of traditional Mexican foods.

Supplementary Materials: The following are available online at www.mdpi.com/2071-1050/9/11/2095/s1, Figure S1: Viability of colon cancer cells (Caco-2) after treatment with extracts obtained after 48, 96 or 192 h of fermentation with isolates (50 µg/mL), Figure S2: Viability of liver cancer cells (HepG2) after treatment with extracts obtained after 48, 96 or 192 h of fermentation with isolates (50 µg/mL).

Acknowledgments: We appreciate financial support from CONACYT CVU 490590, NutriOmics Chair from Tecnológico de Monterrey, Nutrigenomics Research Chair from Fundación FEMSA, PAPIIT DGAPA project IN 207914 from IBT UNAM, as well as Carlos Rodríguez-López for his knowledge and assistance in metabolomics analysis and Albino Vargas from AGMEL S.A de C.V for providing agave sap and agave sap concentrate.

Author Contributions: Luis M. Figueroa carried out all the experiments related to microorganism isolation and identification, compiled and analyzed all the data, and drafted the manuscript. Liliana Santos-Zea obtained the data for saponin profile, collaborated in saponin identification and participated in the writing process. Janet A. Gutiérrez Uribe and Adelfo Escalante obtained the funding for the project and advised throughout the design and experimental phases of the study, as well as through the writing process.

Conflicts of Interest: The authors declare no conflict of interest.

References

1. Lappe-Oliveras, P.; Moreno-Terrazas, R.; Arrizón-Gaviño, J.; Herrera-Suárez, T.; García-Mendoza, A.; Gschaedler-Mathis, A. Yeasts associated with the production of Mexican alcoholic nondistilled and distilled Agave beverages. *FEMS Yeast Res.* **2008**, *8*, 1037–1052. [CrossRef] [PubMed]
2. Escalante, A.; López Soto, D.R.; Velázquez Gutiérrez, J.E.; Giles-Gómez, M.; Bolívar, F.; López-Munguía, A. Pulque, a traditional Mexican alcoholic fermented beverage: Historical, microbiological, and technical aspects. *Front. Microbiol.* **2016**, *7*, 1026. [CrossRef] [PubMed]
3. Gutiérrez-Uribe, J.A.; Serna-Saldívar, S. Agave Syrup Extract Having Anticancer Activity. U.S. Patent 8,470,858, 25 June 2013.
4. Chen, P.-Y.; Chen, C.-H.; Kuo, C.-C.; Lee, T.-H.; Kuo, Y.-H.; Lee, C.-K. Cytotoxic steroidal saponins from *Agave Sisalana*. *Plant. Med.* **2011**, *77*, 929–933. [CrossRef] [PubMed]
5. Santos-Zea, L.; Fajardo-Ramírez, O.R.; Romo-López, I.; Gutiérrez-Uribe, J.A. Fast centrifugal partition chromatography fractionation of concentrated agave (*Agave salmiana*) sap to obtain saponins with apoptotic effect on colon cancer cells. *Plant Foods Hum. Nutr.* **2016**, *71*, 57–63. [CrossRef] [PubMed]
6. Santos-Zea, L.; Rosas-Pérez, A.M.; Leal-Díaz, A.M.; Gutiérrez-Uribe, J.A. Variability in saponin content, cancer antiproliferative activity and physicochemical properties of concentrated agave sap. *J. Food Sci.* **2016**, *81*, H2069–H2075. [CrossRef] [PubMed]
7. Leal-Díaz, A.M.; Santos-Zea, L.; Martínez-Escobedo, H.C.; Guajardo-Flores, D.; Gutiérrez-Uribe, J.A.; Serna-Saldivar, S.O. Effect of *Agave americana* and *Agave salmiana* ripeness on saponin content from aguamiel (agave sap). *J. Agric. Food Chem.* **2015**, *63*, 3924–3930. [CrossRef] [PubMed]
8. Gao, P.; Xu, G. Mass-spectrometry-based microbial metabolomics: Recent developments and applications. *Anal. Bioanal. Chem.* **2015**, *407*, 669–680. [CrossRef] [PubMed]
9. Wishart, D.S. Metabolomics: Applications to food science and nutrition research. *Trends Food Sci. Technol.* **2008**, *19*, 482–493. [CrossRef]
10. Settachaimongkon, S.; Nout, M.J.R.; Antunes Fernandes, E.C.; Hettinga, K.A.; Vervoort, J.M.; van Hooijdonk, T.C.M.; Zwietering, M.H.; Smid, E.J.; van Valenberg, H.J.F. Influence of different proteolytic strains of *Streptococcus thermophilus* in co-culture with *Lactobacillus delbrueckii* subsp. *bulgaricus* on the metabolite profile of set-yoghurt. *Int. J. Food Microbiol.* **2014**, *177*, 29–36. [CrossRef] [PubMed]
11. Escalante, A.; Giles-Gómez, M.; Hernández, G.; Córdova-Aguilar, M.S.; López-Munguía, A.; Gosset, G.; Bolívar, F. Analysis of bacterial community during the fermentation of pulque, a traditional Mexican alcoholic beverage, using a polyphasic approach. *Int. J. Food Microbiol.* **2008**, *124*, 126–134. [CrossRef] [PubMed]
12. He, Y.; Xu, M.; Qi, Y.; Dong, Y.; He, X.; Li, J.; Liu, X.; Sun, L. Differential responses of soil microbial community to four-decade long grazing and cultivation in a semi-arid grassland. *Sustainability* **2017**, *9*, 128. [CrossRef]

13. Antunes-Ricardo, M.; Moreno-García, B.E.; Gutiérrez-Uribe, J.A.; Aráiz-Hernández, D.; Alvarez, M.M.; Serna-Saldivar, S.O. Induction of apoptosis in colon cancer cells treated with isorhamnetin glycosides from *Opuntia ficus-indica* pads. *Plant Foods Hum. Nutr.* **2014**, *69*, 331–336. [CrossRef] [PubMed]

14. Escalante, A.; Rodríguez, M.E.; Martínez, A.; López-Munguía, A.; Bolívar, F.; Gosset, G. Characterization of bacterial diversity in pulque, a traditional Mexican alcoholic fermented beverage, as determined by 16S rDNA analysis. *FEMS Microbiol. Lett.* **2004**, *235*, 273–279. [CrossRef] [PubMed]

15. Kumar, S.; Stecher, G.; Tamura, K. MEGA7: Molecular volutionary Genetics Analysis Version 7.0 for bigger datasets. *Mol. Biol. Evol.* **2016**, *33*, 1870–1874. [CrossRef] [PubMed]

16. Pérez, A.J.; Calle, J.M.; Simonet, A.M.; Guerra, J.O.; Stochmal, A.; Macías, F.A. Bioactive steroidal saponins from *Agave offoyana* flowers. *Phytochemistry* **2013**, *95*, 298–307. [CrossRef] [PubMed]

17. METLIN: A metabolite Mass Spectral Database. Available online: http://metlin.scripps.edu (accesed on 1 May 2017).

18. Yoo, S.-H.; Weon, H.-Y.; Noh, H.-J.; Hong, S.-B.; Lee, C.-M.; Kim, B.-Y.; Kwon, S.-W.; Go, S.-J. *Roseomonas aerilata* sp. nov., isolated from an air sample. *Int. J. Syst. Evol. Microbiol.* **2008**, *58*, 1482–1485. [CrossRef] [PubMed]

19. Lopes, A.; Santo, C.E.; Grass, G.; Chung, A.P.; Morais, P.V. *Roseomonas pecuniae* sp. nov., isolated from the surface of a copper-alloy coin. *Int. J. Syst. Evol. Microbiol.* **2011**, *61*, 610–615. [CrossRef] [PubMed]

20. Kuroda, M.; Notaguchi, E.; Sato, A.; Yoshioka, M.; Hasegawa, A.; Kagami, T.; Narita, T.; Yamashita, M.; Sei, K.; Soda, S.; et al. Characterization of *Pseudomonas stutzeri* NT-I capable of removing soluble selenium from the/phase under aerobic conditions. *J. Biosci. Bioeng.* **2011**, *112*, 259–264. [CrossRef] [PubMed]

21. Zhang, J.; Cao, X.; Xin, Y.; Xue, S.; Zhang, W. Purification and characterization of a dehalogenase from *Pseudomonas stutzeri* DEH130 isolated from the marine sponge *Hymeniacidon perlevis*. *World J. Microbiol. Biotechnol.* **2013**, *29*, 1791–1799. [CrossRef] [PubMed]

22. Romanenko, L.A.; Uchino, M.; Falsen, E.; Lysenko, A.M.; Zhukova, N.V.; Mikhailov, V.V. *Pseudomonas xanthomarina* sp. nov., a novel bacterium isolated from marine ascidian. *J. Gen. Appl. Microbiol.* **2005**, *51*, 65–71. [CrossRef] [PubMed]

23. Fan, F.; Ghanem, M.; Gadda, G. Cloning, sequence analysis, and purification of choline oxidase from *Arthrobacter globiformis*: A bacterial enzyme involved in osmotic stress tolerance. *Arch. Biochem. Biophys.* **2004**, *421*, 149–158. [CrossRef] [PubMed]

24. Mogensen, G.; Salminen, S.; O'Brien, J.; Ouwenhand, A.; Holzapfel, W.; Shortt, C.; Fonden, R.; Miller, G.D.; Donohue, D.; Playne, M.; et al. Inventory of microorganisms with a documented history of use in food. *Bull. Int. Dairy Fed.* **2002**, *377*, 10–19.

25. Roes, M.; Goodwin, C.M.; Meyers, P.R. *Gordonia lacunae* sp. nov., isolated from an estuary. *Syst. Appl. Microbiol.* **2008**, *31*, 17–23. [CrossRef] [PubMed]

26. Teng, W.H.; Sun, W.J.; Yu, B.; Cui, F.J.; Qian, J.Y.; Liu, J.Z.; Wang, L.; Qi, X.H.; Wei, H. Continuous conversion of rice starch hydrolysate to 2-keto-D-gluconic acid by *Arthrobacter globiformis* C224. *Biotechnol. Bioprocess Eng.* **2013**, *18*, 709–714. [CrossRef]

27. Takaichi, S.; Maoka, T.; Akimoto, N.; Carmona, M.L.; Yamaoka, Y. Carotenoids in a Corynebacterineae, *Gordonia terrae* AIST-1: Carotenoid glucosyl mycoloyl esters. *Biosci. Biotechnol. Biochem.* **2008**, *72*, 2615–2622. [CrossRef] [PubMed]

28. Canela, N.; Herrero, P.; Mariné, S.; Nadal, P.; Ras, M.R.; Rodríguez, M.Á.; Arola, L. Analytical methods in sphingolipidomics: Quantitative and profiling approaches in food analysis. *J. Chromatogr. A* **2016**, *1428*, 16–38. [CrossRef] [PubMed]

29. Aida, K.; Kinoshita, M.; Susgawara, T.; Ono, J.; Miyazawa, T.; Ohnishi, M. Apoptosis Inducement by Plant and Fungus Sphingoid Bases in Human Colon Cancer Cells. *J. Oleo Sci.* **2004**, *53*, 503–510. [CrossRef]

sustainability

MDPI

Communication

Carbohydrates Components of Some Italian Local Landraces: Garlic (*Allium sativum* L.)

Silvia Lisciani [1], Loretta Gambelli [1], Alessandra Durazzo [1,*], Stefania Marconi [1], Emanuela Camilli [1], Cecilia Rossetti [1], Paolo Gabrielli [1], Altero Aguzzi [1], Olindo Temperini [2] and Luisa Marletta [1]

[1] Consiglio per la Ricerca in Agricoltura e L'analisi Dell'economia agraria—Centro di ricerca CREA-Alimenti e Nutrizione, Via Ardeatina 546, 00178 Rome, Italy; silvia.lisciani@crea.gov.it (S.L.); loretta.gambelli@crea.gov.it (L.G.); stefania.marconi@crea.gov.it (S.M.); emanuela.camilli@crea.gov.it (E.C.); rossetti.cecilia@gmail.com (C.R.); paolo.gabrielli@crea.gov.it (P.G.); altero.aguzzi@crea.gov.it (A.A.); luisa.marletta@crea.gov.it (L.M.)

[2] Dipartimento di Scienze Agrarie e Forestali (DAFNE), Università degli Studi della Tuscia, 01100 Viterbo, Italy; olindotemperini@gmail.com

* Correspondence: alessandra.durazzo@crea.gov.it; Tel.: +39-06-5149-4430

Received: 11 September 2017; Accepted: 22 October 2017; Published: 24 October 2017

Abstract: Garlic is one of the most widespread and ancient medicinal plants. Its health benefits are due to its chemical components, and among these is carbohydrate, whose characteristics have been so far little investigated. The aim of this study is to typify the various components of carbohydrate (starch, individual sugars, fructans, and total dietary fibre) in four commonly consumed "Italian local landraces": Bianco Piacentino, Rosso di Castelliri, Rosso di Sulmona, Rosso di Proceno, which are grown in two different geographical areas—Viterbo and Alvito—under the same agronomic conditions. This study will also evaluate how genotype and the cultivation area can affect the profile of the carbohydrate components of these landrace strains. Regarding unavailable carbohydrates, all of the varieties showed appreciable contents of fructans, the most representative component, which ranged from 45.8 to 54.4 g/100 g d.w. In contrast, total dietary fibre values varied from 9.1 to 13.1 g/100 g d.w. in Rosso di Castelliri and Bianco Piacentino, respectively, which are both grown in Viterbo. As for starch, only some traces were found, while the amount of total sugars ranged between 2.12 and 3.27 g/100 g d.w., with higher levels of sucrose. Our findings could provide important information that may be adopted to enhance and promote the quality of some local Italian garlic landraces through highlighting the influence that the cultivar and the environmental conditions can have on carbohydrates components.

Keywords: Italian garlic; carbohydrates; fructans; dietary fibre; soluble sugars; cultivar; environmental conditions

1. Introduction

Garlic (*Allium Savitum* L.) is a vegetable bulb native to central Asia and belonging to the plant family Amaryllidaceae [1]. It is widely used in gastronomy for dressing and spicing dishes. Moreover, it has been used as a medicinal plant since ancient times, and it is still being employed in folk medicine all over the world. According to Food and Agriculture Organization of the United Nations (FAO) [2], its global production is close to 15 million tons, with China as world's first export and production country. Between 2008 and 2013, China's shipments increased from 13 to 18 million tons; in second place, just behind, are Spain and Argentina, while the Netherlands play a redistribution function of the product from other countries (China and Argentina), and Italy ranks fifth [3]. It has been observed that garlic production has not significantly decreased over the last decades, thanks to the availability of several new garlic-based products on the gastronomic market (dressings, pickled garlic, seasonings, powders, oil, etc.). Despite this, a lot of Italian varieties of this bulb, although known and appreciated in many national and international dishes, risk disappearing.

Although Italian cuisine is known for its variety and regional diversity, it includes garlic among its main ingredients, which is used to enhance the flavour of many foods and traditional and local recipes. Garlic is grown all over Italy, and its most common landraces mainly belong to two types: white garlic and pink garlic. They are characterised, respectively, by white-silvered and pinkish-purple skin. In particular, some Italian local landraces originally produced in a limited geographical area have spread throughout the country because of their organoleptic properties and long tradition [4].

Numerous studies have investigated the health benefits related to garlic consumption, which include [5–12]: the reduction of risk factors for cardiovascular diseases and cancer, the stimulation of the immune response, the antimicrobial effects [13,14], the invigorating action, the resistance to various stresses, and the potential anti-aging effects [13,15–17]. These physiological properties depend both on the characteristic profile of the bioactive components [18] of garlic and their combined action; the peculiar components of garlic are volatile sulphur compounds, such as sulfoxides and thiosulfinates, which are known not only for their beneficial properties, but also mostly for their typical smell and taste [19–22].

Among the numerous health benefits associated with the consumption of garlic, our attention in particular, is towards the characterization of carbohydrates components, which are related to the prebiotic activity of this bulb. This function is due to its high content of fructans and dietary fibre [8,23]. Several researchers have investigated the physiological role of fructans in plants [24,25]; fructans are fructose polymers that provide the short-term energy storage in many plant species [26]. The global distribution of fructan-accumulating plants shows that they are particularly copious in temperate zones, while they are almost absent in tropical regions, or during seasonal drought or frost [27,28]. The amount of fructans, as for other metabolites, is strongly influenced by genetic and environmental factors, including: growth factors such as light, temperature, humidity, and fertilizers; damages caused by microorganisms and insects, stress induced by UV radiation, heavy metals, and pesticides [29].

According to the type of linkage, fructans are classified into three groups of compounds: inulin, levan, and graminan [8,30,31]. Inulin-type compounds include fructans with a short chain composed by 2–9 units or degree of polymerization (DP), which are generally named fructo-oligosaccharides (FOS), and with long-chain ones (DP > 10), called inulins [4,32,33]. In this study, we include both FOS and inulins within the term "fructans".

Because of the β (2 → 1) linkage of fructose monomers, fructans cannot be digested by the intestinal enzymes; they are defined, indeed, as "non-digestible/unavailable" carbohydrates. However, a regular consumption of such non-digestible polysaccharides has positive effects on human health [34]; in fact, their fermentation takes place in the colon, and this activity is well recognized to play an essential role in the improvement of health and, as a consequence, in the reduction of the risk of diseases. Even though fructans belong to the dietary fibre complex group because they contribute to increasing the water content of the stools and to improving bowel habits [35], they also, however, show some specific fermentative properties that are different from those attributed to dietary fibre. Therefore, fructans may contribute significantly to a well-balanced diet by intensifying and amplifying the beneficial effects of dietary fibre and by influencing several gastrointestinal and systemic functions [36]. In the colon, the fermentation of inulin-type fructans is fast, and based on three main sequential phases: the production of short-chain fatty acids (SCFA) that reduce the intestinal pH; the reduction of putrefactive substances in the intestine, and finally, the increase of the bifidobacteria population, which does not alter the total bacteria levels [8,37–39].

Inulin and FOS are the most studied and well-known prebiotics, because they offer an interesting and remarkable combination of nutritional and technological properties for the food industry. Indeed, interest in these carbohydrates is growing worldwide, because it has been recognized that they can change the composition of the gut microflora by increasing the number of health-promoting bacteria [39].

Although not all of the fibre is prebiotic, some of its digestive functions that produce health benefits have been well defined and characterised: it increases the viscosity of the stomach contents, reduces the intestinal transit time, and enhances the bulk of the food mass [40]. These compounds are catching researchers' attention for their health-promoting effects, to the point that they are gradually being added to a wider variety of food products as fortification agents, and are being increasingly used as the main components in dietary supplements [28,41]. However, recent researches have emphasized that particular attention must be paid to consumers with problems of dietary intolerances to fructans and fermentable oligosaccharides, disaccharides, monosaccharides, and polyols (FODMAPs) [42].

Currently, few studies have provided data about the content of fructans in garlic and, in particular, specific information on the chemical characteristics of Italian varieties of garlic are needed. Some researchers have evaluated the phytochemical compounds of these garlic cultivars by determining the content of carotenoids, flavonoids, vitamin C, and the total antioxidant capacity [43].

At present, the main Italian Food Composition Database [44] mostly contains average values on the nutritional composition of plant foods; while information available for single varieties is currently lacking. The characterisation of the nutrient profile of the different landraces is reported in scientific publications or in limited and specific databases created to promote and preserve the nutritional characteristics of some national traditional foodstuffs [45].

This study is aimed at estimating the carbohydrate components of four Italian landraces of garlic, which are all qualified on the national market, and some are enrolled in the Italian Register of the varieties. This work is addressing to: valorise and promote local products with a long tradition by preserving and protecting their identity; analyse some chemical characteristics for a better understanding of their nutritional role; and verify possible influences exerted by the cultivar and environmental factors on the qualitative properties of garlic.

Furthermore, we propose and suggest the identification of potential genotypes, for more interesting uses for prebiotic sources addressing nutraceutical needs.

2. Materials and Methods

2.1. Samples

Four Italian landraces of *Allium sativum* L.—Bianco Piacentino, Rosso di Castelliri, Rosso di Sulmona, and Rosso di Proceno—that were cultivated at two locations (Viterbo and Alvito) under the same growing conditions, were investigated. In Table 1, we reported the basic information and the main descriptors of morphological characteristics.

Table 1. Main descriptors of morphological characteristics.

	Bianco Piacentino	*Rosso di Castelliri*	*Rosso di Sulmona*	*Rosso di Proceno*
outer skin colour of compound bulb	white silvered	white	cream	cream
skin colour of the cloves	white	red purple	dark red	red
shape of mature dry bulbs	flat globe	ovate	broad oval	broad oval
bulb structure type	regular two-fan groups	regular multi-cloved radial	irregular	regular multi-cloved radial
flavour	pungent	pungent and spicy	spicy	persistent, strong, and spicy

Two of them enrolled in the "National Register of horticultural varieties" (Rosso di Sulmona e Bianco Piacentino) and the ecotype Bianco Piacentino was the only garlic waiting for the Protected Geographical Indication (PGI) trademark; Rosso di Castelliri and Rosso di Proceno cultivars were part of the "Regional Register of indigenous genetic resources of agricultural interest at risk of erosion". They were cultivated and harvested in the period between June and early July.

For Bianco Piacentino, a medium–late cultivar, maturity starts between 20 June and 15 July, and the harvest from 15 to 30 July, while the harvest time of Rosso di Castelliri takes place in the first fifteen days of June. Rosso di Sulmona is an early–medium maturing garlic (around 20–30 June), and for Rosso di Proceno, harvest time is in July (Figure 1).

Currently, all of the four varieties, which were identified for their geographic origin, are the most widespread in Italy, and grown in many geographic areas throughout the country. All of the samples studied were cultivated in three parcels of two different areas in the Lazio region using the same technical–agronomic trail, as shown in Table 2.

Crop density was 18.18 (plants/m^2). The spacing between rows was 0.55 m, and the distance between plants within the row was 0.20 m. The experimental protocol consisted of split parcels with four replicates. Each elemental parcel was 4 m^2, with six four-meter rows containing 240 cloves. The linear meters for each parcel were 24. The total number of bulbs and cloves per cultivar were 240 and 1960, respectively. The soil in Viterbo was muddy and sandy with the following characteristics: pH 6.6, total N 0.09% d.w. soil, exchangeable P 33 ppm, exchangeable K 421 ppm, CaCO$_3$ 5.1% d.w. soil; the soil of Alvito was loamy and clayey, with following parameters: pH 7.7, Total N 0.20% d.w. soil, exchangeable P 78 ppm, exchangeable K 342 ppm, CaCO$_3$ 11.2% d.w. soil.

Figure 1. Photo of Bianco Piacentino, Rosso di Castelliri, Rosso di Sulmona, and Rosso di Proceno samples.

The termo-pluviometric parameters during the whole garlic production were shown in Figure 2. The climatic conditions observed in the graphs showed a similar trend; in the area of Alvito, a greater rainfall was reported, but its effects were not so significant to the point of affecting the production or growth of the garlic.

The biometric parameters of 12 samples for each ecotype and for each cultivation area were examined for a total of 96 bulbs. Samples, once in the lab, were weighed, measured, and deprived of the outer skin; then, individual cloves, evaluated randomly, were counted, weighed, and peeled. For each ecotype and for every cultivated area, a representative sample of garlic was prepared; each pool, then, was homogenised and analysed for the moisture content, frozen at −30 °C and then lyophilised, for subsequent analyses.

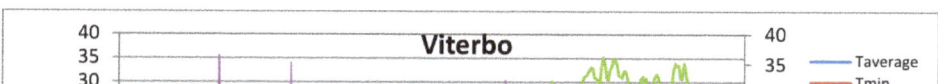

Table 2. Agronomic/Technical trail of garlic production.

Agronomic Trail	
Crop density (plants/m^2)	18.18
Spacing between rows (m)	0.55
Distance (between plants) within the row (m)	0.10
Experimental protocol/design	Split parcels
Replicates	4
Elemental parcel (m^2) - raw (number) - raw (lenght in m)	11.00 (4 × 2.75) 6 4
Linear meters for parcel	24
Cloves for elemental parcel	240
Total cloves for each variety (number)	1920
Total bulbs for each variety (number)	240
Implant date	26–27/11
Planting	The cloves must be buried 4–5 cm deep
Basal dressing (kg/ha)	250 (potassium sulphate) 250 (DAP 18/46)
Top dressing (kg/ha) - in the first decade of March - in the second decade of April	150 (ammonium nitrate) 250 (ammonium nitrate)
Irrigation (400 m^3/intervention)	no irrigation was carried out
Antiparasitic treatment (number)	1 (melody compact -Iprovalicarb + oxychloride-)
Antiparasitic inspection (number)	6 (copper-based products -bordeaux mixture and copper oxychloride- in spring)
Pest check (L/ha)	3—Stomp—Pendimetalin based in pre-emergency 1.5—Setossidim in post-emergency
Fertilisation Unit for proposed technique	143 N 115 P 125 K 44 S 3 Mg
Fertilisation Unit for conventional technique	97 N 115 P 44 S

The biometric parameters of 12 samples for each ecotype and for each cultivation area were examined for a total of 96 bulbs. Samples, once in the lab, were weighed, measured, and deprived of the outer skin; then, individual cloves, evaluated randomly, were counted, weighed, and peeled. For each ecotype and for every cultivated area, a representative sample of garlic was prepared; each pool, then, was homogenised and analysed for the moisture content, frozen at −30 °C and then lyophilised, for subsequent analyses.

The termo-pluviometric parameters during the whole garlic production were shown in Figure 2.

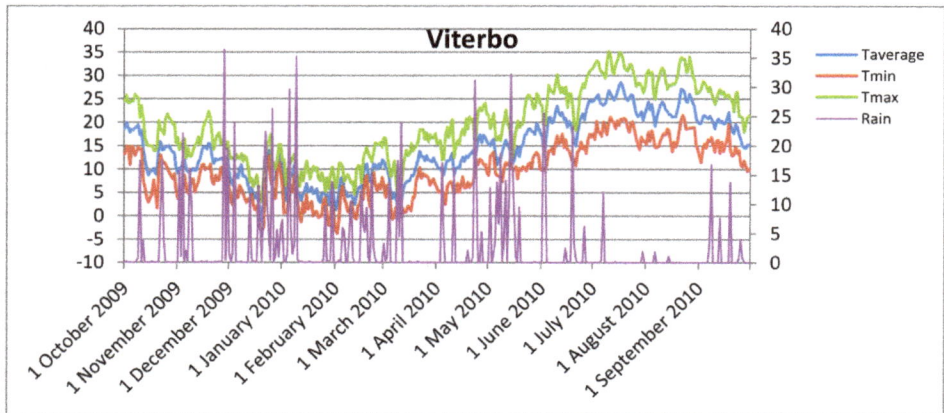

(a)

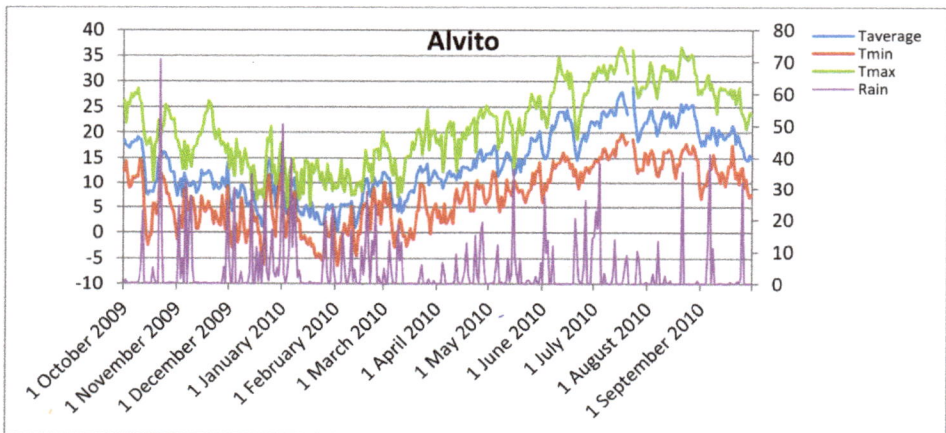

(b)

Figure 2. The termo-pluviometric parameters for the two geographical areas of (**a**) Viterbo and (**b**) Alvito, monitored during the year of production.

2.2. Chemicals and Standards

Reagents and standards (fructose, glucose, and sucrose) were purchased from Sigma-Aldrich Srl (Milan, Italy), Extrasynthese (Genay, France), Carlo Erba (Milan, Italy), J.T. Baker (Deventer, Holland) and BDH Laboratory Supplies (Poole, UK), and were of the analytical grade purity. Double-distilled water (Millipore, Milan, Italy), purified with a Milli-QTM system, was used throughout the study.

2.3. Analyses

All analyses were carried out in triplicate according to official methods [46], and the results were expressed as mean ± SD.

2.3.1. Available Carbohydrates

Starch

Total starch was analysed using the Megazyme Total Starch Assay kit (AOAC Method 996.11 and AACC Method 76.13; Megazyme International Ireland Ltd., Wicklow, Ireland), based on the enzymatic hydrolysis method described by McClearly et al. [47]. Samples were pre-extracted with 80% of ethanol to remove glucose; the complete starch solubilisation was achieved by heating up the samples in presence of thermostable alpha-amylase, followed by hydrolysis to glucose with amyloglucosidase. Maltodextrins were hydrolysed to glucose with glucoamylase and the glucose produced was oxidised to d-gluconate with the release of one mole of hydrogen peroxide (H_2O_2); this was quantitatively measured in a colourimetric reaction employing peroxidase and the production of a quinoneimine dye. The starch content was finally determined as liberated glucose chemically transformed to exhibit absorbance at 510 nm. The analysis procedure was allowed by standard control: regular maize starch.

Soluble Sugars

Soluble sugars (glucose, fructose, and sucrose) were determined by anion exchange chromatography technique using a DIONEX (ion chromatograpy system, mod ICS 5000, Sunnyvale, CA, USA) equipped with GP50 gradient pump, amperometric detector (HPAE-PAD) and column Carbopac PA1 (250 × 4 mm) (Dionex corporation, Sunnyvale, CA, USA); the mobile phase was 160 mm NaOH, and the flow rate was kept constant at 1.0 mL/min [48,49]. Sugars were extracted from garlic using distilled water (20 min in sonicator). Certified material LGC-7103 was analysed as the control of accuracy of the analysis of soluble sugars.

2.3.2. Unavailable Carbohydrates

Fructans (FOS)

The fructan content was determined using the Megazyme Fructan HK Assay kit (AOAC Method 999.03 and AACC Method 32.32; Megazyme International Ireland Ltd., Wicklow, Ireland), based on the enzymatic hydrolysis method described by McClearly and Blakeney [50]. The samples were treated with hot water to dissolve the fructan components. Aliquots, instead, were treated with specific enzymes to hydrolyse the sucrose to glucose, and the fructose and the starch to glucose: sucrose was hydrolysed by a specific sucrase enzyme that has no action on the lower degree of polymerisation [51]. Starch and maltodextrins were hydrolysed to maltose and maltotriose by pullulanase and β-amylase, and these oligosaccharides were then hydrolysed to d-glucose by maltase.

Then, all of the reducing sugars were reduced to sugar alcohols using a treatment based on alkaline borohydride. After this procedure, fructans were hydrolysed to fructose and glucose ultrapure exo-inulinase and endoinulinase [51], then were measured by para-hydroxybenzoic acid hydrazide method for reducing sugars [52]. The analysis procedure was allowed by standard control: dahlia fructan freeze-dried in the presence of α-cellulose.

Total Dietary Fibre (TDF)

The total dietary fibre was evaluated using the enzymatic–gravimetric method described by AOAC 985.29 [47]. Duplicates of freeze-dried samples were treated with a sequential enzymatic digestion by heat stable (α-amylase, protease, and amyloglucosidase) to remove starch and protein. Samples were heated at 100 °C with heat stable α-amylase to give gelatinisation, hydrolysis, and depolymerisation of starch; then, samples were incubated at 60 °C with protease to solubilise and depolymerise proteins, and with amyloglucosidase to hydrolyse starch fragments to glucose. Ethanol was added to precipitate the soluble dietary fibre; the residue was then filtered, washed with ethanol, dried, and weighed. One duplicate was analysed for protein, and the other was incubated at 550 °C to determine ash. Total dietary fibre was calculated as a weight of the residue less the weight of the protein and ash.

2.4. Quality Assurance Procedures

The quality control of analytical procedures was performed through a precise and accurate execution of the methods. Appropriate reagents, equipment, and suitable tests (e.g., system suitability testing) were used during the analysis to check the analytic reproducibility and the validity of the result; the standard deviation (SD) of all of the analytical values was calculated; the accuracy of the procedure was established by the analysis of certified material.

2.5. Statistical Analysis

All the analyses were performed in triplicate, and data were expressed as means $\pm$ standard deviation (SD). The statistical analysis was performed using Statistica for Windows (Statistical package; release 4.5; StatSoft Inc., Vigonza, PD, Italy). One-way Analysis of Variance (ANOVA) was done to determine the statistical significance. A *p*-value of less than 0.05 was taken as significant. The Student's *t*-test was used to compare the cultivation areas for each cultivar.

3. Results and Discussion

An analysis of the biometric parameters was carried out on each variety in order to define the morphological differences of local landraces and possible differences due to the cultivation area. The results showed significant differences between the average weights of bulbs from the two geographic areas, except for the Rosso Proceno. The average weights of samples from Alvito (40 g $\pm$ 6.2) were mostly higher than those of samples grown in Viterbo (33 g $\pm$ 2.9); Rosso di Castelliri landrace showed the largest bulbs (weight 47.1 g and size 5.7 cm), and of those grown in Alvito, the highest values were for cloves weight, as shown in Table 3. The number of cloves in a garlic bulb resulting from both cultivar and cultivation area ranged from 9 to 16. The weights of bulb and cloves are similar to the those reported by Fanaei et al. [53], who analysed different garlic genotypes in the range of 24–50 g and 2.93–5.2 g, respectively; the same authors also observed a number of cloves for bulb from 6 to 11. These variations might be due to the genetic variations among garlic cultivars and their ability for exploiting the environmental sources, particularly light, CO_2, water and nutrients as underlined by several authors [53,54], as well as cultivation conditions [55,56]. Moreover, Dhakulkar et al. [55] studied the effect of spacing and clove size on the yield and quality of garlic, and reported that the weight of bulbs varied from 12.05 to 15.66 g, and the number of cloves per bulb varied from 16.70 to 21.50. A recent study of Sachin et al. [56] marked how the application of organic and inorganic sources of nitrogen significantly affected the quality attributes of garlic.

Table 3. Biometric parameters of four garlic ecotypes.

		Bulbs			Cloves		
		Horizontal Section (cm)	Weight (g)	Waste (%)	Number n°	Weight (g)	Waste (%)
Bianco Piacentino	Viterbo	4.0–5.0	26.6–50.9	5–11	12–21	1.8–4.1	5–11
	Alvito	4.0–6.0	25.7–62.0	1–3	7–20	1.9–5.6	1–9
Rosso di Sulmona	Viterbo	3.8–6.3	17.2–47.7	5–8	6–13	2.0–5.0	3–12
	Alvito	4.5–6.0	26.8–54.3	2–4	10–15	2.6–3.7	2–3
Rosso di Castelliri	Viterbo	4.0–5.5	25.3–38.6	3–12	7–11	3.1–6.3	3–10
	Alvito	5.1–6.4	33.8–62.0	2–3	9–15	3.5–7.3	2–7
Rosso di Proceno	Viterbo	3.9–5.5	22.7–42.5	3–15	6–10	2.7–4.7	5–9
	Alvito	3.5–5.7	18.0–47.6	2–4	7–15	2.0–5.9	2–7

Rosso di	Viterbo	3.8–6.3	17.2–47.7	5–8	6–13	2.0–5.0	3–12
Sulmona	Alvito	4.5–6.0	26.8–54.3	2–4	10–15	2.6–3.7	2–3
Rosso di	Viterbo	4.0–5.5	25.3–38.6	3–12	7–11	3.1–6.3	3–10
Castelliri	Alvito	5.1–6.4	33.8–62.0	2–3	9–15	3.5–7.3	2–7
Rosso di	Viterbo	3.9–5.5	22.7–42.5	3–15	6–10	2.7–4.7	5–9
Proceno	Alvito	3.5–5.7	18.0–47.6	2–4	7–15	2.0–5.9	2–7

The total carbohydrate content in the four local landraces grown in the two selected production areas and analysed in this study showed values ranging from 22.8 to 26.2 g/100 g of fresh edible portion. The percentage distribution of the different carbohydrate fractions was reported in Figure 3: on average, 78% were represented by fructans, 18% by dietary fibre, and 4% by total soluble sugars, with 0% by starch content.

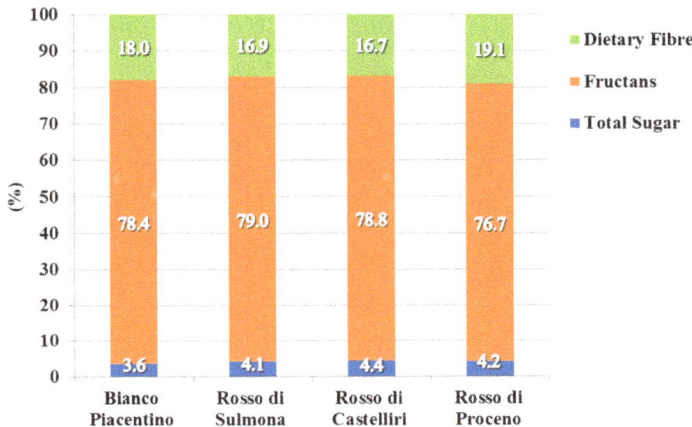

Figure 3. Percentage distribution of the different carbohydrate fractions in four garlic landraces.

The characterisation of the individual soluble sugars of four Italian local landraces cultivated at two locations was shown in Figure 4, where glucose, fructose, and sucrose contents were reported. It appeared clear that the major fraction was represented by sucrose, which comprised 94% of total sugars; furthermore, for each sugar, cultivars grown in Viterbo showed slightly higher values than those of samples coming from Alvito. However, the differences found were not statistically significant.

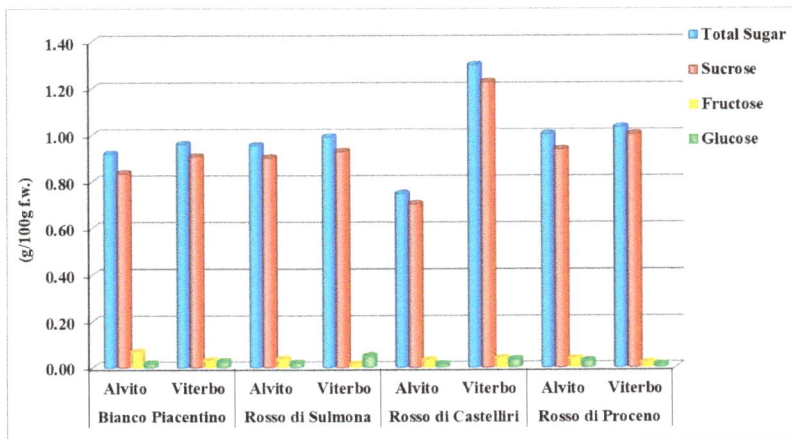

Figure 4. Total and individual sugar content in four garlic landraces.

The results obtained from the measurements of the total carbohydrate components (mean and standard deviation) present in each of the four local garlic landraces were reported in Table 4, taking into account both the variety and the cultivation area.

Table 4. Available and unavailable carbohydrates (g/100 g d.w.) in four garlic landraces from the

The results obtained from the measurements of the total carbohydrate components (mean and standard deviation) present in each of the four local garlic landraces were reported in Table 4, taking into account both the variety and the cultivation area.

Table 4. Available and unavailable carbohydrates (g/100 g d.w.) in four garlic landraces from the Viterbo and Alvito areas *.

Carbohydrates		Area	Bianco Piacentino	Rosso di Sulmona	Rosso di Castelliri	Rosso di Proceno
Available	Starch	Viterbo	tr	tr	tr	tr
		Alvito	tr	tr	tr	tr
	Total Sugars	Viterbo	2.42 ± 0.20 [a]	2.53 ± 0.20 [a]	3.27 ± 0.43 [b]	2.68 ± 0.34 [ab]
		Alvito	2.66 ± 0.29 [b]	2.47 ± 0.25 [ab]	2.12 ± 0.11 [a]	2.49 ± 0.18 [ab]
	p-value [§]		n.s.	n.s.	*p* < 0.05	n.s.
Unavailable	Fructans	Viterbo	54.4 ± 11.19	49.7 ± 8.40	45.8 ± 3.70	49.6 ± 2.80
		Alvito	51.2 ± 4.50	47.2 ± 6.10	49.5 ± 4.10	46.3 ± 2.80
	p-value [§]		n.s.	n.s.	*p* < 0.05	n.s.
	Total dietary Fibre	Viterbo	13.1 ± 1.50 [c]	10.5 ± 0.40 [ab]	9.1 ± 0.90 [a]	12.1 ± 0.30 [bc]
		Alvito	11.7 ± 2.00	10.0 ± 1.40	11.4 ± 1.20	9.8 ± 1.20
	p-value [§]		n.s.	n.s	n.s	*p* < 0.05

* Data are expressed as Mean ± Standard Deviation. Anova, Tukey's honestly significant difference (HSD) Test: by row, means followed by different letters are significantly different (*p* < 0.05). [§] Student's *t*-test; n.s. = not significant.

This includes available carbohydrates (starch and total sugars) and unavailable carbohydrates (fructans and dietary fibre). The values of available carbohydrates, when necessary to know the energy intake, were calculated as the sum of total soluble sugars and starch, and expressed as monosaccharides (conversion factor for starch = 1.1) [40].

Starch was present, but in very low contents; its values were detected, indeed, but identified only in traces (amount < 0.06 g/100 g f.w.) according to Darbyshire and Henry [57]. Although starch is generally the most widespread carbohydrate reserve in the plant kingdom, a storage function is attributed to the fructans present in the bulb tissues; these compounds may have some advantages compared with starch since they are resistant to cold, while the starch biosynthesis dramatically decreases when the temperature drops below 10 °C [58].

Table 4 shows fructan contents to be the most representative component in all the garlic samples, comprising about 78% of the total carbohydrates, and varying from 45.8 to 54.4 g/100 g d.w. These data were in agreement with values reported by Muir et al. [32], but higher than those reported by Peshev and Van den Ende [59], where the dietary fibre content accounted for 16–19% of total carbohydrates.

Statistical analysis revealed that the fructan contents were never influenced by cultivar. No significant differences were observed regarding the effect caused by the cultivation area. Although samples from Viterbo exhibited higher values, only Rosso di Castelliri garlic showed a fructan content higher in samples coming from Alvito than in those grown in Viterbo. Ritsema and Smeekens [60] described how changes in fructan accumulation appear to have a role in the tolerance of some plant species exposed to environmental stresses; moreover, Van den Ende et al. [61] underline that the fructan content seems to be involved in osmoregulation in plants exposed to water restriction or low temperatures during certain periods of the year.

Regarding fructan, it was remarkable to consider that the Megazyme Assay Kit, which was used to quantify the total levels of fructan, does not provide information on the percentage of FOS and inulin, as reported by studies using other methods [32].

All of the samples were found to be a good source of total dietary fibre (Table 4), which ranged from 9.1 to 13.1 g/100 g d.w.; in particular, in bulbs produced in Viterbo, the dietary fibre content varied among cultivars, whereas no significant difference was found between the bulbs from Alvito. The cultivation area did not affect the dietary fibre content except for Rosso di Proceno ($p < 0.05$). The dietary fibre content was higher than the average values observed in Italian commercial samples (about 7 g/100 g d.w.) present in CREA's Food Composition Database [62], and those reported in USDA [63] (3.4/100 g d.w.); moreover, they are in line with values reported by Haciseferogullary et al. [64] for raw garlic grown in Turkey. These results highlight the added value of these Italian local garlic landraces.

Total sugars contents showed variability by cultivar: total sugars ranged from 2.42 to 3.27 g/100 g d.w. in Viterbo samples, and from 2.12 to 2.66 g/100 g d.w. in those coming from Alvito, which are values that are comparable with the average data present in the United States Department of Agriculture (USDA) Nutrient Database [63]. A significant difference ($p < 0.05$) was observed for the total sugar content in Rosso di Castelliri between Viterbo and Alvito samples.

In Table 5, the distribution of the sugar profile (glucose, fructose, sucrose) of the four local garlic landraces from two growing locations is reported.

Table 5. Profiles of total sugar (glucose, fructose, and sucrose) (g/100 g d.w.) in four garlic local landraces from the Viterbo and Alvito cultivation areas *.

	Area	Bianco Piacentino	Rosso di Sulmona	Rosso di Castelliri	Rosso di Proceno
Glucose	Viterbo	0.06 ± 0.01 [b]	0.13 ± 0.02 [d]	0.09 ± 0.01 [c]	0.03 ± 0.00 [a]
	Alvito	0.05 ± 0.01 [a]	0.05 ± 0.01 [a]	0.04 ± 0.01 [a]	0.07 ± 0.01 [b]
	p-value [§]	n.s.	$p < 0.01$	$p < 0.01$	$p < 0.01$
Fructose	Viterbo	0.07 ± 0.01 [b]	0.03 ± 0.00 [a]	0.10 ± 0.02 [c]	0.05 ± 0.01 [ab]
	Alvito	0.20 ± 0.03 [b]	0.09 ± 0.02 [a]	0.09 ± 0.01 [a]	0.09 ± 0.01 [a]
	p-value [§]	$p < 0.01$	$p < 0.05$	n.s.	$p < 0.01$
Sucrose	Viterbo	2.18 ± 0.24 [a]	2.37 ± 0.20 [a]	3.08 ± 0.42 [b]	2.60 ± 0.34 [ab]
	Alvito	2.42 ± 0.26	2.34 ± 0.26	1.99 ± 0.10	2.33 ± 0.18
	p-value [§]	n.s.	n.s.	$p < 0.05$	n.s.

* Data are expressed as Mean ± Standard Deviation. Anova, Tukey's honestly significant difference (HSD) Test: by row, means followed by different letters are significantly different ($p < 0.05$); [§] Student's *t*-test; n.s. = not significant.

In all of the samples, total sugars were represented almost exclusively by sucrose; its concentration (Table 5) showed the lowest (1.99 g/100 g d.w.) and the highest values (3.08 g/100 g d.w.) in the Rosso Castelliri cultivar, coming respectively from Alvito and Viterbo. A similar trend was reported by Cardelle-Cobas et al. [65] for dehydrated commercial samples. In most studies on garlic, the content of sugars was not reported, and the lack of data mainly concerns sucrose.

Glucose and fructose were found at lower quantities than those found for sucrose, with a range of 0.03–0.13 g/100 g d.w. and 0.03–0.20 g/100 g d.w., respectively. In garlic grown in Viterbo, the amount of glucose content varied in the four local landraces, as shown in the following decreasing order: Rosso Sulmona > Rosso di Castelliri > Bianco Piacentino > Rosso di Proceno. However, among the samples from Alvito, the Rosso di Proceno variety achieved the highest values. Fructose content in Bianco Piacentino bulbs was higher in both growing locations.

Moreover, as shown in Table 5, glucose and fructose were greatly affected by the cultivation area for all cultivars, except the Bianco Piacentino for glucose and the Rosso di Castelliri for fructose. On the contrary, sucrose content did not appear to be influenced by the cultivation area, except in the Rosso di Castelliri cultivar ($p < 0.05$).

4. Conclusions

Our findings have revealed that even the carbohydrates of the four Italian varieties chosen for the study can be influenced, like other nutrients, by the genotype and the cultivation area, and that their individual components are differently affected by these factors. These traditional varieties can be considered a good source of dietary fibre and fructans, especially the Bianco Piacentino ecotype grown in Viterbo. Fructans, in particular, are a good opportunity to add value to the production of foods both in terms of innovation—combined with the protection of consumers' interest—and profitability for the food industry.

This research provides evidence for the importance of studying the interaction between genotype, environmental conditions, and nutrient composition, in order to define the specific properties of Italian garlic local landraces and their total quality.

Since the four garlic ecotypes are registered in the list of traditional Italian agri-food products published by the Italian Ministry of Agricultural, Food, and Forestry Policies (MiPAAF) [66] in 2016, our results wish to contribute to raising knowledge and awareness about the quality of local products—some of which are at risk of erosion—by means of highlighting their nutritional characteristics, as well as contributing to their rural development and the protection of biodiversity.

Acknowledgments: This research was supported by the Ministry of Agricultural, Food and Forestry Policies—"BIOVITA" Project. The authors thank Annalisa Lista for the linguistic revision and editing of manuscript.

Author Contributions: Luisa Marletta and Olindo Temperini conceived and designed the experiments; Silvia Lisciani, Loretta Gambelli, Cecilia Rossetti, Paolo Gabrielli, Altero Aguzzi performed the experiments; Silvia Lisciani, Alessandra Durazzo, Loretta Gambelli, Stefania Marconi and Emanuela Camilli analyzed the data; Silvia Lisciani, Loretta Gambelli, Cecilia Rossetti, Paolo Gabrielli, Altero Aguzzi contributed reagents/materials/analysis tools; Silvia Lisciani, Loretta Gambelli, Alessandra Durazzo and Luisa Marletta wrote the paper.

Conflicts of Interest: The authors declare no conflict of interest.

References

1. Angiosperm Phylogeny Group (APG). An update of the Angiosperm Phylogeny Group classification for the orders and families of flowering plants: AGP III. *Biol. J. Linn. Soc. Lond.* **2009**, *161*, 105–121.

2. Food and Agriculture Organization of the United Nations (FAO). *The State of Food and Agriculture. Agricultural Trade and Poverty*; FAO: Rome, Italy, 2005.

3. ISMEA. Il Mercato Dell'aglio Tendenze Recenti e Dinamiche Attese. 2014. Available online: www.ismeamercati.it/flex/cm/pages/ServeBLOB.php/L/IT/IDPagina/3977 (accessed on 18 June 2017).

4. Brandolini, V.; Tedeschi, P.; Cereti, E.; Maietti, A.; Barile, D.; Coisson, J.D.; Mazzotta, D.; Arlorio, M.; Martelli, A. Chemical and genomic combined approach applied to the characterization and identification of Italian *Allium savitum* L. *J. Agric. Food Chem.* **2005**, *53*, 7–14. [CrossRef] [PubMed]

5. Ali, M.; Thomson, M.; Afzal, M. Garlic and onions: Their effect on eicosanoid metabolism and its clinical relevance. *Prostaglandins Leukot. Essent. Fatty Acids* **2000**, *62*, 55–73. [CrossRef] [PubMed]

6. Wilson, E.A.; Demmig-Adams, B. Antioxidant, anti-inflammatory, and antimicrobial properties of garlic and onions. *Nutr. Food Sci.* **2007**, *37*, 178–183. [CrossRef]

7. Corzo-Martinez, M.; Corzo, N.; Villaiel, M. Biological properties of onions and garlic. *Trends Food Sci. Technol.* **2007**, *18*, 609–625. [CrossRef]

8. Choque Delgado, G.T.; Tamashiro, W.M.S.C.; Pastore, G.M. Immunomodulatory effects of fructans. *Food Res. Int.* **2010**, *43*, 1231–1236. [CrossRef]

9. Zhou, X.F.; Ding, Z.S.; Liu, N.B. Allium vegetables and risk of prostate cancer: Evidence from 132,192 subjects. *Asian Pac. J. Cancer Prev.* **2013**, *14*, 4131–4134. [CrossRef] [PubMed]

10. Nair, S.S.; Gaikwad, S.S.; Kulkarni, S.P.; Mukne, A.P. Allium sativum constituents exhibit anti-tubercular activity in vitro and in RAW 264.7 mouse macrophage cells infected with *Mycobacterium tuberculosis* H37Rv. *Pharmacogn. Mag.* **2017**, *13*, S209–S215. [PubMed]

11. Nasiri, A.; Ziamajidi, N.; Abbasalipourkabir, R.; Goodarzi, M.T.; Saidijam, M.; Behrouj, H.; SolemaniAsl, S. Beneficial effect of aqueous garlic extract on inflammation and oxidative stress status in the kidneys of type 1 diabetic rats. *Indian J. Clin. Biochem.* **2017**, *32*, 329–336. [CrossRef] [PubMed]

12. Ziamajidi, N.; Behrouj, H.; Abbasalipourkabir, R.; Lotfi, F. Ameliorative effects of *Allium sativum* extract on iNOS gene expression and NO production in liver of Streptozotocin + Nicotinamide-induced diabetic rats. *Indian J. Clin. Biochem.* **2017**, 1–7. [CrossRef]

13. Banerjee, S.K.; Maulik, S.K. Effect of garlic on cardiovascular disorder—A review. *Nutr. J.* **2002**, *1*, 4–14. [CrossRef] [PubMed]

14. Peinado, M.J.; Ruiz, R.; Echavarii, A.; Aranda-Olmedo, I.; Rubio, L.A. Garlic derivative PTS-O modulates intestinal microbiota composition and improves digestibility in growing broiler chickens. *Anim. Feed Sci. Technol.* **2013**, *181*, 87–92. [CrossRef]

15. Amagase, H.; Petesch, B.L.; Matsuura, H.; Kasuga, S.; Itakura, Y. Intake of garlic and its bioactive components. *J. Nutr.* **2001**, *131*, 955–962.

16. Banerjee, S.K.; Mukherjee, P.K.; Maulik, S.K. Garlic as an antioxidant: The good, the bad and the ugly. *Phytother. Res.* **2003**, *17*, 97–106. [CrossRef] [PubMed]

17. Santhosha, S.G.; Jamuna, P.; Prabhavathin, S.N. Bioactive components of garlic and their physiological role in health maintenance: A review. *Food Biosci.* **2013**, *3*, 59–74. [CrossRef]

18. Diretto, G.; Rubio-Moraga, A.; Argandoña, J.; Castillo, P.; Gómez-Gómez, L.; Ahrazem, O. Tissue-specific accumulation of sulfur compounds and saponins in different parts of garlic cloves from purple and white ecotypes. *Molecules* **2017**, *22*, 1359. [CrossRef] [PubMed]

19. González, R.; Soto, V.; Sance, M.; Camargo, A.; Galmarini, C.R. Variability of solids, organosulfur compounds, pungency and health-enhancing traits in garlic (*Allium sativum* L.) cultivars belonging to different ecophysiological groups. *J. Agric. Food Chem.* **2009**, *57*, 10282–10288. [CrossRef] [PubMed]

20. Touloupakis, E.; Ghanotakis, D.F. Nutraceutical use of garlic sulfur-containing compounds. *Adv. Exp. Med. Biol.* **2010**, *698*, 110–121. [PubMed]

21. Lee, S.H.; Liu, Y.T.; Chen, K.M.; Lii, C.K.; Liu, C.T. Effect of garlic sulfur compounds on neutrophil infiltration and damage to the intestinal mucosa by endotoxin in rats. *Food Chem. Toxicol.* **2012**, *50*, 567–574. [CrossRef] [PubMed]

22. Liang, D.; Wang, C.; Tocmo, R.; Wu, H.; Deng, L.-W.; Huang, D. Hydrogen sulphide (H$_2$S) releasing capacity of essential oils isolated from organosulphur rich fruits and vegetables. *J. Funct. Foods* **2015**, *14*, 634–640. [CrossRef]

23. Koruri, S.S.; Banerjee, D.; Chowdhury, R.; Bhattacharya, P. Studies on prebiotic food additive (inulin) in Indian dietary fibre sources—Garlic (*Allium sativum*), wheat (*Triticum* spp.), oat (*Avena sativa*) and dalia (Bulgur). *Int. J. Pharm. Sci.* **2014**, *6*, 278–282.

24. Vijn, I.; Smeekens, S. Fructan: More than a reserve carbohydrate. *Plant Physiol.* **1999**, *120*, 351–359. [CrossRef] [PubMed]

25. Gupta, A.K.; Kaur, N. *Carbohydrate Reserves in Plants—Synthesis and Regulation*, 1st ed.; Elsevier Science: Amsterdam, The Netherlands, 2000; ISBN 0444502696.

26. De Oliveira, A.J.B.; Gonçalves, R.A.C.; Chierrito, T.P.C.; dos Santos, M.M.; de Souza, M.S.; Gorin, P.A.J.; Sassak, G.L.I.; Iacomini, M. Structure and degree of polymerization of fructooligosaccharides present in roots and leaves of *Stevia rebaudiana* (Bert.) Bertoni. *Food Chem.* **2011**, *129*, 305–311. [CrossRef]

27. Hendry, G.A.F.; Wallace, R.K. The origin, distribution, and evolutionary significance of fructans. In *Science and Technology of Fructans*; Suzuki, M., Chatterton, N.J., Eds.; CRC Press: Boca Raton, FL, USA, 1993; pp. 119–139.

28. Van den Ende, W. Multifunctional fructans and raffinose family oligosaccharides. *Front. Plant Sci.* **2013**, *4*, 247. [PubMed]

29. Orcutt, D.M.; Nilsen, E.T. *The Physiology of Plants under Stress: Soil and Biotic Factors*; John Wiley & Sons: New York, NY, USA, 2000.

30. Lewis, D.H. Nomenclature and diagrammatic representation of oligomeric fructans—A paper for discussion. *New Phytol.* **1993**, *124*, 583–594. [CrossRef]

31. Baumgartner, S.; Dax, T.G.; Praznik, W.; Falk, H. Characterisation of the high-molecular weight fructan isolated from garlic (*Allium sativum* L.). *Carbohydr. Res.* **2000**, *328*, 177–183. [CrossRef]

32. Muir, J.G.; Shepherd, S.J.; Rosella, O.; Rose, R.; Barrett, J.Q.; Gibson, P.R. Fructan and free fructose content of common Australian vegetables and fruit. *J. Agric. Food Chem.* **2007**, *55*, 6619–6627. [CrossRef] [PubMed]

33. Singh, R.S.; Singh, R.P.; Kennedy, J.F. Recent insights in enzymatic synthesis of fructooligo-saccharides from inulin. *Int. J. Biol. Macromol.* **2016**, *85*, 565–572. [CrossRef] [PubMed]

34. Shoaib, M.; Shehzad, A.; Omar, M.; Rakha, A.; Raza, H.; Sharif, H.R.; Niazi, S. Inulin: Properties, health benefits and food applications. *Carbohydr. Polym.* **2016**, *147*, 444–454. [CrossRef] [PubMed]

35. Roberfroid, M.B. Introducing inulin-type fructans. *Br. J. Nutr.* **2005**, *93*, S13–S25. [CrossRef] [PubMed]

36. Roberfroid, M.B. Prebiotics: The concept revisited. *J. Nutr.* **2007**, *137*, 830S–837S. [PubMed]
37. Seifert, S.; Watzl, B. Inulin and oligofructose: review of experimental data on immune modulation 1–4. *J. Nutr.* **2007**, *137*, 2563–2567.
38. Roberfroid, M.; Gibson, G.R.; Hoyles, L.; McCartney, A.L.; Rastall, R.; Rowland, I.; Wolvers, D.; Watztl, B.; Szajewska, H.; Stahl, B.; et al. Prebiotic effects: Metabolic and health benefits. *Br. J. Nutr.* **2010**, *104*, S1–S63. [CrossRef] [PubMed]
39. Zhang, N.; Huang, X.; Zeng, Y.; Wu, X.; Peng, X. Study on prebiotic effectiveness of neutral garlic fructan in vitro. *Food Sci. Hum. Wellness* **2013**, *2*, 119–123. [CrossRef]
40. Slavin, J. Why whole grains are protective: Biological mechanisms. *Proc. Nutr. Soc.* **2003**, *62*, 129–134. [CrossRef] [PubMed]
41. Leenen, C.H.M.; Dieleman, L.A. Inulin and oligofructose in chronic inflammatory bowel disease. *J. Nutr.* **2007**, *137*, 2572–2575.
42. Fedewa, A.; Rao, S.S.C. Dietary fructose intolerance, fructan intolerance and FODMAPs. *Curr. Gastroenterol. Rep.* **2014**, *16*, 370. [CrossRef] [PubMed]
43. Azzini, E.; Durazzo, A.; Foddai, M.S.; Temperini, O.; Venneria, E.; Valentini, S.; Maiani, G. Phytochemicals content in Italian garlic bulb (*Allium sativum* L.) varieties. *J. Food Res.* **2014**, *3*, 26–31.
44. Carnovale, E.; Marletta, L. *Tabelle di Composizione Degli Alimenti—Aggiornamento 2000 Inran*; EDRA: Milan, Italy, 2000.
45. Marletta, L.; Camilli, E. *Biodiversità e Agroalimentare: Strumenti per Descrivere la Realtà Italiana*; Banca Dati BIOVITA; Casa Editrice CRA-ex INRAN: Rome, Italy, 2013.
46. Greenfield, H.; Southgate, D.A.T. *Food Composition Data: Production, Management, and Use*, 2nd ed.; FAO: Rome, Italy, 2003.
47. McCleary, B.V.; Gibson, T.S.; Mugford, D.C. Measurement of total starch in cereal products by amyloglucosidase—α-Amylase method: Collaborative study. *J. AOAC Int.* **1997**, *80*, 571–579.
48. Lee, Y.C. Carbohydrate Analyses with High-Performance Anion-Exchange Chromatography. *J. Chromatogr. A* **1996**, *720*, 137–149. [CrossRef]
49. Ruggeri, S.; Cappelloni, M.; Gambelli, L.; Nicoli, S.; Carnovale, E. Chemical composition and nutritive value of nuts grown in Italy. *Ital. J. Food Sci.* **1998**, *10*, 243–252.
50. McCleary, B.V.; Blakeney, A.B. Measurement of inulin and oligofructan. *Cereal Foods World* **1999**, *44*, 398–406.
51. McCleary, B.V. Measuring Dietary Fibre. *World Ingred.* **1999**, 50–53.
52. Lever, M. Colorimetric and fluorimetric carbohydrate determination with *p*-hydroxybenzoic acid hydrazide. *Biochem. Med.* **1973**, *7*, 247–281. [CrossRef]
53. Fanaei, H.; Narouirad, M.; Farzanjo, M.; Ghasemi, M. Evaluation of Yield and Some Agronomical Traits in Garlic Genotypes (*Allium sativum* L.). *Annu. Res. Rev. Biol.* **2014**, *4*, 3386–3391. [CrossRef]
54. Noorbakhshian, S.J.; Mousavi, S.A.; Bagheri, H.R. Evaluation of agronomic traits and path coefficient analysis of yield for garlic cultivars. *Pajouhesh Sazandegi* **2008**, *77*, 10–18.
55. Dhakulkar, N.D.; Ghawade, S.M.; Dalal, S.R. Effect of spacing and clove size on growth and yield of garlic under Akola conditions. *Int. J. Chem. Stud.* **2017**, *5*, 559–562.
56. Sachin, A.J.; Bhalerao, P.P.; Patil, S.J. Effect of organic and inorganic sources of nitrogen on growth and yield of garlic (*Allium sativum* L.) var. GG-4. *Int. J. Chem. Stud.* **2017**, *5*, 559–562.
57. Darbyshire, B.; Henry, R.J. Differences in fructan content and synthesis in some allium species. *New Phytol.* **1981**, *87*, 249–256. [CrossRef]
58. Pollock, C.J.; Jones, T. Seasonal patterns of fructan metabolism in forage grasses. *New Phytol.* **1979**, *83*, 9–15. [CrossRef]
59. Peshev, D.; Van den Ende, W. Fructans: Prebiotics and immunomodulators. *J. Funct. Foods* **2014**, *8*, 348–357. [CrossRef]
60. Ritsema, T.; Smeekens, S. Fructans: Beneficial for plants and humans. *Curr. Opin. Plant Biol.* **2003**, *6*, 223–230. [CrossRef]
61. Van den Ende, W.; De Coninck, B.; Van Laere, A. Plant fructanexohydrolases: A role in signaling and defense? *Trends Plant Sci.* **2004**, *9*, 523–528. [CrossRef] [PubMed]
62. Marletta, L.; Camilli, E. Aggiornamento 2008 Food Composition Database of CREA. Available online: http://nut.entecra.it/646/Tabelle_di_composizione_degli_alimenti.html (accessed on 5 June 2017).

63. U.S. Department of Agriculture National Agricultural Research Service. *USDA Food Composition Database for Standard Reference*; Software developed by the National Agricultural Library: Beltsville, MD, USA, 2016.

64. Hacıseferoğulları, H.; Özcanb, M.; Demira, F.; Çalışıra, S. Some nutritional and technological properties of garlic (*Allium sativum* L.). *J. Food Eng.* **2005**, *68*, 463–469. [CrossRef]

65. Cardelle-Cobas, A.; Costo, R.; Corzo, N.; Villamiel, M. Fructo-oligosaccharide changes during the storage of dehydrated commercial garlic and onion samples. *Int. J. Food Sci. Technol.* **2009**, *44*, 947–952. [CrossRef]

66. Ministero Politiche Agricole Alimentari e Forestali (MiPAAF). Prodotti Agroalimentari Tradizionali. 2016. 16° Revisione. Available online: https://www.politicheagricole.it/flex/cm/pages/ServeBLOB.php/L/IT/IDPagina/10241 (accessed on 10 June 2017).

sustainability

MDPI

Communication

Alternative Use of Extracts of Chipilín Leaves (*Crotalaria longirostrata* Hook. & Arn) as Antimicrobial

Johana Miranda-Granados [1], Cesar Chacón [1], Nancy Ruiz-Lau [2], María Elena Vargas-Díaz [3], L. Gerardo Zepeda [3], Peggy Alvarez-Gutiérrez [2], Rocio Meza-Gordillo [1] and Selene Lagunas-Rivera [2,*

[1] Instituto Tecnológico de Tuxtla Gutiérrez, Carretera Panamericana km. 1080, Tuxtla Gutiérrez 29050, Chiapas, Mexico; jmirandagranados@gmail.com (J.M.-G.); ingecesarfigueroa10@hotmail.com (C.C.); romego71@yahoo.com (R.M.-G.)

[2] CONACyT, Tecnológico Nacional de México/Instituto Tecnológico de Tuxtla Gutiérrez, Carretera Panamericana km. 1080, Tuxtla Gutiérrez 29050, Chiapas, Mexico; nruizla@conacyt.mx (N.R.-L.); pealvarezgu@conacyt.mx (P.A.-G.)

[3] Departamento de Química Orgánica, Escuela Nacional de Ciencias Biológicas, Instituto Politécnico Nacional, Prol. de Carpio y Plan de Ayala, Ciudad de Mexico 11340, Mexico; evargasvd@yahoo.com.mx (M.E.V.-D.); luisgzepeda@gmail.com (L.G.Z.)

* Correspondence: slagunari@conacyt.mx; Tel.: +521-777-1034945

Received: 24 December 2017; Accepted: 18 March 2018; Published: 20 March 2018

Abstract: The genus Crotalaria comprises about 600 species that are distributed throughout the tropics and subtropical regions of the world; they are antagonistic to nematodes in sustainable crop production systems, and have also shown antimicrobial capacity. Chipilín (*C. longirostrata*), which belongs to this genus, is a wild plant that grows in the state of Chiapas (Mexico) and is traditionally is used as food. Its leaves also have medicinal properties and are used as hypnotics and narcotics; however, the plant has received little research attention to date. In the experimental part of this study, dried leaves were macerated by ethanol. The extract obtained was fractionated with ethyl ether, dichloromethane, ethyl acetate, 2-propanone, and water. The extracts were evaluated against three bacteria—namely, *Escherichia coli* (Ec), *Citrobacter freundii* (Cf), and *Staphylococcus epidermidis* (Se)—and three fungi—*Fusarium oxysporum* A. comiteca (FoC), *Fusarium oxysporum* A. tequilana (FoT), and *Fusarium solani* A. comiteca (FSC). During this preliminary study, a statistical analysis of the data showed that there is a significant difference between the control ciprofloxacin (antibacterial), the antifungal activity experiments (water was used as a negative control), and the fractions used. The aqueous fraction (WF) was the most active against FoC, FsC, and FoT (30.65, 20.61, and 27.36% at 96 h, respectively) and the ethyl ether fraction (EEF) was the most active against Se (26.62% at 48 h).

Keywords: traditional food; antimicrobial; bioassay; PIRG; fractions

1. Introduction

The number of plant diseases caused by pests attacking crops has increased the need for new antimicrobials to eliminate the pathogens. This need has led to a renewed focus on natural extracts from plants, fungi, bacteria, algae, etc. [1]. Every year, plant diseases cause an estimated 40 billion dollars in losses worldwide [2]. Chemical fungicides are not readily biodegradable and tend to persist for years in the environment. As a result, the use of natural products for the management of fungal diseases in plants is considered a reasonable substitute for synthetic fungicides [3]. The genus Crotalaria includes around 600 species distributed throughout the tropics and subtropical regions of the world, which have been used as antagonists to nematodes in sustainable crop production systems [4,5].

There are also previous studies showing the anti-inflammatory [6], anthelmintic [7], antitumoral capacity [8] and antimicrobial activity of *C. madurensis* [9] and *C. burhia* [10,11], which showed activity against *Bacillus subtilis* and *Staphylococcus aureus*, while *C. pallida* demonstrated that it has an effect on *Escherichia coli* and *Pseudomonas* sp. [12–14]. The species of this genus contain alkaloids, saponins, and flavonoids to which biological activity is attributed [4]. Chipilín (*Crotalaria longirostrata*) belongs to this genus; it is a wild plant that grows in the state of Chiapas, Mexico that is used traditionally food [15], and also has ethnobotanical properties as hypnotics and narcotics [16]. Since there are few reports of the biological activity of the species *C. longirostrata*, this study fractionates the crude extract from Chipilín (*C. longirostrata*) leaves, obtaining ethyl ether (FEE), dichloromethane (FDM), ethyl acetate (FEA), 2-propanone (FAO), and aqueous fractions (FW), as a preliminary measure in order to evaluate its potential as an antimicrobial.

2. Materials and Methods

2.1. Plant Material

The leaves of *C. longirostrata* were collected in Ocozocoautla, Chiapas, México, geographic location: latitude 16°45′32″ N and longitude 93°21′53″ O.

2.2. Extraction

The plant material were shade-dried for seven days. The dried leaves were grounded to a fine texture, then soaked (0.15 g of dry matter/mL of solvent) in EtOH (96%) (Meyer, CDMex, Mexico) for 15 days. After filtration, the extract was evaporated to obtain the crude extract. About half of the crude extract was suspended in distilled water (H_2O) (Sigma-Aldrich-Merck, Darnstadt, Germany) and separately partitioned with ethyl ether (Et_2O) (Meyer, CDMex, Mexico), followed by dichloromethane (CH_2Cl_2) (Meyer, CDMex, Mexico), ethyl acetate (AcOEt) (Meyer, CDMex, Mexico), and 2-propanone (C_3H_6O) (Meyer, CDMex, Mexico), respectively. The organic layer of each solvent was concentrated to dryness under reduced pressure, and dried over anhydrous sodium sulfate to afford Et_2O, CH_2Cl_2, AcOEt, C_3H_6O, and H_2O fractions. These fractions were stored at 4 °C until use. Each fraction was dissolved in dimethyl sulfoxide (DMSO) (Sigma-Aldrich-Merck, Darnstadt, Germany), and prepared at a concentration of 200 mg/mL in all bioassays [17].

2.3. Cultivation of Microorganism

The microorganisms used were: *Escherichia coli* (Ec) (ITTG-1879), *Citrobacter freundii* (Cf) (Cf-ITTG), and *Staphylococcus epidermidis* (Se) (ITTG-850), which were inoculated on Tryptone-Soya-Agar TSA and incubated at 33 ± 2 °C for 24 h [18]; and *Fusarium oxysporum* A. comiteca (FoC) (FoC-ITTG), *Fusarium oxysporum* A. tequilana (FoT) (FoT-ITTG), and *Fusarium solani* A. comiteca (FSC) (FsC-ITTG), which were inoculated in potato dextrose agar (PDA) at 28 ± 2 °C. For the bioassays, Whatman No. 1 paper discs of 6 mm diameter were placed on the periphery of the Petri dishes [18].

2.4. Evaluation of Antifungal Activity

In order to evaluate the effect by direct contact of the fractions on the microorganisms, Whatman No. 1 paper discs were impregnated with 10 μL of the corresponding fraction. Later, discs were placed with the microorganism on the disks with the fraction. In the second bioassay, the effect of the fractions that showed antimicrobial activity in the first bioassay was evaluated. A 5 mm paper disc with the fraction was placed in the Petri dishes colonized by the microorganism. Microbial growth was measured every 24 h until 96 h, as a positive control sterile distilled water (H_2O) was employed. A solvent test was also performed using a filter paper disc treated with sterile DMSO [19]. Negative control test wity DMSO were performed (data provided in the Supplementary Materials). The diameters for the inhibition zones were measured in millimeters. The percentage inhibition of radial growth (PIRG) was calculated using the Abbott formula: PIRG (%) = [(RC − RT)/RC] × 100, where RC is the radius of the control, and RT the radius of the treatment [20].

2.5. Evaluation of Antibacterial Activity

A volume of 0.1 mL of inoculated cell suspension broth was placed on each Petri dish (Ec 2.95 × 10³ CFU/mL; Cf 8.47 × 10³ CFU/mL and 2.86 × 10⁶ CFU/mL for Se) [12]. Then, four Whatman No. 1 paper discs were impregnated with 10 μL of the corresponding fraction. The diameter of the growth inhibition zone was measured at 15 h, 24 h, 40 h, and 48 h. Ciprofloxacin 125 mg/mL for Ec and Cf, and chloramphenicol at 5 mg/mL for Se [20] were used as a positive control. A solvent test was also performed using a filter paper disc treated with sterile DMSO [19]. DMSO tests were performed, observing growth on the whole plate identical to the control (water) Percentage inhibition (PI) was calculated using a modified expression of the Abbott formula: PI (%) = [DT/DC] × 100, where DC is the diameter of the inhibition halo of the control, and DT is the diameter of the inhibition halo of the treatment [20].

2.6. Experimental Design

A completely randomized experimental design with three replicates was used for each microorganism, taking as a response variable to PIRG or PI. A simple ANOVA was performed with a comparison of means using the Tukey test at 95% confidence.

3. Results

3.1. Evaluation of Antifungal Activity

The bioassays were carried out to know the possible antimicrobial activity of the different fractions, and directed towards the most promising fraction after other specific bioassays.

In the first bioassay, the fractions showed a fungistatic effect on the three fungi. For each fungus, the most effective fraction at 24 h was different. In the case of FoC, it was the aqueous fraction with a PIRG of 50.00%. For FoT, the highest value of PIRG was obtained with the dichloromethane fraction (FDM, 61.76%), and for FsC, it was the 2-propanone fraction that obtained higher values of inhibition, with 35.00% (Table 1). However, for the three species fungi, the aqueous fraction (FW) was the one with the highest percentage of inhibition (PIRG) at 48 h, 72 h, and 96 h (Table 1).

For the second bioassay, the aqueous fraction was employed. For FoC and FsC, a mycelial growth-promoting effect was observed at the end of the test time. For FoT, the aqueous fraction showed a value of 27.94% of inhibition in the first 24 h; however, this effect did not last after 72 h (Table 2).

3.2. Evaluation of Antibacterial Activity

The fractions for 2-propanone (AF) and ethyl acetate (EAF) had a low antimicrobial activity at 15 h (17.62% and 18.10%, respectively). The lowest percentage inhibition was observed in the dichloromethane fraction (5.71% at 15 h); while the ethyl ether fraction (EEF) showed better antimicrobial activity than the other fractions at 48 h (26.62%) against Se (Table 3).

Table 1. Percent inhibition by direct contact of Chipilín (C. longirostrata) active fractions on phytopathogenic fungal species.

Treatment	Strain											
	FoC				FoT				FsC			
	Time (h)											
	24	48	72	96	24	48	72	96	24	48	72	96
(SDW)	0.00 [a]	0.00 [a]	0.00 [a]	0.00 [a]	0.00 [a]	0.00 [a]	0.00 [a]	0.00 [a]	0.00 [ab]	0.00 [ab]	0.00 [a]	0.00 [a]
AF	25.00 [ab]	14.13 [b]	4.91 [ab]	9.61 [bc]	49.02 [bc]	8.99 [ab]	6.92 [b]	18.66 [b]	35.00 [b]	11.61 [ab]	7.29 [a]	8.65 [ab]
EAF	40.00 [b]	28.26 [c]	17.89 [c]	13.25 [c]	57.84 [bc]	12.70 [b]	9.34 [b]	15.42 [b]	7.50 [ab]	12.26 [b]	10.42 [ab]	12.98 [bc]
WF	50.00 [b]	41.85 [d]	26.67 [d]	30.65 [d]	40.20 [b]	16.40 [b]	18.34 [c]	27.36 [c]	2.50 [ab]	15.48 [b]	20.83 [b]	20.61 [c]
DMF	42.50 [b]	20.65 [bc]	8.77 [b]	12.99 [c]	61.76 [c]	12.70 [b]	8.30 [b]	18.16 [b]	−17.50 [a]	9.68 [ab]	8.68 [a]	10.69 [bc]
EEF	42.50 [b]	18.48 [bc]	10.18 [bc]	3.38 [ab]	55.88 [bc]	2.12 [a]	3.46 [ab]	15.92 [b]	22.50 [b]	−5.16 [a]	10.76 [ab]	13.99 [bc]

The data are given in percentage inhibition of radial growth (PIRG). FoC (Fusarium oxysporum A. comiteca), FoT (Fusarium oxysporum A. tequilana), and FSC (Fusarium solani A. comiteca). AF: 2-propanone fraction; EAF: ethyl acetate fraction; WF: aqueous fraction; DMF: dichloromethane fraction; EEF: ethyl ether fraction and SDW: sterile distilled water. Latin letters a,b and c,d (superscript) in the same column indicates significant differences. Tukey 95% $p \leq 0.0046$.

Table 2. Volatility test in the aqueous fraction of Chipilín (*C. longirostrata*) on phytopathogenic fungal species.

Treatment	Strain								
	FoC			FoT			FsC		
	Time (h)								
	24	48	72	24	48	72	24	48	72
(SDW)	0.00 [a]	0.00 [a]	0.00 [a]	0.00 [a]	0.00 [a]	0.00 [a]	0.00 [a]	0.00 [a]	0.00 [a]
WF	−2.00 [a]	−7.09 [b]	−5.53 [b]	27.94 [b]	4.44 [a]	−4.78 [a]	3.77 [a]	−6.47 [b]	−5.63 [b]

The data are given in PIRG. FoC (*Fusarium oxysporum* A. comiteca), FoT (*Fusarium oxysporum* A. tequilana), and FSC (*Fusarium solani* A. comiteca). WF: aqueous fraction and SDW: sterile distilled water. Latin letters a,b (superscript) in the same column indicates significant differences. Tukey 95% $p \le 0.0529$.

Table 3. Percent inhibition of Chipilín (*C. longirostrata*) fractions on pathogenic bacteria.

Treatment	Strain											
	Ec				Cf				Se			
	Time (h)											
	15	24	40	48	15	24	40	48	15	24	40	48
Positive control		100 [a]			100 [c]	100 [b]	100 [b]	100 [c]	100 [c]	100 [d]	100 [d]	100 [d]
AF		NI				NI			17.62 [b]	13.78 [b]	13.97 [b]	14.39 [b]
EAF		NI			5.43 [a]	3.43 [a]	1.43 [a]	0.00 [a]	18.10 [b]	5.61 [a]	6.15 [ab]	5.04 [a]
WF		NI				NI			10.48 [a]	8.67 [ab]	8.38 [ab]	5.04 [a]
DMF		NI				NI			5.71 [a]	4.08 [a]	3.91 [a]	2.16 [a]
EEF		NI			10.29 [b]	4.00 [a]	4.00 [a]	2.59 [b]	24.29 [c]	22.45 [c]	22.91 [c]	26.62 [c]
MSD	-				3.82523	3.46982	2.57792	1.50361	7.07667	7.63128	7.9051	7.60449

The data are given in percentage inhibition (PI). Ec (*Escherichia coli*), Cf (*Citrobacter freundii*), and Se (*Staphylococcus epidermidis*). AF: 2-propanone fraction; EAF: ethyl acetate fraction; WF: aqueous fraction; DMF: fraction of dichloromethane; EEF: ethyl ether fraction; and positive control ciprofloxacin at 125 mg/mL for *E. coli* and *C. freundii*, and chloramphenicol at 5 mg/mL for *S. epidermidis*. NI: not inhibited. Latin letters a,b,c,d (superscript) in the same column indicates significant differences. Tukey 95% $p \le 0.0000$. Minimum Significant Difference (MSD).

4. Discussion

In the last two decades, there has been growing interest in research for extracts for medicinal plants as sources of new antimicrobial agents [12–14]. Recent findings about species of the genus *Crotalaria* describe their biological activity, by example the ethanolic fractions of *C. retusa*, the chloroform fraction of *C. prostrate*, the ethanolic extract of *C. medicaginea*, the ethanolic extract of *C. pallida*, the methanolic extract of *C. burhia*, and the fractions of *C. bernieri* and *C. madurensis* showed inhibitory capacity against *E. coli* [4,9,10,13,21]. The species *C. longirostrata* has been reported to have ethnobotanical activity, but an antimicrobial evaluation had not yet been done, and the chemical compounds responsible remained unknown. We evaluated its antimicrobial potential, finding that the fraction of dichloromethane is more effective in inhibiting the growth of FoT (61.7%) in comparison with the aqueous extract of *C. medicagenina* (33%) and the methanolic extract of *C. filipes* (55%). However, it was less efficient than the isolated peptide of *C. pallida* (70%) [1,12,22]. Further, the aqueous fraction showed low inhibition values (2.5%) against *F. solani*, while the 2-propanone fraction (35%) revealed activity for the aqueous extract of *C. juncea* [23]. Subsequently, the antibacterial evaluation of the fractions of *C. longirostrata* found statistically significant differences between the fractions and the control in *C. freundii* and *S. epidermidis*; however, the percentage of inhibition was lower than that of the antibiotic (control).

The results obtained of this preliminary study show the fungistatic capacity, but not fungicidal capacity, of the fractions obtained from the Chipilín (*C. longirostrata*). Some phenolic compounds, alkaloids, essential oils, and glycosides have shown to be responsible for antifungal activity [1]. This suggests the presence of these compounds in the fractions that were analyzed, which proved to be more effective against fungi than bacteria. The effect of substances of plant origin is due to mechanisms of direct fungitoxic action [21], while the bactericidal potential is associated with anthraquinones and flavonoids of a catechinic nature [24]. There is a great diversity in the forms of action of secondary metabolites that have been reported as antifungal [25–27]. However, each extract showed a spectrum of specific activity that could be due to the difference between the chemical nature and the concentration of bioactive compounds in extracts [21]. For example, the EEF fraction against FsC showed low inhibition or equal to the growth of the control, so its PIRG value was diminished. Therefore, it is necessary to carry out further phytochemical studies for the identification of the secondary metabolites of the Chipilín (*C. longirostrata*) leaves that are responsible for its antimicrobial activity.

Supplementary Materials: The following are available online at http://www.mdpi.com/2071-1050/10/3/883/s1, Table S1. Solvent test on the mycelial growth of the phytopathogenic fungi evaluated.

Acknowledgments: J.M.-G. and C.C. thanks Consejo Nacional de Ciencia y Tecnologia (CONACyT)-Mexico (597369 and 594550 respectively) postgraduate fellowship.

Author Contributions: Selene Lagunas-Rivera conceived and designed the experiment and all authors were involved in analyzing the data; Johana Miranda-Granados and Cesar Chacon performed the experiments and analyzed all the samples; Nancy Ruiz-Lau, Peggy Alvarez-Gutierrez and Rocio Meza-Gordillo contributed analysis tools; Maria Elena Vargas-Diaz and Gerardo Zepeda-Vallejo contributed with the reagents/materials/analysis tools, all authors were involved and contributed to writing the paper.

Conflicts of Interest: The authors declare no conflict of interest.

References

1. Akarsh, S.; Prashith, T.R.; Ranjitha, M.C.; Vidya, P.; Firdos, G.F. Inhibitory activity of some plants against *Colletotrichum capsici* and *Fusarium oxysporum* f. sp. zingiberi. *J. Med. Plants Stud.* **2016**, *4*, 165–168.

2. Ab Rahman, S.F.S.; Singh, E.; Pieterse, C.M.J.; Schenk, P.M. Emerging microbial biocontrol strategies for Plant Pathogens. *Plant Sci.* **2017**, *267*, 102–111. [CrossRef] [PubMed]

3. Kumar, R.A.; Kumar, H.G.R. In vitro antifungal activity of some plant extracts against *Fusarium oxysporum* f. sp. *lycopersici. Asian J. Plant Sci. Res.* **2015**, *5*, 22–27.

4. Devendra, B.N.; Srinivas, N.; Solmon, K. A comparative pharmacological and phytochemical analysis of in vivo and in vitro propagated. *Crotalaria* species. *Asian Pac. J. Trop. Med.* **2012**, *5*, 37–41. [CrossRef]

5. Casimiro, D.; Fechine, J.T.; Ferreira dos Santos, P.; Vieira Sobral, C.M.; Agra, M.; Lima Subrinho, F.; Braz-Filhob, R.; Sobral da Silva, M. Structural elucidation and NMR assignments of a new pyrrolizidine alkaloid from *Crotalaria vitellina* Ker Gawl. *Magn. Reson. Chem.* **2013**, *51*, 497–499. [CrossRef] [PubMed]

6. Ahmed, B.; Al-Howiriny, T.A.; Mossa, J.S. Crotalic and emarginellic acids: Two triterpenes from *Crotalaria emarginella* and anti-inflammatory and antihepatotoxic activity of crotalic acid. *Phytochemistry* **2006**, *67*, 956–964. [CrossRef] [PubMed]

7. Panda, S.K.; Debajyoti, D.; Tripthathy, N.K. Phytochemical investigation and anthelmintic activity of various leaf extracts of *Crotalaria pallida* aiton. *World J. Pharm. Pharm. Sci.* **2015**, *4*, 336–342.

8. Belal, A.; El-Dien, B. Pyrrolizines: Promising scaffolds for anticancer drugs. *Bioorg. Med. Chem.* **2014**, *22*, 46–53. [CrossRef] [PubMed]

9. Bhakshu, L.M.; Venkata, R.K.; Venkataraju, R.R. Medicinal properties and antimicrobial activity of *Crotalaria madurensis* Var. Kurnoolica. *Ethnobot. Leafl.* **2008**, *12*, 758–762.

10. Sandeep, K.; Shrivastava, B.; Khajuria, R.K. Antimicrobial activity of *Crotalaria burhia* Buch.-Ham. *Indian J. Nat. Prod. Resour.* **2010**, *1*, 481–484.

11. Mansoor, H.; Muhammad, A.; Al-Quriany, F.; Tahira, N.; Muhammad, S. Medicinal flora of the Cholistan desert. *Pak. J. Bot.* **2011**, *43*, 39–50.

12. Pelegrini, P.B.; Farias, L.R.; Saude, A.C.; Costa, C.J.; Silva, L.P.; Oliveira, A.S.; Gomes, C.E.; Sales, M.P.; Franco, O.L. A novel antimicrobial peptide from *Crotalaria pallida* seeds with activity against human and phytopathogens. *Curr. Microbiol.* **2009**, *59*, 400–404. [CrossRef] [PubMed]

13. Aravinthan, K.; Kiruthiga, R.; Rakkimuthu, R. Antibacterial activity of *Crotalaria pallida* Aiton. (Fabaceae). *Indian J. Pharm. Biol. Res.* **2014**, *2*, 82–85.

14. Ukil, S.; Laskar, S.; Roy, R.N. Physicochemical characterization and antibacterial activity of the leaf oil of *Crotalaria pallida* Aiton. *J. Taibah Univ. Sci.* **2016**, *10*, 490–496. [CrossRef]

15. Caballero, R.A.; Talavera, A.; Dumani, M.; Escobar, D. Los recursos vegetales en la alimentación de mujeres tsotsiles de la Selva El Ocote, Chiapas, México. *Lacandonia* **2011**, *5*, 141–147.

16. Morton, J.F. Pito (*E. berteroana*) and Chipilín (*Crotalaria longirostrata*), (*Fabaceae*), two soporific vegetables of central América. *Econ. Bot.* **1994**, *48*, 130–138. [CrossRef]

17. Al Akeel, R.; Mateen, A.; Janardhan, K.; Gupta, V. Analysis of anti-bacterial and antioxidative activity of *Azadirachta indica* bark using various solvents extracts. *Saudi J. Biol. Sci.* **2017**, *24*, 11–14. [CrossRef] [PubMed]

18. Sánchez, Y.; Pino, O.; Correa, T.; Naranjo, E.; Iglesia, A. Chemical and microbiological study of the essential oil of *Piper auritum* Kunth (Caisimón de anís). *Rev. Prot. Veg.* **2009**, *24*, 39–46.

19. Sánchez, Y.; Pino, O.; Lazo, F.; Abreu, Y.; Naranjo, E.; Iglesia, A. Promisory activity of essential oils from species belonging to *Pipereae* tribu against *Artemia salina* and *Xanthomonas albilineans*. *Rev. Prot. Veg.* **2011**, *26*, 45–51.

20. Moo, F.; Alejo, J.C.; Reyes-Ramírez, A.; Tun-Suárez, J.M.; Sandoval-Luna, R.; Ramírez-Pool, J.A. In vitro activity of an aqueous extract of *Bonellia flammea* against phytopathogenic fungi. *Agrociencia* **2014**, *48*, 833–845.

21. Andriamampianina, H.; Rakoto, D.D.; Petit, T.; Ramanankierana, H.; Randrianarivo, H.; Jeannoda, V. Antimicrobial activity of extracts from *Crotalaria bernieri* Baill. (Fabaceae). *Afr. J. Microbiol. Res.* **2016**, *10*, 1229–1239.

22. Kumar, A.; Tripathi, S.C. Evaluation of the leaf juice of some higher plants for their toxicity against soil borne pathogens. *Plant Soil* **1991**, *32*, 297–301. [CrossRef]

23. Hussain, F.; Abid, M.; Shaukat, S.; Akbar, M. Anti-fungal activity of some medicinal plants on different pathogenic fungi. *Pak. J. Bot.* **2015**, *47*, 2009–2013.

24. Celis, Á.; Mendoza, C.; Pachón, M.; Cardona, J.D.; Cuca, L.E. Plant extracts used as biocontrol with emphasis on *Piperaceae* family—A review. *Agron. Colomb.* **2008**, *26*, 97–106.

25. Lucini, E.; Zunino, M.P.; López, M.L.; Zygadlo, J.A. Effect of monoterpenes on lipid composition and sclerotial development of *Sclerotium cepivorum* Berk. *J. Phytopathol.* **2006**, *154*, 441–446. [CrossRef]

26. Cruz-Rodríguez, R.I.; Meza-Gordillo, R.; Rodríguez-Mendiola, M.A.; Arias-Castro, C.; Mancilla-Margalli, N.A.; Ávila-Miranda, M.E.; Culebro-Ricaldi, J.M.; Gutiérrez-Miceli, F.A.; Ruiz-Valdiviezo, V.M.; Ayora-Talavera, T.R. Antifungal activity of *Crotalaria longirostrata* Hook. & Arn. extracts against phytopathogen fungi from maize. *Gayana Bot.* **2017**, *74*, 167–175.

27. Montes, B.R. Chemical diversity in plants against phytopathogenic fungi. *Rev. Mex. Micol.* **2009**, *29*, 73–82.

sustainability

MDPI

Article

Characterization and Antimicrobial Activity of Alkaloid Extracts from Seeds of Different Genotypes of *Lupinus* spp.

Flora Valeria Romeo [1,*], Simona Fabroni [1], Gabriele Ballistreri [1], Serena Muccilli [1], Alfio Spina [2] and Paolo Rapisarda [1]

[1]   Consiglio per la ricerca in agricoltura e l'analisi dell'economia agraria (CREA), Centro di Ricerca Olivicoltura, Frutticoltura e Agrumicoltura, Corso Savoia, 190-95024 Acireale, CT, Italy; simona.fabroni@crea.gov.it (S.F.); gabriele.ballistreri@crea.gov.it (G.B.); serena.muccilli@crea.gov.it (S.M.); paolo.rapisarda@crea.gov.it (P.R.)

[2]   Consiglio per la ricerca in agricoltura e l'analisi dell'economia agraria (CREA), Centro di Ricerca Cerealicoltura e Colture Industriali—Laboratorio di Acireale, Corso Savoia, 190-95024 Acireale, CT, Italy; alfio.spina@crea.gov.it

*   Correspondence: floravaleria.romeo@crea.gov.it; Tel.: +39-095-765-3136

Received: 22 February 2018; Accepted: 9 March 2018; Published: 13 March 2018

Abstract: Alkaloid profiles of 22 lupin genotypes belonging to three different cultivated species, *Lupinus albus* L., *Lupinus luteus* L., and *Lupinus angustifolius* L., collected from different Italian regions and grown in Sicily, were studied by gas chromatography mass spectrometry (GC-MS) to determine alkaloid composition. More than 30 alkaloids were identified. The lowest alkaloid concentration was observed in the *L. albus* Luxor, Aster, and Rosetta cultivars, and in all the varieties of *L. luteus* and *L. angustifolius*. The highest content was observed in all the landraces of *L. albus*. Surprisingly, the white lupin Lublanc variety and the commercial seeds of cv Multitalia had a high alkaloid content. The tested species and the different genotypes exhibited different alkaloid profiles: lupanine, 13α-hydroxylupanine, and albine were the main alkaloids in the analyzed *L. albus* seeds; angustifoline and 13α-tigloyloxylupanine were well-represented in *L. albus* landraces; sparteine and lupinine were typical of *L. luteus*; and lupanine, 13α-hydroxylupanine, and angustifoline were the main alkaloids in *L. angustifolius* seeds. The samples with the highest amounts of total alkaloids proved to be interesting from a pharmaceutical viewpoint. The alkaloid extracts showed significant activity on *Klebsiella pneumoniae* and *Pseudomonas aeruginosa* clinical isolates.

Keywords: alkaloids; antimicrobial activity; germplasm; *Klebsiella*; landraces; lupanine; *Pseudomonas*; varieties

---

1. Introduction

The *Lupinus* genus belongs to the Fabaceae family, subfamily Papilionoideae, and includes about 170 species [1], but only four species are cultivated, three of which originate from the Mediterranean area: *Lupinus albus* L. (white lupin, chromosomally, $2n = 50$), *L. angustifolius* L. (narrow-leafed lupin, $2n = 40$), and *L. luteus* L. (yellow lupin, $2n = 52$). One, *L. mutabilis* Sweet, commonly known as pearl lupin or Tarwi ($2n = 48$) [2], originates from the Andean mountain. Narrow-leafed lupins are important for both animal feed and human foodstuff for the production of lupin flour and isolate proteins, whereas yellow lupin is only used in the livestock chain [3,4].

Lupinus species are mainly grown in Australia, Chile, and Eastern and Central Europe, but it is almost absent in the Mediterranean basin because it requires acid or sub-acid soils [5].

All *Lupinus* species produce quinolizidine and bipiperidine alkaloids, but the former are the main lupin alkaloids [2,6–10]. Lupin alkaloids are secondary metabolites that the plant stores in its organs, including seeds, likely as chemical agents against insects, microorganisms, and herbivores [11]. Quinolizidine alkaloids displayed oral toxicity due to neurological effects. Therefore, minimizing the risk of high levels of alkaloid uptake is important. For this reason, the health authorities of some countries, such as Great Britain, France, Australia, and New Zealand, have fixed the maximum alkaloid content in lupin food and flour marketing at 200 mg/kg [12,13].

The pharmacological benefits of alkaloids have been reported, with activity on the circulatory system, metabolism against obesity, cardiac dysfunction, and skin disease. In some cases, they act as hypoglycemic and hypolipidemic agents, as well as antibiotic, antivirus, anti-hepatitis, anti-inflammatory, anti-oxidant, anti-cancer, and neuroprotective agents. Moreover, alkaloids have a sedative effect on the central nervous system [14–17]. Therefore, a complete knowledge of lupin alkaloid patterns is important, not only because of their potential toxicity but also for their potential pharmacological benefits.

The genetic breeding program, conducted for nearly a century mainly by Sengbusch in Germany, Gladstones in Australia, and Baer in Chile, has led to the selection of sweet mutants with low or no alkaloid content (0.01–0.05% versus 1–8% of landraces) [4,18]. However, bitter lupins are still used in some parts of the world where the new sweet varieties are not well suited for the climate [17]. A renewed interest in lupin has grown in relation to its interesting nutritional properties and potential health benefits [19–21]. Lupin seeds are a good source for animal feeding and human nutrition due to their high protein content of 40–48% [18,22]. Lupin-based foods, such as ice cream, baking products, snacks, and meat-free products, including steaks, chops, and cutlets, as well as food supplements such as flour added to bread, have been developed in which the *L. albus* is one of the main ingredients [23]. In addition, the increase in demand for Genetically Modified Organisms-free materials in livestock chains has led to the reconsideration of the national grain protein-species, such as lupin, as an alternative to soybean as a source of protein [4].

As the application of antibiotics is limited in some contexts, the pharmaceutical industry and researchers have evaluated alternative antibacterial and antifungal agents of natural origin for use against pathogens [24–26]. The quinolizidine alkaloids of lupin are included in these natural agents. In addition, the extensive use of antibiotics has rapidly increased bacterial antibiotic resistance.

The aim of this work was to analyze 22 lupin seed samples corresponding to different genotypes of three *Lupinus* species to assess their total alkaloid content and the differences in the alkaloid pattern of the analyzed genotypes. This characterization was aimed at identifying and selecting the landraces and, eventually, the cultivars with the highest alkaloid content to be used as a source of bioactive compounds for pharmaceutical applications. This work also aimed to test the antibacterial and antifungal activities of *Lupinus* spp. alkaloid extracts against type strains and clinical isolate strains of Gram positive and negative bacteria and yeasts, considering the need for new antimicrobial natural agents against foodborne and clinical pathogens.

2. Materials and Methods

2.1. Plant Material and Sampling

The seeds of 10 sweet lupin cultivars: 6 white lupin (*L. albus*: Aster, Lublanc, Lutteur, Luxor, Rosetta, and Multitalia 1), 3 yellow lupin (*L. luteus*: Dukat, Mister, and Taper), and 1 narrow-leafed lupin (*L. angustifolius*: Sonet), together with 9 Italian landraces of *L. albus*, were collected from different Italian regions and tested in Sicily, Italy. Note that Multitalia 1 is a historically bitter variety less selected by geneticists [27]. The samples were also compared with three other Italian Multitalia seed samples. Two of these were cultivated in two other areas in Southern Italy: Multitalia 2 and Multitalia 3 from Battipaglia, Campania and Acireale, Sicily, respectively; and one commercial certified seed, Multitalia 4, reproduced in Northern Italy. The names of the Italian landraces are the same as the region from where they were harvested, with the exception of Modica and Scicli, which were from Sicily. The trial was conducted in 2012–2013 on volcanic soil in East Sicily, Giarre, Italy. All seed samples were sowed in duplicated plots of 5 m^2 (2.5 × 2 m). Manual seeding was completed on

November 24, 2012. Fertilization was applied during sowing with 30 kg/ha of ammoniacal nitrogen (ammonium sulfate) and 60 kg/ha of mineral perphosphate (P_2O_5). Chemical weed control was applied at 200 mL/hL of Pendimetalin (Stomp 330, Basf, Ludwigshafen, Germany) pre-emergence and mechanical post-emergence. Flood irrigation was required from March to June. Aphicide treatment with 50 mL/hL of Imidacloprid (Confidor, Bayer CropScience, Milan, Italy) was applied in late March. The crop was harvested on 20 June 2013. Seeds of all studied genotypes were deposited within the germplasm collection of the Research Centre for Cereal and Industrial Crops (CREA), Laboratory of Acireale (Italy).

2.2. Alkaloid Extraction

Alkaloid extraction was performed as described by Erdemoglu et al. [18] with modifications. Lupin seeds were finely ground and 1 g of each sample was suspended in 10 mL of 0.5 N HCl. After stirring for 30 min at room temperature, the homogenate was centrifuged for 10 min at 4 °C and 10,000 rpm. For quantitative analysis, the pellet was suspended in 0.5 N HCl and centrifuged again. Both supernatants were then pooled and adjusted to pH 12 with 5 N NaOH. Alkaloids were extracted by solid phase extraction using Extrelut columns (NT20 Merck, Darmstadt, Germany). Total alkaloids were eluted after 20 min with CH_2Cl_2 (3 × 20 mL) and the solvent evaporated until dry under vacuum at 40 °C. The residue was diluted in 1.5 mL dichloromethane and analyzed by gas chromatography-mass spectrometry (GC-MS) apparatus. Each sample was independently extracted and analyzed at least three times.

2.3. GC-MS Analysis

The analyses were performed on an Agilent 6890 gas chromatograph equipped with an Agilent 5973 Network quadrupole mass selective spectrometer and an Agilent 7683B Series autosampler. The separation was achieved using a VF-5ms 5% phenyl, 95% methylpolysiloxane capillary column (30 m × 0.25 mm, 0.25 μm film thickness, Varian). The GC-MS analysis was performed under the following conditions: the ion source temperature was 220 °C in EI mode at 70 eV, injector temperature was 250 °C, interface was 270 °C, carrier gas helium at 1 mL/min, split ratio 1/10, injection volume 1 μL, and mass range of 50 to 450 m/z. GC oven temperature was kept at 70 °C for 1 min, and programmed to 150 °C, heating at a rate of 40 °C/min, then to 300 °C at a rate of 6 °C/min, and kept constant for 1 min. Ionization was kept off during the first 3 min to avoid solvent overloading. The analyses were performed in full-scan mode. The retention index (RI) of the alkaloids was determined according to the Kovats method [13] by injecting a mixture of linear C8–C20 (cod. 04070, Sigma-Aldrich, Milan, Italy) and C21–C40 (cod. 04071, Sigma-Aldrich, Milan, Italy) alkanes. The compounds were identified by comparing their mass spectra with data in the NIST 05 MS Library Database [28] and the literature [11,29,30].

2.4. Alkaloid Quantification

The alkaloid quantification was performed using the external standard method, using (−)-sparteine (Sigma-Aldrich, St. Louis, MO, USA) as the standard. The calibration curve was prepared by injecting six solutions of sparteine at different concentrations in the range of 10 to 1000 mg/L and then a known amount (100 mg/L) of caffeine (Sigma-Aldrich, St. Louis, MO, USA) was added to each solution to check the response of the instrument. Since tetracyclic lupanine and sparteine are the most representative alkaloids in the seeds of lupins [6], and standards of most compounds are not commercially available, in the present work, the reported quantitative results are expressed as sparteine. The precision of the method was assessed by analyzing sparteine solutions within the same day and on different days, obtaining relative standard deviations (RSD %) of 2–3% and 4–5%, respectively. The limit of detection (LOD) of sparteine in the standard solutions was 0.1 mg/L.

2.5. Antimicrobial Activity

2.5.1. Microorganisms

The strains used in this study were type strains: *Candida albicans* DSM 1386, *Saccharomyces cerevisiae* DSM 1333, *Pseudomonas aeruginosa* DSM 1117, *Escherichia coli* DSM 1103, and *Trichophyton interdigitale* DSM 4870. Clinical isolates used were: *Candida krusei* (wound and vaginal tampon), *Staphylococcus aureus* (skin), *P. aeruginosa* (skin and ulcer), *Klebsiella pneumoniae* (skin and inguinal skin site), and *Proteus mirabilis* (skin). Clinical isolates were identified with vitek 2 (bioMerieux, Florence, Italy) in the Laboratory of Microbiology of University Hospital Policlinico Vittorio Emanuele (Catania, Italy).

2.5.2. Antibacterial and Antifungal Tests

The minimum inhibitory concentrations (MICs) of the extracts were determined following the Broth Microdilution Techniques according to the National Committee for Clinical Laboratory Standards [31,32]. Mueller-Hinton Broth (Merck, Darmstadt, Germany) and Sabouraud Broth (Oxford, UK) were used for growing and diluting the bacteria and fungi, respectively. Alkaloid extracts, obtained as per Section 2.2 from the two samples showing the highest alkaloid amount (Multitalia 4 and Calabria 2, as reported in Figure 1), after the evaporation to dryness under vacuum at 40 °C, were dissolved in dimethylsulfoxide (DMSO). Then the extracts, ranging from 3.75 to 1,000 µg/mL, were prepared for the test. Before the broth microdilution procedure, microorganism inocula were standardized to a turbidity of 0.5 McFarland standard (10^6 yeasts or 10^8 bacteria cells/mL). Final concentrations were approximately 10^3 cells/mL for yeasts and 10^4 cells/mL for bacteria. The microorganisms and pure media were placed in the wells of a microtiter plate together with the different concentration extracts. Proper blanks were tested simultaneously. Microtiter plates were incubated under atmospheric conditions at 37 °C for 24 h for bacteria, and at 25 °C for 48 h for the yeasts. Spectrophotometric lectures were performed at 600 nm. Each extract was tested in triplicate.

2.6. Statistical Analysis

SPSS software (version 21.0, IBM Statistics Corp., New York, NY, USA) was used for data processing. One-way analysis of variance (ANOVA) was used to test the effects of the different genotypes on the measured factors (total and each single alkaloid). Duncan's multiple range test was used to compare means when a significant variation was highlighted by the analysis of variance. The total amount of alkaloids and all single compounds were also analyzed using principal component analysis (PCA).

3. Results and Discussion

The mass spectral data of seed extracts from the different lupin samples, shown in Table 1, revealed and confirmed the presence of quinolizidine alkaloids previously reported by other authors [12,29,30,33].

Table 1. Mass spectral data of alkaloids and their distribution along the three *Lupinus* species. The alkaloids in bold letters are those shared by all the analyzed species. RI, retention index.

Peak n.	Alkaloid	RI	M+	Characteristic Ions	*Lupinus albus*	*Lupinus luteus*	*Lupinus angustifolius*
1	lupinine	1464	169	83-152-138		X	
2	unknown_a	1624	208	121-175		X	
3	gramine	1679	174	130-103-77		X	
4	genisteine (α-isosparteine)	1759	234	98-137		X	
5	sparteine	1827	234	137-98	X	X	
6	unknown_b	1879	232	134-232-98	X		
7	β-isosparteine	1883	234	137-98	X	X	

Table 1. *Cont.*

Peak n.	Alkaloid	RI	M+	Characteristic Ions	*Lupinus albus*	*Lupinus luteus*	*Lupinus angustifolius*
8	11,12-dehydrosparteine	1893	232	134	X	X	X
9	ammodendrine	1932	208	165-110	X	X	X
10	unknown_c	1951	208	166-136-110		X	
11	albine	1984	232	191-110	X	X	X
12	unknown_d	2030	232	191-110	X		
13	isoangustifoline	2127	234	193-112	X		X
14	tetrahydrhorombifoline	2135	248	207	X		X
15	angustifoline	2178	234	193-112	X		X
16	α-isolupanine	2206	248	136-248	X		X
17	5,6-dehydrolupanine	2225	246	98	X		X
18	unknown_e	2235	246	150-136-110	X		
19	lupanine	2273	248	136-149	X	X	X
20	11,12-dehydrolupanine	2296	246	134-246	X		
21	unknown_f	2308	248	110-191-149	X		
22	11,12-seco-12,13-didehydromultiflorine	2330	246	58-205	X		
23	unknown_g	2349	246	134	X		
24	3β-hydroxylupanine	2361	264	136-44	X		
25	unknown_h	2429	262	150-164-96	X		
26	multiflorine	2441	246	134-246	X	X	X
27	unknown_i	2460	264	134-152-246	X		
28	17-oxolupanine	2482	262	150-110-262	X		
29	*N*-formylangustifoline	2502	262	193-112-221	X		
30	13α-hydroxylupanine	2534	264	152-246	X		X
31	unknown_l	2570	264	152-246-134	X		
32	unknown_m	2581	262	245-150	X		
33	unknown_n	2609	260	260-148-112	X		
34	*N*-formylalbine	2688	260	219-96	X		
35	unknown_o	2712	262	149-150-148	X		
36	13α-hydroxymultiflorine	2727	262	150	X		X
37	13α-isovaleroyloxylupanine	2779	348	246-134-112	X		
38	13α-angeloyloxylupanine	2858	346	246	X		
39	13α-tigloyloxylupanine	2879	346	246	X		X
40	unknown_p	2912	348	246-134-231	X		
41	unknown_q	2930	348	246-134-112	X		
42	3β-tigloyloxylupanine	2944	346	134	X		
43	unknown_r	3073	344	132-244-149	X		
44	feruloyllupinine	3097	345	152-151		X	
45	13α-tigloyloxymultiflorine	3112	344	132	X		
46	unknown_s	3262	394	246-134-112	X		

The estimated total amount of alkaloids from the seeds of the 22 genotypes is shown in Figure 1. The post-hoc test revealed 11 homogeneous groups. Among the 10 selected cultivars belonging to the three species, Luxor, Aster, Dukat, Rosetta, Taper, Mister, and Sonet exhibited low and similar total alkaloid content, whereas Multitalia 1 and Lutteur contained higher amounts of alkaloids.

The commercial certified seed, Multitalia 4, was directly produced from the certified seed of the first generation, which was established and officially controlled with the purpose of producing certified second generation seed, marked with a red card, as provided by the Italian seed legislation [34]. Despite this fact, Multitalia 4 was one of the most alkaloid-rich samples along with Calabria 2 landrace, with 18,979 and 19,340 mg/kg, respectively. This seed was probably selected for genetic purity without a genetic selection for alkaloid content. The results obtained for Multitalia 2 and 3, and especially for Multitalia 4 and Lublanc, which were formerly sweet cultivars, indicated a genetic contamination occurred during the reproduction of those varieties. In fact, *L. albus* species demonstrated prevalent cross pollination, and for this reason, the alkaloid content increased over the years if the seeds with low alkaloid content were not selected and reproduced. However, Multitalia 1 showed levels of alkaloids in accordance with those reported by Calabrò et al. [27] and Gresta et al. [4] for cv

Multitalia. The amount of quinolizidine alkaloids can differ between years due to differing weather [35] or agronomic conditions [36]. Conversely, this problem was not observed in *L. angustifolius* and *L. luteus* since they are prevalently autogamous species. The most alkaloid-rich landraces may be of interest because they may be used for extracting pure alkaloids, as analytical standards of the majority of quinolizidine alkaloids are not commercially available [12]. Moreover, as mentioned above, the alkaloids possess several favorable pharmacological properties [37]. Therefore, the landraces could be practically and conveniently used in further studies of their properties and potential future applications in medicine and phytotherapy.

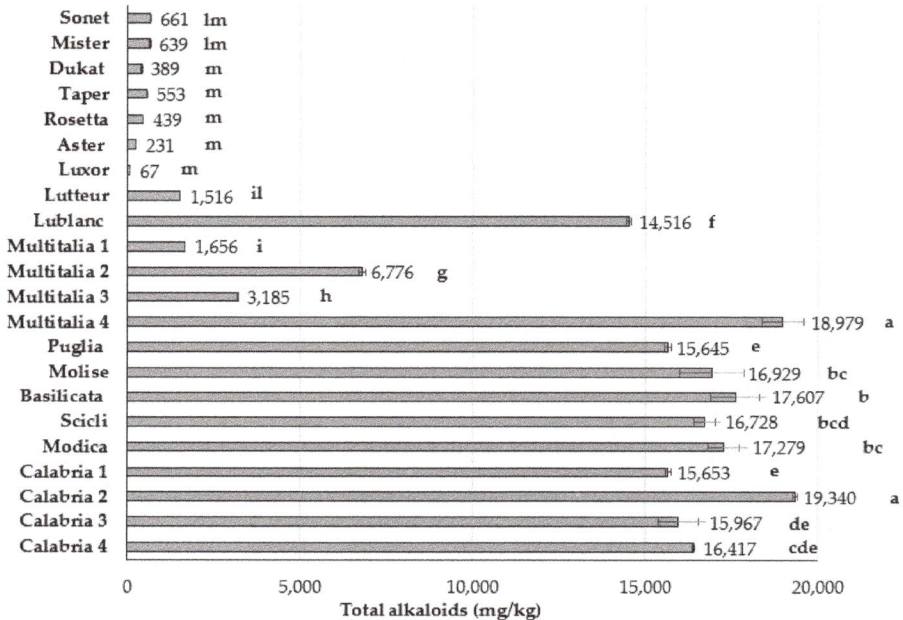

Figure 1. Total alkaloids in seed samples of the 22 genotypes belonging to the three analyzed lupin species. Mean values followed by different letters are significantly different (Duncan's post-hoc test).

In Table 2, the most representative alkaloids of the *L. albus* seed samples are shown. The seven compounds were also the only alkaloids shared by all *L. albus* samples, and the amounts of these compounds were significantly different among the genotypes ($p < 0.01$). Lupanine, albine, and 13α-hydroxylupanine were the main alkaloids were found in the highest amounts, but angustifoline and 13α-tigloyloxylupanine were also well-represented in *L. albus* landraces. Moreover, two landraces, Molise and Calabria 2, showed the presence of sparteine (data not shown), the main alkaloid found in *L. luteus* species [12]. Among the selected varieties, Lublanc had the largest number of alkaloids (36/46) and a different alkaloid pattern, possessing more than six compounds unshared with the other cultivars: β-isosparteine, 11,12-dehydrosparteine, 11,12-dehydrolupanine, 17-oxolupanine, *N*-formylangustifoline, *N*-formylalbine, and some unknown alkaloids (data not shown).

Table 2. Alkaloids shared by all the analyzed *Lupinus albus* genotypes.

Genotype	Albine	Tetrahydro-Rhombifoline	Angustifoline	Lupanine	13α-Hydroxy-Lupanine	13α-Tigloyloxy-Lupanine	3β-Tigloyloxy-Lupanine
Aster	30.2 ± 1.3 [n]	12.8 ± 0.1 [g]	12.1 ± 0.5 [h]	60.8 ± 0.3 [m]	17.4 ± 1.2 [g]	37.4 ± 2.9 [l]	6.2 ± 0.2 [c,d]
Lublanc	1012.4 ± 38.5 [g]	24.1 ± 0.8 [c,d]	268.4 ± 6.1 [e]	11,218.6 ± 82.5 [e,f]	491.7 ± 3.8 [c,d]	102.0 ± 1.7 [f,g]	16.6 ± 1.1 [a]
Luxor	11.5 ± 0.2 [n]	1.0 ± 0.4 [n]	3.6 ± 0.3 [h]	30.4 ± 1.2 [m]	3.1 ± 0.3 [g]	6.7 ± 0.5 [m]	2.4 ± 0.1 [e]
Lutteur	162.0 ± 5.7 [m]	8.8 ± 0.6 [i]	65.0 ± 1.2 [g]	863.2 ± 6.1 [i,l]	157.4 ± 1.8 [f]	44.4 ± 2.2 [i,l]	7.5 ± 1.4 [c]
Rosetta	29.3 ± 0.2 [n]	1.1 ± 0.2 [m,n]	6.4 ± 0.2 [h]	361.8 ± 5.4 [l,m]	4.7 ± 0.1 [g]	6.4 ± 0.1 [m]	1.5 ± 0.1 [e]
Calabria 1	1199.7 ± 10.5 [f]	23.1 ± 1.0 [d]	265.2 ± 8.4 [e]	12,363.6 ± 138.0 [c,d]	510.3 ± 15.5 [c]	75.4 ± 5.8 [h]	5.2 ± 1.4 [d]
Calabria 2	1436.2 ± 53.5 [d]	19.4 ± 0.3 [e,f]	466.8 ± 9.3 [b]	14,480.3 ± 10.0 [a]	687.2 ± 4.5 [b]	227.8 ± 1.1 [d]	12.1 ± 1.4 [b]
Calabria 3	1767.4 ± 17.6 [b]	35.9 ± 2.5 [b]	431.1 ± 19.0 [b]	11,165.4 ± 157.3 [e,f]	893.4 ± 66.7 [a]	98.7 ± 2.3 [g]	8.5 ± 1.0 [c]
Calabria 4	1232.3 ± 4.2 [f]	44.1 ± 0.1 [a]	444.2 ± 1.8 [b]	12,935.3 ± 16.2 [b,c]	413.9 ± 2.2 [d]	87.6 ± 0.3 [g,h]	11.1 ± 0.1 [b]
Modica	2596.2 ± 2.2 [a]	18.8 ± 0.8 [d]	518.0 ± 15.0 [a]	11,358.3 ± 140.4 [e,f]	747.4 ± 25.0 [b]	124.5 ± 7.6 [e]	12.4 ± 0.6 [b]
Scicli	1367.6 ± 60.6 [e]	44.1 ± 1.6 [a]	382.2 ± 13.6 [c]	11,916.5 ± 130.1 [d,e]	506.2 ± 41.9 [c]	308.4 ± 1.6 [c]	8.5 ± 0.1 [c]
Basilicata	1451.2 ± 55.9 [d]	23.5 ± 0.4 [d]	454.0 ± 34.6 [b]	11,583.0 ± 170.8 [e,f]	864.6 ± 66.5 [a]	419.6 ± 38.1 [a]	10.7 ± 0.3 [b]
Molise	668.2 ± 30.1 [i]	10.1 ± 0.5 [h,i]	315.8 ± 13.0 [d]	13,243.8 ± 161.9 [b]	562.3 ± 36.7 [c]	315.0 ± 15.3 [c]	12.0 ± 1.0 [b]
Puglia	845.0 ± 1.2 [h]	21.1 ± 1.1 [e]	322.6 ± 7.7 [d]	10,919.9 ± 6.6 [f]	548.5 ± 70.1 [c]	347.8 ± 2.2 [b]	17.0 ± 2.4 [a]
Multitalia 1	136.8 ± 12.3 [m]	2.9 ± 0.2 [m]	20.7 ± 1.8 [h]	1377.9 ± 7.0 [i]	12.4 ± 0.3 [g]	11.8 ± 1.1 [m]	1.5 ± 0.3 [e]
Multitalia 2	838.1 ± 10.4 [h]	11.1 ± 0.3 [g,h]	124.1 ± 4.2 [f]	4744.2 ± 88.0 [g]	305.3 ± 4.1 [e]	65.5 ± 0.4 [h,i]	10.9 ± 0.5 [b]
Multitalia 3	375.1 ± 3.0 [l]	6.5 ± 0.3 [l]	41.9 ± 50.3g [h]	2161.2 ± 11.8 [h]	157.5 ± 1.2 [f]	51.4 ± 0.1 [i,l]	6.5 ± 1.0 [c,d]
Multitalia 4	1647.4 ± 12.3 [c]	25.9 ± 0.4 [c]	431.5 ± 15.7 [b]	13,588.0 ± 188.3 [b]	730.9 ± 10.7 [b]	121.1 ± 4.5 [e,f]	18.0 ± 0.7 [a]
Sig.	**	**	**	**	**	**	**

Data are expressed as mg/kg of means ± SD. Different superscript letters indicate statistical differences within the same column (** Significance at $p < 0.01$).

For this reason, Lublanc seems to be the most interesting, together with the landraces, for further pharmaceutical studies. Table 3 shows the alkaloids observed in the three *L. luteus* seed samples.

Table 3. Alkaloids shared by the three analyzed *Lupinus luteus* cultivars.

Alkaloid	Dukat	Mister	Taper	Sig.
Lupinine	177.1 ± 30.5 [b]	281.4 ± 24.5 [a]	194.6 ± 21.0 [b]	*
Sparteine	139.0 ± 1.3 [b]	233.9 ± 9.5 [a]	120.0 ± 0.8 [c]	**
β-Isosparteine	4.5 ± 0.1	5.7 ± 0.9	4.2 ± 0.4	n.s.
Ammodendrine	26.8 ± 0.4 [a,b]	31.2 ± 0.3 [a]	22.7 ± 2.5 [c]	*
Unknown_c	7.6 ± 0.2	8.5 ± 0.7	7.6 ± 0.7	n.s.
Lupanine	8.7 ± 0.5 [b]	36.6 ± 4.7 [a]	6.4 ± 0.6 [b]	**
Feruloyllupinine	12.8 ± 0.8 [a]	13.1 ± 0.5 [a]	9.1 ± 0.2 [b]	*

Data are expressed as mg/kg of means ± SD. Different superscript letters indicate statistical differences within the same row (** Significance at $p < 0.01$; * Significance at $p < 0.05$; n.s., not significant).

Lupinine and sparteine were the main quinolizidine alkaloids, as confirmed by Aniszewski [6]. The Mister cultivar showed the highest amount of each of the seven shared alkaloids. Lupanine, the main alkaloid of *L. albus* species, was also observed in *L. luteus*, but at much lower concentrations, as reported by other authors [4]. Although gramine and other indole alkaloids are not usually detected in lupin seeds, in the present work, gramine was detected in the Dukat and Taper samples at 7 and 190 mg/kg, respectively (data not shown). This result agrees with previous findings [7].

The principal component analysis score plot was completed using all alkaloids and their total amounts as measured factors (Figure 2a). The first two principal components (PCs) explained 63.9% of the total variability (PC1 = 46.18%; PC2 = 17.70%). The score plot highlighted three sample sets, the first of which included Lublanc and Multitalia 4. Those two samples were different from each other and they were positively related to both the PCs. Lublanc and Multitalia 4 appeared to be related to the largest number of unknown alkaloids and to the identified compounds located in the upper right side of the loading plot (Figure 2b), such as N-formylalbine, N-formylangustifoline, 11,12-dehydrosparteine, and 17-oxolupanine. The second cluster included all nine *L. albus* landraces, all samples were positively related to PC2 and negatively related to PC1. The landraces were principally related to unknown_d, isoangustifoline, multiflorine, 13α-hydroxymultiflorine, 13α-angeloyloxylupanine, and 13α-tigloyloxymultiflorine, which were observed in all landrace profiles. Moreover, the landraces were related to the total alkaloid content, as confirmed by Figure 1, in which all landraces showed a similar total alkaloid amount. The third sample set highlighted the genotypes that were positively related to PC1 and negatively to PC2. That group should be sub-clustered into two groups: the *L. luteus* group (Mister, Dukat, and Taper), and the *L. albus* varieties and cv Sonet (*L. angustifolius*). All *L. albus* varieties and cv Sonet had almost a zero PC1 value and positive PC2 values, whereas the PC2 value of the *L. luteus* was higher than that of the *L. albus* group. All compounds located in the upper left side of the loading plot (Figure 2b) were typical of *L. luteus* samples, with the exception of sparteine, which was also found in some *L. albus* landraces. The position of *L. albus* cultivars in the plot discriminated them from the landraces in terms of typical alkaloids and their total amount.

Table 4 shows the antimicrobial results of the two tested extracts. Two species were affected differently by the alkaloid extracts. Among the tested strains, only the two clinical isolates *Pseudomonas aeruginosa* (skin and ulcer) and *Klebsiella pneumoniae* (skin and inguinal skin site) were inhibited by the lupin alkaloids. Antibacterial activity is considered by Erdemoglu et al. [18,38] as significant when the MIC is less than or equal to 100 μg/mL, and moderate when the MIC is 100–500 μg/mL. According to this classification, both alkaloid extracts showed significant activity on *K. pneumoniae* and significant moderate activity on *P. aeruginosa*. The extract of the landrace Calabria 2 had the highest activity.

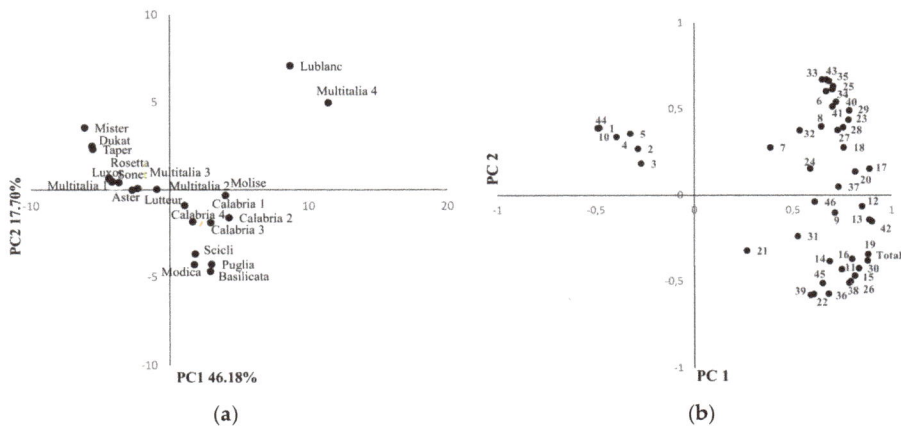

Figure 2. Score plot of (**a**) Principal Component Analysis showing the distribution of lupin genotypes along principal components PC1 and PC2, and (**b**) loading plot showing the distribution of the variables (total amount of alkaloids and all individual alkaloid compounds; each variable number refers to Table 1).

Table 4. Antimicrobial activity of alkaloid extracts of Multitalia 4 and Calabria 2 (*Lupinus albus*) seeds.

Microorganism	Multitalia 4	Calabria 2
Pseudomonas aeruginosa (from skin)	128	67
Klebsiella pneumoniae (from inguinal skin)	16	67

Data are expressed as minimum inhibitory concentrations (MICs, µg/mL).

The opportunistic pathogen *K. pneumoniae* can cause severe nosocomial infections, such as septicemia, pneumonia, urinary tract infections, and soft tissue infections. The indiscriminate use of antibiotics has resulted in a considerable increase in outbreaks caused by microorganisms resistant to antimicrobial drugs. Moreover, an alert was recently released due to the increasing evidence of the ability of *K. pneumoniae* to form biofilm, mostly on medical devices, thus new approaches are needed to control infection [39]. *P. aeruginosa* is a notoriously difficult organism to control with antibiotics or disinfectants due to its low antibiotic susceptibility. This resistance is attributable to a concerted action of multidrug efflux pumps with chromosomally-encoded antibiotic resistance genes and resistance developed due to mutation of chromosomally-encoded genes [40].

The present results about the antibacterial property of *Lupinus* extracts indicate that compounds from the plants of this genus could be used against common pathogens, as previously reported by other authors [18,41]. However, these extracts must be studied in animal models to determine their in vivo efficacy and potential toxicity, and to elucidate their mechanisms of action, as in vitro activity does not necessarily correspond to in vivo efficacy.

4. Conclusions

The three tested lupin species exhibited different alkaloid profiles, with some typical alkaloids present as the main compounds of each species. Lupanine, 13α-hydroxylupanine, and albine were determined as the main alkaloids in the *L. albus* seed samples; angustifoline and 13α-tigloyloxylupanine were also well-represented in *L. albus* landraces; whereas sparteine and lupinine were typical of *L. luteus*.

Finally, lupanine, 13α-hydroxylupanine, and angustifoline were the main alkaloids of *L. angustifolius* seeds. Some alkaloids were shared by all three species: 11,12-dehydrosparteine, ammodendrine, albine, lupanine, and multiflorine.

The *L. luteus* and *L. angustifolius* samples, together with most of the *L. albus* varieties, had a lower alkaloid amount, thus supporting their use as a human foodstuff and/or animal feed (white lupin and narrow-leafed lupin), and yellow lupin for the livestock chain only. Conversely, all landraces and the Lublanc cultivar (*L. albus*) proved to be interesting from a pharmaceutical viewpoint due to their several unknown alkaloids and for having the highest total amount of alkaloids. The alkaloid extracts from landrace Calabria 2 showed high activity on *K. pneumoniae* and moderate activity on *P. aeruginosa* clinical isolates.

Additional studies are needed to test the potential pharmacological effects and the in vivo antibacterial properties of lupin extracts.

Acknowledgments: Dedicated to the memory of Carmela Spatafora. The authors would also like to thank Guido Carpinteri and Agata Sciacca for their technical support. This work was supported by the ALI.FU.I.DE.A. Project (Functional foods and nutraceutical supplements based on white lupin and citrus derivatives), action 4.1.1.1 of PO FESR Sicilia 2007–2013.

Author Contributions: F.V.R. and P.R. conceived and designed the experiments; F.V.R. and S.M. performed the experiments; S.F. and G.B. analyzed the data; S.M. and A.S. contributed materials and analysis tools; F.V.R. and A.S. wrote the paper.

Conflicts of Interest: The authors declare no conflict of interest.

References

1. Gresta, F.; Wink, M.; Prins, U.; Abberton, M.; Capraro, J.; Scarafoni, A.; Hill, G. Lupins in European cropping systems. In *Legumes in Cropping Systems*; Murphy-Bokern, D., Stoddard, F., Watson, C., Eds.; CABI: Wallingford, Oxfordshire, UK, 2017; pp. 88–108, ISBN 9781780644981.

2. Carvajal-Larenas, F.E.; Van Boekel, M.J.A.S.; Koziol, M.; Nout, M.J.R.; Linnemann, A.R. Effect of processing on the diffusion of alkaloids and quality of *Lupinus mutabilis* sweet. *J. Food Process. Preserv.* **2014**, *38*, 1461–1471. [CrossRef]

3. Annicchiarico, P.; Manunza, P.; Arnoldi, A.; Boschin, G. Quality of *Lupinus albus* L. (White lupin) seed: Extent of genotypic and environmental effects. *J. Agric. Food Chem.* **2014**, *62*, 6539–6545. [CrossRef] [PubMed]

4. Gresta, F.; Abbate, V.; Avola, G.; Magazzù, G.; Chiofalo, B. Lupin seed for the crop-livestock food chain. *Ital. J. Agron.* **2010**, *4*, 333–340. [CrossRef]

5. FAOSTAT. Statistics Database of the Food and Agriculture Organization of the United Nations. Food and Agriculture Organization of the United Nations: Rome. Available online: http://faostat3.fao.org/home/E (accessed on 18 September 2017).

6. Aniszewski, T. *Alkaloids: Chemistry, Biology, Ecology, and Applications*, 2nd ed.; Elsevier: Amsterdam, The Netherlands, 2015; ISBN 9780444594334.

7. Magalhães, S.C.; Fernandes, F.; Cabrita, A.R.; Fonseca, A.J.; Valentão, P.; Andrade, P.B. Alkaloids in the valorization of European *Lupinus* spp. seeds crop. *Ind. Crops Prod.* **2017**, *95*, 286–295. [CrossRef]

8. Musco, N.; Cutrignelli, M.I.; Calabrò, S.; Tudisco, R.; Infascelli, F.; Grazioli, R.; Lo Presti, V.; Gresta, F.; Chiofalo, B. Comparison of nutritional and antinutritional traits among different species (*Lupinus albus* L.; *Lupinus luteus* L.; *Lupinus angustifolius* L.) and varieties of lupin seeds. *J. Anim. Physiol. Anim. Nutr.* **2017**, *101*, 1227–1241. [CrossRef] [PubMed]

9. Muzquiz, M.; Cuadrado, C.; Ayet, G.; de la Cuadra, C.; Burbano, C.; Osagie, A. Variation of alkaloid components of Lupin seeds in 49 genotypes of *Lupinus albus* L. from different countries and locations. *J. Agric. Food Chem.* **1994**, *42*, 1447–1450. [CrossRef]

10. Reinhard, H.; Rupp, H.; Sager, F.; Streule, M.; Zoller, O. Quinolizidine alkaloids and phomopsins in lupin seeds and lupin containing food. *J. Chromatogr. A* **2006**, *1112*, 353–360. [CrossRef] [PubMed]

11. Wink, M.; Meibner, C.; Witte, L. Pattern of quinolizidine alkaloids in 56 species of the genus *Lupinus*. *Phytochemistry* **1995**, *38*, 139–153. [CrossRef]

12. Boschin, G.; Annicchiarico, P.; Resta, D.; D'Agostina, A.; Arnoldi, A. Quinolizidine Alkaloids in Seeds of Lupin Genotypes of Different Origins. *J. Agric. Food Chem.* **2008**, *56*, 3657–3663. [CrossRef] [PubMed]

13. Resta, D.; Boschin, G.; D'Agostina, A.; Arnoldi, A. Evaluation of total quinolizidine alkaloids content in lupin flours, lupin-based ingredients, and foods. *Mol. Nutr. Food Res.* **2008**, *52*, 490–495. [CrossRef] [PubMed]

14. Canu Boido, C.; Tasso, B.; Boido, V.; Sparatore, F. Cytisine derivatives as ligands for neuronal nicotine receptors and with various pharmacological activities. *Farmaco* **2003**, *58*, 265–277. [CrossRef]

15. Guo, C.; Zhang, C.; Li, L.; Wang, Z.; Xiao, W.; Yang, Z. Hypoglycemic and hypolipidemic effects of oxymatrine in high-fat diet and streptozotocin-induced diabetic rats. *Phytomedicine* **2014**, *21*, 807–814. [CrossRef] [PubMed]

16. Lin, Z.; Huang, C.F.; Liu, X.S.; Jiang, J. In Vitro Anti-Tumour Activities of Quinolizidine Alkaloids Derived from *Sophora flavescens* Ait. *Basic Clin. Pharmacol.* **2010**, *108*, 304–309. [CrossRef] [PubMed]

17. Trugo, L.C.; von Baer, E.; von Baer, D. Breeding of grains: Lupin Breeding. *Ref. Mod. Food Sci.* **2016**, 1–8. [CrossRef]

18. Erdemoglu, N.; Ozkan, S.; Tosun, F. Alkaloid profile and antimicrobial activity of *Lupinus angustifolius* L. alkaloid extract. *Phytochem. Rev.* **2007**, *6*, 197–201. [CrossRef]

19. Arnoldi, A.; Boschin, G.; Zanoni, C.; Lammi, C. The health benefits of sweet lupin seed flours and isolated proteins. *J. Funct. Foods* **2015**, *18*, 550–563. [CrossRef]

20. Fontanari, G.G.; Batistuti, J.P.; da Cruz, R.J.; Saldiva, P.H.N.; Arêas, J.A.G. Cholesterol-lowering effect of whole lupin (*Lupinus albus*) seed and its protein isolate. *Food Chem.* **2012**, *132*, 1521–1526. [CrossRef] [PubMed]

21. Sirtori, C.R.; Lovati, M.R.; Manzoni, C.M.; Castiglioni, S.; Duranti, M.; Magni, C.; Moranti, S.; D'Agostina, A.; Arnoldi, A. Proteins of white lupin seed, a naturally isoflavone-poor legume, reduce cholesterolemia in rats and increase LDL receptor activity in HepG2 cells. *J. Nutr.* **2004**, *134*, 18–23. [CrossRef] [PubMed]

22. Nasar-Abbas, S.M.; Jayasena, V. Effect of lupin flour incorporation on the physical and sensory properties of muffins. *Qual. Assur. Saf. Crops Foods* **2012**, *4*, 41–49. [CrossRef]

23. Spina, A.; Cambrea, M.; Platania, A.; Roccasalva, D.; Rapisarda, P. Germoplasma di lupino (*Lupinus albus* L., *Lupinus angustifolius* L. e *Lupinus luteus* L.) in collezione presso il CRA-ACM di Acireale. In *Conservazione, Biodiversità, Gestione Banche Dati e Miglioramento Genetico*; D'Andrea, F., Ed.; Edizioni Nuova Cultura: Roma, Italy, 2013, pp. 271–279. [CrossRef]

24. Campolo, O.; Romeo, F.V.; Malacrinò, A.; Laudani, F.; Carpinteri, G.; Fabroni, S.; Rapisarda, P.; Palmeri, V. Effects of inert dusts applied alone and in combination with sweet orange essential oil against *Rhyzopertha dominica* (Coleoptera: Bostrichidae) and wheat microbial population. *Ind. Crop. Prod.* **2014**, *61*, 361–369. [CrossRef]

25. Li Destri Nicosia, M.G.; Pangallo, S.; Raphael, G.; Romeo, F.V.; Strano, M.C.; Rapisarda, P.; Droby, S.; Schena, L. Control of postharvest fungal rots on citrus fruit and sweet cherries using a pomegranate peel extract. *Postharvest Biol. Technol.* **2016**, *114*, 54–61. [CrossRef]

26. Romeo, F.V.; Ballistreri, G.; Fabroni, S.; Pangallo, S.; Li Destri Nicosia, M.G.; Schena, L.; Rapisarda, P. Chemical characterization of different sumac and pomegranate extracts effective against *Botrytis cinerea* rots. *Molecules* **2015**, *20*, 11941–11958. [CrossRef] [PubMed]

27. Calabrò, S.; Cutrignelli, M.; Lo Presti, V.; Tudisco, R.; Chiofalo, V.; Grossi, M.; Infascelli, F.; Chiofalo, B. Characterization and effect of year of harvest on the nutritional properties of three varieties of white lupine (*Lupinus albus* L.). *J. Sci. Food Agric.* **2015**, *95*, 3127–3136. [CrossRef] [PubMed]

28. NIST Chemistry WebBook 2015. Available online: http://webbook.nist.gov/ (accessed on 05 May 2016).

29. El-Shazly, A.; Ateya, A.M.M.; Wink, M. Quinolizidine Alkaloid Profiles of *Lupinus varius orientalis*, *L. albus albus*, *L. hartwegii*, and *L. densiflorus*. *Z. Naturforsch. C* **2001**, *56*, 21–30. [CrossRef] [PubMed]

30. Torres, K.B.; Quintos, N.R.; Herrera, J.M.; Tei, A.; Wink, M. Alkaloid profiles of leaves and seeds of *Lupinus aschenbornii* Schauer from Mexico. In *Lupin, an Ancient Crop for the New Millennium, Proceedings of the 9th International Lupin Conference, Klink/Müritz, Germany, 20–24 June 1999*; van Santen, E., Wink, M., Weissmann, S., Roemer, P., Eds.; International Lupin Association Publisher: Canterbury, New Zealand, 1999; pp. 301–304, ISBN 0-86476-123-6.

31. CLSI. Methods for Dilution Antimicrobial Susceptibility Tests for Bacteria That Grow Aerobically; Approved Standard. In *Clinical and Laboratory Standards Institute Document M7-A7*, 7th ed.; CLSI: Wayne, PA, USA, 2006; ISBN 1-56238-587-9.

32. CLSI. Reference method for broth dilution antifungal susceptibility testing of yeasts; Approved standard. In *Clinical and Laboratory Standards Institute document M27-A3*, 3rd ed.; CLSI: Wayne, PA, USA, 2008; ISBN 1-56238-666-2.

33. Hernández, E.M.; Rangel, M.L.C.; Corona, A.E.; Cantor del Angel, A.E.; Sánchez López, J.A.; Sporer, F.; Wink, M.; Torres, K.B. Quinolizidine alkaloid composition in different organs of *Lupinus aschenbornii*. *Rev. Bras. Farmacogn.* **2011**, *21*, 824–828. [CrossRef]

34. Law 1096 of the 25 November 1971 and Subsequent Amendments and Additions. pp. 1–28. Available online: http://scs.entecra.it/leggiEdisposizioni/NORM-NAZIONALI/Legge-1096-del-25-11-71.pdf (accessed on 18 September 2017).

35. Jansen, G.; Jürgens, H.U.; Ordon, F. Effects of temperature on the alkaloid content of seeds of *Lupinus angustifolius* cultivars. *J. Agron. Crop Sci.* **2009**, *195*, 172–177. [CrossRef]

36. Barlóg, P.K. Effect of magnesium and nitrogenous fertilisers on the growth and alkaloid content in *Lupinus angustifolius* L. *Aust. J. Agric. Res.* **2002**, *53*, 671–676. [CrossRef]

37. Jayasena, V.; Nasar-Abbas, S.M. Development and quality evaluation of high-protein and high-dietary-fiber pasta using lupin flour. *J. Texture Stud.* **2012**, *43*, 153–163. [CrossRef]

38. Erdemoglu, N.; Ozkan, S.; Duran, A.; Tosun, F. GC-MS analysis and antimicrobial activity of alkaloid extract from *Genista vuraiii*. *Pharm. Biol.* **2009**, *47*, 81–85. [CrossRef]

39. Vuotto, C.; Longo, F.; Balice, M.P.; Donelli, G.; Varaldo, P.E. Antibiotic resistance related to biofilm formation in *Klebsiella pneumoniae*. *Pathogens* **2014**, *3*, 743–758. [CrossRef] [PubMed]

40. Lambert, P.A. Mechanisms of antibiotic resistance in *Pseudomonas aeruginosa*. *J. R. Soc. Med.* **2002**, *95*, 22–26. [PubMed]

41. Khan, M.K.; Karnpanit, W.; Nasar-Abbas, S.M.; Huma, Z.E.; Jayasena, V. Phytochemical composition and bioactivities of lupin: A review. *Int. J. Food Sci. Techol.* **2015**, *50*, 2004–2012. [CrossRef]

sustainability

MDPI

Article

Variability and Site Dependence of Grain Mineral Contents in Tetraploid Wheats

Laura Del Coco [1], Barbara Laddomada [2], Danilo Migoni [1,3], Giovanni Mita [2], Rosanna Simeone [4] and Francesco Paolo Fanizzi [1,3,*]

[1]   Department of Biological and Environmental Sciences and Technologies (Di.S.Te.B.A.), University of Salento, via Monteroni, 73100 Lecce, Italy; laura.delcoco@unisalento.it (L.D.C.); danilo.migoni@unisalento.it (D.M.)

[2]   Institute of Sciences of Food Production (ISPA), National Research Council (CNR), via Monteroni, 73100 Lecce, Italy; barbara.laddomada@ispa.cnr.it (B.L.); giovanni.mita@ispa.cnr.it (G.M.)

[3]   Inter-University Consortium for Research on Chemistry of Metals in Biological Systems (C.I.R.C.M.S.B.), Lecce Research Unit, 73100 Lecce, Italy

[4]   Department of Soil, Plant & Food Sciences, Genetics and Plant Breeding Section, University Aldo Moro, via G. Amendola, 165/a 70126 Bari, Italy; rosanna.simeone@uniba.it

*   Correspondence: fp.fanizzi@unisalento.it; Tel.: +39-0832-299265

Received: 10 December 2018; Accepted: 28 January 2019; Published: 31 January 2019

Abstract: Crop production and natural resource use, especially in developing countries, represents one of the most important food sources for humans. In particular, two wheat species (tetraploid, which is mostly used for pasta and hexaploid, which is primarily used for bread) account for about 20% of the whole calories consumed worldwide. In order to assess the mineral accumulation capability of some popular tetraploid wheat genotypes, a metabolomic (metallomic) approach was used in this study. The metallomic profile related to micro- (Zn, Fe, Cu, Mn, Ni and Cr), macro- (Ca, Mg and K) and toxic trace elements (Cd and Pb) was obtained by ICP-AES analysis in a large set of tetraploid wheat genotypes (*Triticum turgidum* L.) that were grown in two different experimental fields. Correlations and multivariate statistical analyses were performed, grouping the samples under two wheat sets, comprising cultivated durum cultivars (*T. turgidum* subsp. *durum*) and wild accessions (*T. turgidum* subsp. *dicoccum* and subsp. *dicoccoides*). The site dependence ranking for the selected genotypes with the highest nutrient accumulation was obtained. The significantly higher content of Mg (among the macronutrients) and the highest levels of Mn, Fe and Zn (among the micronutrients) were found for wild accessions with respect to durum cultivars. Moreover, the former genotypes were also the ones with the lowest level of accumulation of the trace toxic elements, in particular Cd. According to the performed statistical analyses, the wild accessions appeared also to be less influenced by the different environmental conditions. This is in accord with literature data, indicating the superiority of "old" with respect to modern wheat cultivars for mineral content. Although further studies are required on a wider range of genotypes to confirm these findings, the obtained results could be used to better select the less demanding and better performing cultivars in specific target wheat growing environments.

Keywords: tetraploid wheat; metallomics; macronutrients; micronutrients; plants adaptability

---

1. Introduction

Wheat is one of the most important commodities, with a world production of 750 million tons and 220 million ha harvested in 2016, as reported by Faostat [1]. Interestingly, wheat consumption is closely related with the adoption of a "western lifestyle" [2], as well as some countries that are not climatically adapted to wheat cultivation, such as some Sub-Saharan Africa areas [3].

Among the wheat species, two types are prevalent: tetraploid wheat (2n = 28, genome AABB), which is mostly used for pasta, and hexaploid wheat (2n = 42, genome AABBDD), which is primarily used for bread. Together, these wheats account for about 20% of the whole calories that are consumed worldwide. Considering both types, wheat cultivation occurs in a wide range of environmental and soil conditions, and wheat growers face several challenges to reaching high yield and quality standards that are also sustainable and economically feasible. Among the major challenges, there is the identification of agronomical practices or genotypes that may help in reducing the inputs of fertilizers, herbicides and chemical treatments to counteract biotic or abiotic diseases.

Among tetraploid wheat (*Triticum turgidum* L.), the subsp. *durum* is the most cultivated and economically important typically rainfed crop and adapted to the semiarid conditions of the Mediterranean Basin. Indeed, the species *T. turgidum* also includes other subspecies, displaying a large variability for many traits, such as plant adaptability and defence, or quality traits. This variability is still largely unexplored for genes that could meet the needs for a more sustainable management of wheat cultivation. For instance, the subspecies *dicoccoides* (Körn. ex Asch. et Graebner) Thell.), the wild progenitor of durum wheat, is well known to accumulate more proteins and minerals in mature grains [4]. Also, the subspecies *dicoccum*, the direct domesticated progenitor of durum wheat, may carry other interesting traits related to plant adaptability that could be easily transferred into the durum background through breeding programs [5]. The occurrence of such variability could be of particular significance to improve yield, especially in those countries that are undergoing urbanization and industrialization, or to address market expectations with respect to quality, nutrition and health issues.

In general, wheat grains have been considered a primary source of energy (carbohydrates), proteins, vitamins and minerals (especially micronutrients, such as iron and zinc) in human diets [2,3,6–10]. Indeed, the potential benefits of wheat on human health are well known and have been extensively studied [6,11–13]. Besides the major components, several phytochemicals conferring antioxidant properties [14–17] and dietary fibers contribute to reduce the risks for cardiovascular disease and colon cancers [2]. The high variability in those health-promoting components was shown to be influenced by both genetic and environmental factors [2,13,18,19].

To increase the content of minerals in wheat grains, in recent years, several biofortification breeding efforts were undertaken, mainly focused on hexaploid wheat. Those studies led to the release of bred varieties having competitive grain yields and about 30–40% more Zn compared with other varieties [20]. Those results were possibly due to the identification of wheat genotypes with elevated Zn contents [21] that were subsequently crossed with modern elite wheat lines that generally do not vary much in terms of mineral contents [22].

Similar studies are mostly missing for tetraploid wheats [17,23]. So far, investigations are urgent to assess the current mineral levels in durum cultivars and in other tetraploid wheats subspecies to underpin the extent of genetic variability that is available within the primary gene pool. In fact, the assessment of genetic variation for mineral content is essential for the success of breeding activities that are aimed at developing new micronutrient-rich wheat genotypes that also display high grain yields [24,25]. Some literature data [26–28] indicated the superiority of "old" with respect to modern high-yielding wheat cultivars for mineral contents, however a comprehensive survey in tetraploid wheats is still lacking [28]. Also, more efforts devoted to investigate the influence of different growing sites on mineral accumulation in durum wheat grains would be important to identify precise genotype x environment combinations resulting in higher contents of some micronutrients, such as Fe, Zn, Mn and Cu, which have important physiological functions in humans [2,3,12,14,15,18,19,24–26,28–30].

In order to investigate the genetic variability for mineral content in tetraploid wheats, this study was conducted with the following objectives: (1) to evaluate, by ICP-AES analysis, the metallomic profile related to micro- (Zn, Fe, Cu, Mn, Ni and Cr), macro- (Ca, Mg and K) and toxic trace elements (Cd and Pb) in a collection of tetraploid wheats, including durum cultivars and *dicoccoides* and *dicoccum* accessions; (2) to investigate the site dependence features exhibited by durum and wild wheat by both univariate and multivariate statistical analyses, and (3) to evaluate a possible ranking of exhibited site-dependence among the considered subspecies.

2. Materials and Methods

2.1. Data Collection

The wheat collection that was analyzed in this study was composed of 25 tetraploid wheat genotypes comprising *dicoccum*, *dicoccoides* and *durum* subspecies of *T. turgidum* (Table 1). The plant material was grown under conventional farming in the experimental fields of the Department of Soil, Plant and Food Sciences at Valenzano, Italy (site-A) in 2013–2014 [31], and at Policoro, Italy (site B) in 2014–2015 [32]. The plants were grown in a randomized complete block design with three field replicates and plots consisting of 1m rows that were 30 cm apart, with 50 germinating seeds per plot. During the growing season, 120 kg/ha N were applied and standard cultivation practices were adopted. Plots were hand-harvested at maturity.

Table 1. Taxonomic classification of the wheat genotypes considered in the study.

Genotype	Taxonomic Classification
MG5323	*T. turgidum* L. subsp. *dicoccum*
MG4343	*T. turgidum* L. subsp. *dicoccoides*
MG4330	*T. turgidum* L. subsp. *dicoccoides*
MG29896	*T. turgidum* L. subsp. *dicoccoides*
Anco Marzio, Aureo, AC Avonlea, Ciccio, Duilio, Fiore, Grecale, Iride, Isildur, Latino, Latinur, Liberdur, Messapia, Neolatino, Normanno, Preco, Primadur, Saragolla, Svevo, Tiziana, UC1113	*T. turgidum* L. subsp. *durum*

2.2. Determination of Metal Concentration

The concentration of macronutrients (Ca, Mg, K), micronutrients (Zn, Fe, Cu, Mn, Ni, Cr) and toxic trace elements (Cd, Pb) was determined using the Inductively Coupled Plasma Atomic Emission Spectroscopy (ICP-AES) in the whole grains of 25 wheat accessions and cultivars for each site. Wheat samples were mineralized following standard procedures [33]. Briefly, in a Teflon vessel, 0.5 g of each sample were added with 6 mL of super pure HNO_3 69% and 4mL H_2O_2 and were digested at 180 °C for 10 min using a Milestone START D microwave digestion system. After mineralization, the samples were cooled at room temperature, diluted to a final volume of 30 mL with super pure water and filtered. The solutions that were obtained after the mineralization process were subsequently analyzed by a Thermo Scientific iCAP 6000 ICP-AES spectrometer. It should be noted that the obtained metallomic profiles refer to the whole grain wheat mass (upon chemical digestion and ICP analyses). Therefore, no kernel related adjustments were considered since a simple metabolomic approach was used in this study in order to assess the bioaccumulation ability of different wheat species groups.

2.3. Statistical Analysis

Standard analysis of variance (One Way-ANOVA) with Tukey's honestly significant difference (HSD) post hoc test was applied to compare the means between the two cultivation sites and for multiple comparisons of groups (wheat type) using the R statistical environment, Version 3.5.1 on a 64 bit Windows machine [34]. The levels of statistical significance were at least at p-values < 0.05 with a 95% confidence level. Moreover, the correlation matrix based on Pearson's coefficient was calculated for all the measured elements by using MetaboAnalyst 4.0, which is a web-based tool for the visualization of metabolomics [35,36]. This approach is the most widely used in this type of data [31,37] in order to assess the existence of a possible linear relationship between minerals for both the two cultivation sites [38]. Multivariate statistical analyses and graphics were obtained using SIMCA 14 software, (Sartorius Stedim Biotech, Umeå, Sweden) [39]. Specifically, exploratory data analysis was performed using Principal Component Analysis (PCA), while Projection to Latent Structures

(PLS)-based methods were used for discriminant analysis and data set comparison. The models were validated using an internal cross-validation default method (7-fold) and were further evaluated with a permutation test (400 permutations) [40,41]. The quality of the models was described by R^2, Q^2 and p values (p[CV-ANOVA], at a 95.0% confidence level, which were obtained from the analysis of variance testing of cross-validated predictive residuals (CV-ANOVA) [41,42]). To investigate the role of the measured variables in classification, a combination of loadings, variable influence on projection (VIP) parameters and p(corr) were analyzed. Loadings describe the correlations that the Principal or PLS component has with the original variables. VIP parameters summarize the overall contribution of each variable to the model and p(corr) represents the loadings scaled as a correlation coefficient (ranging from −1.0 to 1.0) between the model and the original data [43].

3. Results

Mineral Composition of Durum and Wild Wheat

The 25 wheat genotypes were used to assess the potential similarities and/or differences in the mineral content for the considered two harvesting sites (A, Valenzano, Bari, and B, Policoro, Matera) and among the different cultivars. All samples were grown in both the two harvesting sites and three replicates were collected for each sample, reaching a total of 148 (73 and 75 for site A and B, respectively). According to the wheat type, samples were also classified into durum (63 for each site, for a total of 126 samples) and wild (10 and 12 for site A and B, for a total of 22 samples).

Average and standard deviation values for concentrations of macro- (Ca, K, Mg), micro- (Cr, Cu, Fe, Mn, Zn) and toxic elements (Cd, Pb), which were expressed in part per million (ppm), were measured for all the studied cultivars and sites and are reported in Supplementary material (Tables S1 and S2). A summary clustering all of the studied cultivars according to the different wheat species (durum and wild) and the two cultivation sites (A and B) is reported in Table 2 and Figure 1. Tukey HSD test was applied for multiple comparisons of groups (wheat type) and One-way ANOVA was applied to compare the means between the two cultivation sites. As expected [44,45], among all the measured elements, the highest concentration was found for K, followed by Mg and Ca for all the examined wheat types. In particular, the highest K average value was found for durum wheat (4357.19 ± 618.72 ppm), followed by wild wheat (3864.32 ± 561.39 ppm). On the other hand, the highest Mg content resulted for wild wheat, with a similar content for the two cultivation sites (1665.97 ± 147.10 and 1651.89 ± 203.23 ppm for site A and B, respectively). Moreover, for durum wheat, a significantly lower value for Mg content was found in site B with respect to site A. Average Ca values were comparable in the two wheat types, with the highest content for durum wheat samples (833.08 ± 240.16 ppm) being significantly different from the durum wheat of site A (731.11 ± 295.70 ppm). It should be noted that for each type of tetraploid wheats, the average content of the macro-elements had the expected sequence K > Mg > Ca when considering both the harvesting sites [45]. Regarding micro-elements, in accordance with the literature data [27,37,45,46], higher levels of Mn, Fe and Zn and lower Cu and Cr, with Ni (Ni data not shown), in very low traces for all the samples were found in wild wheat. Interestingly, the highest Mn and Fe values were measured in wild wheat, followed by durum genotypes for both the harvesting sites. Generally higher values of Mn resulted in site A with respect to site B, while similar content of Fe was measured in the two sites. Very significant and site-dependent differences were found only for Zn, with significantly higher levels for the three tetraploid subspecies in site B with respect to site A. Moreover, the highest Zn content was found in wild (32.86 ± 7.50 ppm), followed by durum (30.59 ± 3.69 ppm), all from site B. Finally, higher Cd and Pb levels were found in durum wheat genotypes, with the highest in samples of site A (Cd = 0.10 ± 0.04 ppm, Pb = 0.09 ± 0.04 ppm), followed by wild for both the two harvesting sites.

Table 2. Average and standard deviation (SD) levels of macro-, micro- and toxic trace elements, calculated for the samples of three subspecies of tetraploid wheats (expressed as part per million, ppm) and for each site (A, Valenzano, Bari and B, Policoro, Matera). Tukey HSD test was applied for multiple comparisons of groups (wheat type). Letters ([a, b]) indicate significant differences for Tukey HSD test at least for 5% statistical probability ([a] significant difference between durum and wild in site A; [b] significant difference between durum and wild in site B). One-way ANOVA was applied to compare the means between the two cultivation sites (significant codes: *** p-value < 0.001; ** p-value < 0.01; * p-value < 0.05).

Type of Wheat		Durum	Wild
		Macroelements	
Ca	A	731.11 ± 295.70 *	728.73 ± 181.26
	B	833.08 ± 240.16 *	836.73 ± 232.90
K	A	3458.34 ± 329.72 ***	3362.44 ± 195.09 *
	B	4357.19 ± 618.72 ***,[a]	3864.32 ± 561.39 *,[a]
Mg	A	1510.34 ± 176.55 ***,[a]	1665.97 ± 147.10 [a]
	B	1398.73 ± 184.45 ***,[b]	1651.89 ± 203.23 [b]
		Microelements	
Cr	A	0.05 ± 0.06	0.06 ± 0.05
	B	0.03 ± 0.02	0.04 ± 0.02
Cu	A	4.12 ± 0.57 ***,[a]	5.32 ± 2.11 [a]
	B	5.77 ± 0.80 ***	5.29 ± 0.63
Fe	A	33.31 ± 6.87 [a]	38.51 ± 3.77 [a]
	B	31.96 ± 12.75	41.32 ± 13.79
Mn	A	35.11 ± 3.13 ***,[a]	42.85 ± 6.57 **,[a]
	B	26.93 ± 3.89 ***,[b]	32.66 ± 5.24 **,[b]
Zn	A	14.03 ± 1.75 ***,[a]	16.35 ± 2.34 ***,[a]
	B	30.59 ± 3.69 ***,[b]	32.86 ± 7.50 ***,[b]
		Toxic trace elements	
Cd	A	0.10 ± 0.04 ***	0.08 ± 0.03 ***
	B	0.07 ± 0.04 ***,[a]	0.035 ± 0.01 ***,[a]
Pb	A	0.09 ± 0.04 **	0.08 ± 0.03
	B	0.07 ± 0.02 **,[b]	0.10 ± 0.05 [b]

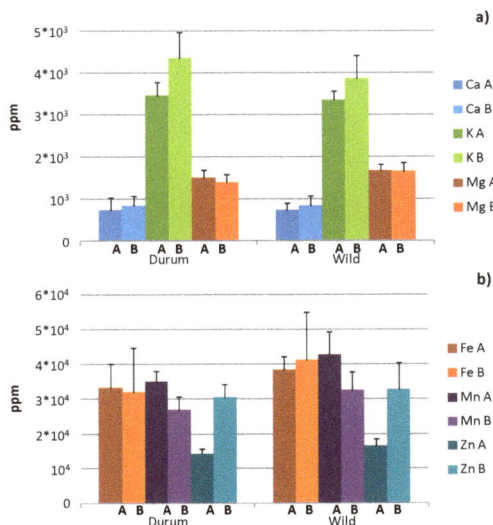

Figure 1. Average and standard deviation content of (**a**) macro- (Ca, K, Mg) and (**b**) higher micro- (Fe, Zn, Mn) elements, calculated for the samples of two types of wheat (expressed as part per million, ppm) and for each site (A, Valenzano, Bari and B, Policoro, Matera).

A further level of investigation was performed by calculating the correlation matrix based on Pearson's coefficient for all the measured elements. An overview about the potential linear relationship between the metals (macronutrients, Ca, Mg, K, micronutrients, Zn, Fe, Cu, Mn and toxic trace elements, Cd, Pb) was obtained for each tetraploid genotype (durum and wild) and for each cultivation site, A and B (Tables 3–6). Although a simple correlation analysis, especially considering that a large data set should be carefully used for expressing nutrient relationships, we could find and will discuss a limited range of significant correlations that were obtained in this study. A high level of correlation was observed for durum wheat for both the two sites: in particular six couples of elements (Cu/Zn, Mg/Zn with significance at $p < 0.001$ and Cu/Mg, Cu/Mn, Zn/Mn, Cr/Fe with significance at $p < 0.01$) for site A and seven couples of elements (K/Cu, K/Mg, K/Mg, Cu/Mg, Cu/Mn, Cr/Fe with significance at $p < 0.001$ and Ca/Cd with significance at $p < 0.01$) for site B (Tables 3 and 5), while wild wheat showed a low number of correlations in both the two sites (Cr/Cu, Fe/Mn with significance at $p < 0.001$ and Cr/Fe, Cr/Mg, Cd/Mn and Ca/Mg with significance at $p < 0.01$ for site A; Cr/Fe with significance at $p < 0.001$ and Ca/Cd, K/Zn, K/Cr, K/Fe, Cr/Fe with significance at $p < 0.01$ for site B, Tables 4 and 6). It should also be noted that in some cases, more significant or positive correlation values for other elements (i.e., Mn/Mg) were expected, although comparable results to those that were obtained in this work were also reported in literature [28,37,47,48].

Table 3. Pearson correlation matrix among all the variables of durum wheat genotypes grown in the A site (Valenzano, Bari). *, **, *** indicate significance at $p < 0.05$, $p < 0.01$ and $p < 0.001$, respectively.

	Ca	K	Cu	Zn	Cr	Fe	Pb	Cd	Mg
K	0.06								
Cu	0.03	0.23							
Zn	−0.03	0.00	0.50 ***						
Cr	0.03	0.07	0.28 *	0.09					
Fe	−0.05	0.08	0.28 *	0.17	0.35 **				
Pb	0.11	0.05	−0.01	0.10	0.18	0.08			
Cd	−0.20	0.21	0.09	0.17	−0.08	0.22	0.03		
Mg	0.06	0.29 *	0.39 **	0.45 ***	0.25 *	0.22	0.13	−0.02	
Mn	−0.04	−0.17	0.36 **	0.36 **	0.20	0.27 *	0.07	0.30 *	0.29 *

Table 4. Pearson correlation matrix among all the variables of wild wheat genotypes grown in the A site (Valenzano, Bari). *, **, *** indicate significance at $p < 0.05$, $p < 0.01$ and $p < 0.001$, respectively.

	Ca	K	Cu	Zn	Cr	Fe	Pb	Cd	Mg
K	0.14								
Cu	−0.49	−0.42							
Zn	0.46	−0.16	−0.41						
Cr	−0.49	−0.46	0.93 ***	−0.20					
Fe	−0.08	−0.41	0.61	0.19	0.80 **				
Pb	0.39	−0.25	−0.18	0.29	0.04	0.33			
Cd	0.44	−0.49	0.40	0.21	0.39	0.62	0.09		
Mg	0.66 **	0.34	−0.45	0.10	−0.66 **	−0.51	−0.38	0.26	
Mn	0.26	−0.36	0.51	0.16	0.62	0.87 ***	0.33	0.80 **	−0.18

Table 5. Pearson correlation matrix among all the variables of durum wheat genotypes grown in the B site (Policoro, Matera). *, **, *** indicate significance at $p < 0.05$, $p < 0.01$ and $p < 0.001$, respectively.

	Ca	K	Cu	Zn	Cr	Fe	Pb	Cd	Mg
K	0.03								
Cu	0.03	0.55 ***							
Zn	−0.01	0.72	0.60						
Cr	−0.17	−0.17	−0.27 *	−0.21					
Fe	−0.14	0.04	−0.06	−0.08	0.82 ***				
Pb	0.03	−0.06	−0.13	0.00	0.06	−0.02			
Cd	0.32 **	0.26 *	0.29 *	0.28 *	0.00	0.05	−0.10		
Mg	0.08	0.86 ***	0.46 ***	0.65	−0.24	−0.05	−0.03	0.29 *	
Mn	0.12	0.46 ***	0.45 ***	0.54	−0.20	−0.05	−0.17	0.21 *	0.24 *

Table 6. Pearson correlation matrix among all the variables of wild wheat genotypes grown in the B site (Policoro, Matera). *, **, *** indicate significance at $p < 0.05$, $p < 0.01$ and $p < 0.001$, respectively.

	Ca	K	Cu	Zn	Cr	Fe	Pb	Cd	Mg
K	0.48								
Cu	0.01	0.10							
Zn	0.62 *	0.78 **	0.15						
Cr	0.09	0.73 **	0.49	0.67 *					
Fe	0.08	0.74 **	0.60 *	0.65 *	0.97 ***				
Pb	−0.16	−0.39	0.19	−0.11	−0.08	−0.09			
Cd	0.74 **	0.35	0.20	0.56	0.26	0.19	−0.01		
Mg	0.20	0.44	−0.08	0.70 *	0.54	0.46	0.09	0.12	
Mn	0.43	0.37	0.46	0.16	0.32	0.35	−0.33	0.50	−0.40

Multivariate statistical analysis (PCA and OPLS-DA) was used to deeply investigate the variation in macronutrients (Ca, Mg, K), micronutrients (Zn, Fe, Cu, Mn) and toxic trace elements (Cd, Pb) for the whole dataset of 25 wheat accessions and cultivars (three replicates for each cultivar studied) that is representative of durum and wild species in the two cultivation sites. The whole data were studied by OPLS-DA in order to evaluate the potential effect of the different pedoclimatic conditions on the wheat species (Figure 2). In particular, two OPLS-DA models were built using the same number of components (OPLS-DA model of site A: $1 + 2 + 0$, $R^2X = 0.48$, $R^2Y = 0.59$, $Q^2 = 0.43$, p[CV-ANOVA] = 2.40612×10^{-6}; for OPLS-DA model of site B: $1 + 2 + 0$, $R^2X = 0.56$, $R^2Y = 0.83$, $Q^2 = 0.78$, p[CV-ANOVA] = 2.10965×10^{-19}). This approach showed that different pedoclimatic conditions characterize the two sites. Samples cultivated in site A (Valenzano, Bari) were homogeneously distributed in the space of the OPLS-DA graph, while those cultivated in site B (Policoro, Matera) appeared well differentiated in two groups (durum, wild).

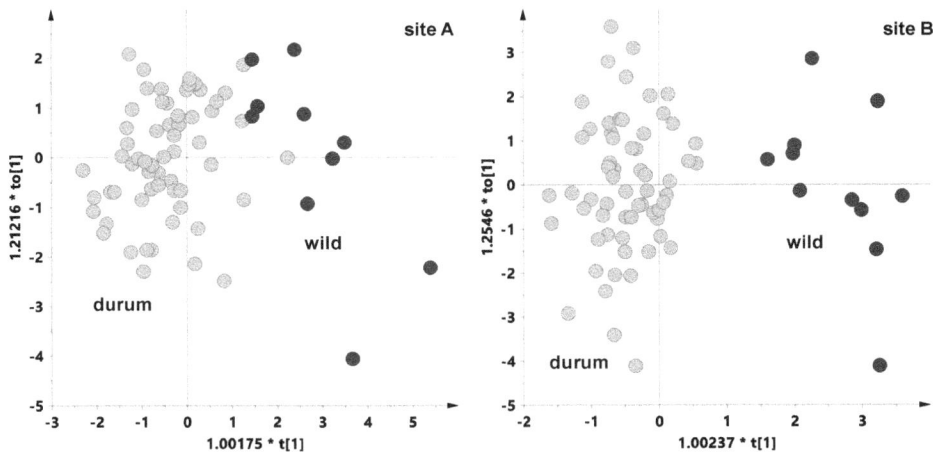

Figure 2. OPLS-DA scoreplots for each site (A, **left**, B, **right**). For OPLS-DA model of site A: $1 + 2 + 0$, $R^2X = 0.48$, $R^2Y = 0.59$, $Q^2 = 0.43$, p[CV-ANOVA] = 2.40612×10^{-6}; for OPLS-DA model of site B: $1 + 2 + 0$, $R^2X = 0.56$, $R^2Y = 0.83$, $Q^2 = 0.78$, p[CV-ANOVA] = 2.10965×10^{-19}.

With the aim to assess the different element uptake ability of the studied tetraploid wheat subspecies, an OPLS-DA analysis was performed for the whole data. In the first place, the dataset was analysed by differentiating in the model two different categories, the durum and the wild wheat types. The resulting OPLS-DA model, which was built with one predictive (t[1]) and three orthogonal components $(1 + 3 + 0)$ beside the observed classification parameters for the two categories ($R^2X = 0.58$, $Q^2 = 0.49$) gave an interesting hint for the sample distribution among the two sites (Figure 3). Indeed, this appears to be the natural discrimination ($R^2Y = 0.53$) observed in the first orthogonal component (to[1]). The colour encoded sample distribution among the two sites is clearly observed in Figure 3. The OPLS-DA scoreplot showed that durum wheat genotypes were mainly distributed at negative values of the predictive component t[1], while the intra-class variation resulted in a large distribution of the data along the predictive component to[1] in a wide range of values (from -7 to 3). On the other hand, wild wheat accessions were clustered both at positive values of t[1] and in a range of -2 and $+2$ of to[1].

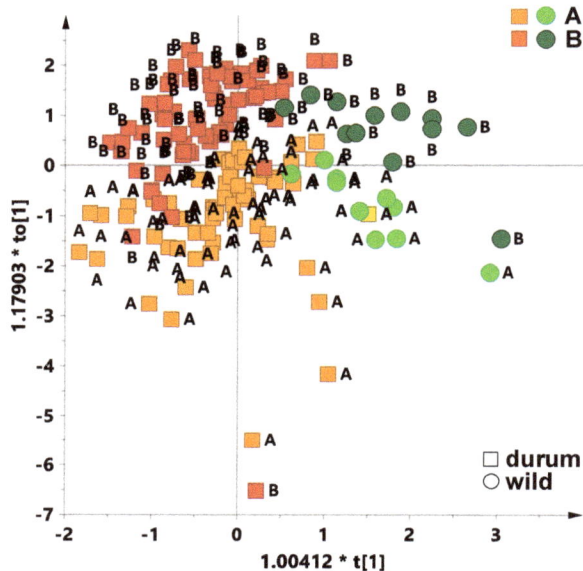

Figure 3. OPLS-DA scoreplot for the whole data considering the two sites (A and B) as discriminating class (1 + 3 + 0, $R^2X = 0.58$, $R^2Y = 0.53$, $Q^2 = 0.49$, p[CV-ANOVA] = 3.9421×10^{-17}).

A further MVA was then performed with the aim to differentiate, according to the cultivation site, durum and wild wheat types. For this purpose, two independent PCA and OPLS-DA models were built using the cultivation site as discrimination class. Both for the PCA (data not shown) and OPLS-DA (Figure 4) models, the two wheat groups resulted differently for the two cultivation sites. According to the OPLS-DA models that are depicted in Figure 4, and relative model parameters (in particular Q^2 values), the more pronounced separation among the two cultivation sites was observed for durum ($Q^2 = 0.92$) with respect to wild type ($Q^2 = 0.75$) wheat. Moreover, due to a different sample size of durum (126 samples) in comparison with wild (22 samples) statistical models, a further cross check was applied to assess the soundness of the obtained parameters for the OPLS-DA models that are reported in Figure 4. By using the Weka open-source data mining software (v. 3.8.3, University of Waikato New Zealand) [49], a filter was applied to randomly remove a given percentage of samples from the durum sample set. For this purpose, 80% of durum samples were randomly excluded in order to obtain a comparable sample size for durum set (25) with respect to wild (22) samples. Subsequently, a Naïve Bayes classification was applied separately for both the durum and wild sets. The models' reliability (indicated with "Correctly Classified Instances" Tables S3 and S4), resulted with 87.5% and 71.4% for durum and wild samples, respectively, confirming our hypothesis that wild wheat is the less site sensitive species. This result suggested that a profitable search for the less site sensitive species had to be performed within the considered wild wheat cultivars.

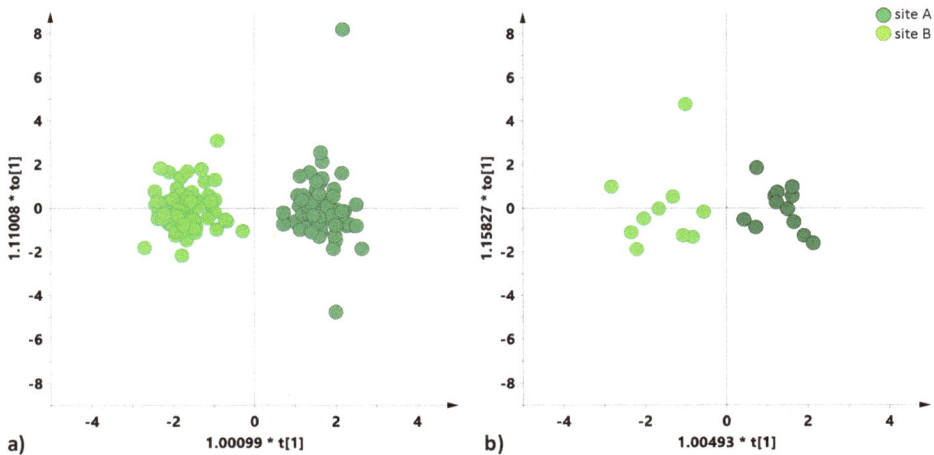

Figure 4. OPLS-DA scoreplots for each of the types of wheat (durum, wild) when they were considered separately in order to evaluate the effect of the site on the wheat species ((**a**). OPLS-DA model for durum wheat: 1 + 1 + 0, R^2X = 0.453, R^2Y = 0.933, Q^2 = 0.921, p[CV-ANOVA] = 0; (**b**). OPLS-DA model for wild wheat: 1 + 1 + 0, R^2X = 0.47, R^2Y = 0.85, Q^2 = 0.75, p[CV-ANOVA] = 6.05943 × 10^{-5}).

Finally, new OPLS-DA models were then calculated by using the four available data sets (three replicates for each of the two cultivation sites) for the four wild wheat accessions. The cultivation site was chosen as the discriminating class for each of the four OPLS-DA models. The obtained calculated quality model parameters are reported in Table 7, showing the lowest (MG29896) and highest (MG5323) cultivation site differentiation. Among the genotypes of wild wheat tetraploids, MG29896 resulted as the least affected from the cultivation site, having the lowest predictive ability, expressed as Q^2 value, followed by MG4330, MG5323.

Table 7. R^2X, R^2Y and Q^2 parameters reported for every OPLS-DA model performed on the single wild tetraploid wheat genotypes (three replicates for each of the two cultivation sites).

Genotype	Model	R^2X	R^2Y	Q^2
MG5323	1 + 1 + 0	0.66	0.99	0.94
MG4343	1 + 1 + 0	0.82	0.98	0.93
MG4330	1 + 1 + 0	0.81	0.95	0.85
MG29896	1 + 1 + 0	0.74	0.99	0.82

4. Discussion

Variation in mineral micronutrient concentrations were found in the grains of wheat lines of diverse origin [50], and significant differences between wheat genotypes were found for grain Fe and Zn, however not Se concentration. Spelt, einkorn and emmer wheat appeared to contain higher Se concentration in their grains than bread and durum wheat. Other studies found that the genetic variability available in the modern wheat pool is moderate, and it would be necessary to use other wheat genetic resources in the breeding process [23]. In this respect, Cakmak et al. and Gomez-Becerra et al. [19,22] have shown that the ssp. *dicoccoides* and ssp. *dicoccum* could be a good source of high micronutrients concentration [51]. Ficco et al. [52] estimated the magnitude of genotype × environment interaction effects in a collection of Italian durum wheat cultivars that were evaluated for the mineral elements concentration and suggested that the breeding activity for Fe and Zn would be difficult because G × E interaction is prevalent, though multi-location evaluation of germplasm collection might help to identify superior genotypes.

Analysis of the contents of bioactive components in a wheat diversity panel grown in six diverse environments showed that the extent of variation due to variety and the environment differed significantly between components [53]. However, significant correlations were found between bioactive components and environmental factors, with even highly heritable components differing in amount between grain samples grown in different years on different sites [10]. Detailed analyses of bioactive components were carried out under the EU FP6 HEALTHGRAIN program on a wheat collection grown on a single site, and principal component analysis allowed us to classify the wheat genotypes on the basis of the bioactive components and to clearly separate wheat species (bread, durum, spelt, emmer and einkorn wheat) from related cereals (barley, rye and oats) [53]. It is also important to consider the evolutionary differences of wheat with different ploidy levels. Indeed, the comparison of photosynthesis and antioxidant defense systems in wheat with different ploidy levels showed significant differences in diploid, tetraploid and hexaploid wheats, suggesting an important regulatory role in photosystems and antioxidative systems of plants [16]. Several investigations have shown wide variation in phytochemical composition between genotypes, with important effects of environmental factors and genotypes x environment interactions, and indicated that ancient wheats have health benefits compared with the current cultivated wheat. However, recent reviews on the health benefits of ancient and modern wheat have ascertained that most studies were not comparable in terms of genetic materials, growth conditions and processing, and reached the conclusion that further studies are required by different research groups and should consider a wider range of genotypes of ancient and modern wheat species grown together in the same field experiments [11,54].

In the present study, the data obtained by ICP-AES (macro-, micronutrients and toxic trace elements) were analyzed for different single wheat genotypes belonging to the three different *T. turgidum* wheat subspecies, under different pedoclimatic conditions (two sites, two growing seasons). The significantly higher content of Mg (among the macronutrients) and the highest levels of Mn, Fe and Zn (among the micronutrients) were found for wild accessions with respect to durum wheat cultivars for both the two pedoclimatic conditions (two sites, two growing seasons). Moreover, wild types were also the wheat ones with the lowest level of accumulation of the trace toxic elements, in particular for Cd concentration. It should be noted that, in all the measured wheat samples, cadmium levels never exceeded the EU maximum allowed level of this contaminant in cereals and cereal products intended for human consumption (European Commission Regulation 2006) [55]. On the other hand, we found that among all the elements, the Zn levels that resulted were mainly affected by the cultivation site in all the measured wheat samples. This is mainly due to specific Zn bioaccumulation capacities of the soil [56] and, therefore, resulted in a very different uptake behaviour in the three wheat species, with the highest level in wild wheat species.

The above reported results, obtained by applying both correlation studies and multivariate analysis, also demonstrated that wild tetraploid wheats appeared the less affected from the pedoclimatic conditions which characterized the two cultivation sites (and their intrinsic mineral bioavailability).

Indeed, according to correlation studies, durum wheat appeared as the "most demanding" species (with respect to wild), showing a high number of correlations among elements for both the two cultivation sites. Therefore, for this intrinsic characteristic, durum wheat genotypes could be less prone to adaptation in different agricultural sites. The obtained results were also confirmed by the evidence that was also found by other authors [57–59], that both the concentrations of the analyzed elements and the relationships between them are species-specific. As reported in literature [55], antagonistic as well as synergistic interactions of elements are determined by the level of each nutrient, both in the soil and plant species, and sometimes even among cultivars of the same species, suggesting that some elements are translocated in plants in a similar way [37]. As a consequence, the hypothesis that the presence of a high level of correlations among elements (with consequent constraints in their uptake) could result in a wheat genotypes with a higher demanding character for simultaneous micro-nutrients availability could be reasonable. Finally, multivariate analysis (OPLS-DA) was also applied in order to establish the site dependence ranking among the considered durum and wild tetraploid wheats. For this reason, the statistic supervised model was built first using the durum and the wild wheat genotypes, and subsequently the cultivation site as the discriminating category. This approach allowed the focusing on the wild wheat species as the less site sensitive in the site dependence ranking. Despite the small sample size, multivariate analysis (OPLS-DA) was then applied in order to establish the existence of a ranking among all the studied wild cultivars. Consequently, OPLS-DA models using the two cultivation sites as a discriminating category were built for each of the wild wheat accessions and the obtained predictive ability (expressed by Q^2 values) that was evaluated. This operating system could be successfully used in order to reveal potential differences between the wild wheat genotypes with respect to their cultivation site-dependence [60]. Among the wild wheat types, the MG29896 accession was the least affected from the cultivation site, having the lowest Q^2, followed by MG4330, MG4343, MG5323. The lowest predictability (Q^2 value) of the MG29896 OPLS-DA model corresponded to the lowest difference of samples, which were cultivated in two different pedoclimatic conditions (two sites, two growing seasons). Among the genotypes of the present study, wild appeared as the most remarkable for higher content of Mg (among the macronutrients), higher levels of Mn, Fe and Zn (among the micronutrients) and lower level of Cd accumulation. According to correlation studies and MVA analyses, wild wheat also appeared the less demanding for simultaneous micro-nutrients availability and environmental conditions sensitivity. Therefore, the wild tetraploid wheats could be recommended for improvement and biofortification breeding programs in order to satisfy human nutrition requirements, as also reported by other authors [25–28,37,61]. Although both macro- and micro-nutrient concentration in wheat are very important for increasing the productivity and efficiency of a crop's yield, it is worth noting that micronutrient balance represents a very crucial point, since micronutrients deficiency in soil is a global phenomenon [62]. Therefore, bioavailability and their correlation should be considered [37]. In general, the present analysis of cultivar differences, and within wheat classes and subclasses, in different environmental conditions gave promising results. The differences are reflected in mineral concentration and absorption capability, and accumulation could also be useful in cross linkage studies. Nevertheless, further and more detailed studies are required, particularly on a wider range of genotypes of ancient and modern wheat species.

Supplementary Materials: The following are available online at http://www.mdpi.com/2071-1050/11/3/736/s1.

Author Contributions: Formal analysis, L.D.C.; Investigation, B.L. and D.M.; Methodology, R.S. and F.P.F.; Resources, F.P.F.; Validation, D.M.; Visualization, L.D.C.; Writing—original draft, L.D.C., B.L. and G.M.; Writing—review & editing, L.D.C. and F.P.F.

Funding: This research received no external funding.

Acknowledgments: This work was supported by PON (PONa3_00334) "Human and Environment Health Research 329 Center (2HE—Research Center). The authors gratefully acknowledge Prof. Antonio Blanco (Department of Soil, Plant & Food Sciences, Genetics and Plant Breeding Section, University Aldo Moro, Bari, Italy) for critical reading of the manuscript.

Conflicts of Interest: The authors declare no conflict of interest.

References

1. Faostat, F. *Statistical Databases*; Food and Agriculture Organization of the United Nations: Rome, Italy, 2018.
2. Shewry, P.R.; Hey, S.J. The contribution of wheat to human diet and health. *Food Energy Secur.* **2015**, *4*, 178–202. [PubMed]
3. Shewry, P.R.; Hey, S. Do "ancient" wheat species differ from modern bread wheat in their contents of bioactive components? *J. Cereal Sci.* **2015**, *65*, 236–243. [CrossRef]
4. Uauy, C.; Distelfeld, A.; Fahima, T.; Blechl, A.; Dubcovsky, J. A NAC gene regulating senescence improves grain protein, zinc, and iron content in wheat. *Science* **2006**, *314*, 1298–1301. [CrossRef] [PubMed]
5. Faris, J.D. Wheat domestication: Key to agricultural revolutions past and future. In *Genomics of Plant Genetic Resources*; Springer: Berlin, Germany, 2014; pp. 439–464.

6. Filipčev, B.; Kojić, J.; Krulj, J.; Bodroža-Solarov, M.; Ilić, N. Betaine in cereal grains and grain-based products. *Foods* **2018**, *7*, 49. [CrossRef]
7. Merah, O.; Deleens, E.; Nachit, M.; Monneveux, P. Carbon isotope discrimination, leaf characteristics and grain yield of interspecific wheat lines and their durum parents under Mediterranean conditions. *Cereal Res. Commun.* **2001**, *29*, 143–149.
8. Merah, O.; Deleens, E.; Al Hakimi, A.; Monneveux, P. Carbon isotope discrimination and grain yield variations among tetraploid wheat species cultivated under contrasting precipitation regimes. *J. Agron. Crop Sci.* **2001**, *186*, 129–134. [CrossRef]
9. Merah, O. Carbon isotope discrimination and mineral composition of three organs in durum wheat genotypes grown under Mediterranean conditions. *Comptes Rendus De L'académie Des Sci.* **2001**, *324*, 355–363. [CrossRef]
10. Shewry, P.R.; Hawkesford, M.J.; Piironen, V.; Lampi, A.-M.; Gebruers, K.; Boros, D.; Andersson, A.A.; Åman, P.; Rakszegi, M.; Bedo, Z. Natural variation in grain composition of wheat and related cereals. *J. Agric. Food Chem.* **2013**, *61*, 8295–8303. [CrossRef] [PubMed]
11. Shewry, P.R. Do ancient types of wheat have health benefits compared with modern bread wheat? *J. Cereal Sci.* **2017**, *79*, 469–476. [CrossRef]
12. Kärkkäinen, O.; Lankinen, M.A.; Vitale, M.; Jokkala, J.; Leppänen, J.; Koistinen, V.; Lehtonen, M.; Giacco, R.; Rosa-Sibakov, N.; Micard, V.; et al. Diets rich in whole grains increase betainized compounds associated with glucose metabolism. *Am. J. Clin. Nutr.* **2018**, *108*, 971–979. [CrossRef]
13. Godfrey, D.; Hawkesford, M.J.; Powers, S.J.; Millar, S.; Shewry, P.R. Effects of crop nutrition on wheat grain composition and end use quality. *J. Agric. Food Chem.* **2010**, *58*, 3012–3021. [CrossRef] [PubMed]
14. Laddomada, B.; Caretto, S.; Mita, G. Wheat bran phenolic acids: Bioavailability and stability in whole wheat-based foods. *Molecules* **2015**, *20*, 15666–15685. [CrossRef] [PubMed]
15. Laddomada, B.; Durante, M.; Minervini, F.; Garbetta, A.; Cardinali, A.; D'Antuono, I.; Caretto, S.; Blanco, A.; Mita, G. Phytochemical composition and anti-inflammatory activity of extracts from the whole-meal flour of Italian durum wheat cultivars. *Int. J. Mol. Sci.* **2015**, *16*, 3512–3527. [CrossRef] [PubMed]
16. Brestic, M.; Zivcak, M.; Hauptvogel, P.; Misheva, S.; Kocheva, K.; Yang, X.; Li, X.; Allakhverdiev, S.I. Wheat plant selection for high yields entailed improvement of leaf anatomical and biochemical traits including tolerance to non-optimal temperature conditions. *Photosynth. Res.* **2018**, *136*, 245–255. [CrossRef] [PubMed]
17. Mao, H.; Chen, M.; Su, Y.; Wu, N.; Yuan, M.; Yuan, S.; Brestic, M.; Zivcak, M.; Zhang, H.; Chen, Y. Comparison on Photosynthesis and Antioxidant Defense Systems in Wheat with Different Ploidy Levels and Octoploid Triticale. *Int. J. Mol. Sci.* **2018**, *19*, 3006. [CrossRef] [PubMed]
18. Laddomada, B.; Durante, M.; Mangini, G.; D'Amico, L.; Lenucci, M.S.; Simeone, R.; Piarulli, L.; Mita, G.; Blanco, A. Genetic variation for phenolic acids concentration and composition in a tetraploid wheat (*Triticum turgidum* L.) collection. *Genet. Resour. Crop Evol.* **2017**, *64*, 587–597. [CrossRef]
19. Gomez-Becerra, H.F.; Yazici, A.; Ozturk, L.; Budak, H.; Peleg, Z.; Morgounov, A.; Fahima, T.; Saranga, Y.; Cakmak, I. Genetic variation and environmental stability of grain mineral nutrient concentrations in Triticum dicoccoides under five environments. *Euphytica* **2010**, *171*, 39–52. [CrossRef]
20. Velu, G.; Singh, R.; Arun, B.; Mishra, V.; Tiwari, C.; Joshi, A.; Virk, P.; Cherian, B.; Pfeiffer, W. Reaching out to farmers with high Zinc wheat varieties through public-private partnerships–An Experience from Eastern-Gangetic Plains of India. *Adv. Food Technol. Nutr. Sci* **2015**, *1*, 73–75. [CrossRef]
21. Guzmán, C.; Medina-Larqué, A.S.; Velu, G.; González-Santoyo, H.; Singh, R.P.; Huerta-Espino, J.; Ortiz-Monasterio, I.; Peña, R.J. Use of wheat genetic resources to develop biofortified wheat with enhanced grain zinc and iron concentrations and desirable processing quality. *J. Cereal Sci.* **2014**, *60*, 617–622. [CrossRef]
22. Cakmak, I.; Pfeiffer, W.H.; McClafferty, B. Biofortification of durum wheat with zinc and iron. *Cereal Chem.* **2010**, *87*, 10–20. [CrossRef]
23. Magallanes-López, A.M.; Hernandez-Espinosa, N.; Velu, G.; Posadas-Romano, G.; Ordoñez-Villegas, V.M.G.; Crossa, J.; Ammar, K.; Guzmán, C. Variability in iron, zinc and phytic acid content in a worldwide collection of commercial durum wheat cultivars and the effect of reduced irrigation on these traits. *Food Chem.* **2017**, *237*, 499–505. [CrossRef] [PubMed]
24. Yan, J.; Xue, W.-T.; Yang, R.-Z.; Qin, H.-B.; Zhao, G.; Tzion, F.; Cheng, J.-P. Quantitative Trait Loci Conferring Grain Selenium Nutrient in Durum Wheat× Wild Emmer Wheat RIL Population. *Czech J. Genet. Plant Breed.* **2018**, *54*, 52–58. [CrossRef]

25. Awan, S.; Ahmad, S.; Ali, M.; Ahmed, M.; Rao, A. Use of multivariate analysis in determining characteristics for grain yield selection in wheat. *Sarhad J. Agric.* **2015**, *31*, 139–150. [CrossRef]

26. Shewry, P.R.; Pellny, T.K.; Lovegrove, A. Is modern wheat bad for health. *Nat. Plants* **2016**, *2*, 16097. [CrossRef] [PubMed]

27. Arzani, A.; Ashraf, M. Cultivated Ancient Wheats (*Triticum* spp.): A Potential Source of Health-Beneficial Food Products. *Compr. Rev. Food Sci. Food Saf.* **2017**, *16*, 477–488. [CrossRef]

28. Murphy, K.M.; Reeves, P.G.; Jones, S.S. Relationship between yield and mineral nutrient concentrations in historical and modern spring wheat cultivars. *Euphytica* **2008**, *163*, 381–390. [CrossRef]

29. Kaur, K.D.; Jha, A.; Sabikhi, L.; Singh, A. Significance of coarse cereals in health and nutrition: A review. *J. Food Sci. Technol.* **2014**, *51*, 1429–1441. [CrossRef]

30. Srinivasa, J.; Arun, B.; Mishra, V.K.; Singh, G.P.; Velu, G.; Babu, R.; Vasistha, N.K.; Joshi, A.K. Zinc and iron concentration QTL mapped in a *Triticum spelta* × *T. aestivum* cross. *Theor. Appl. Genet.* **2014**, *127*, 1643–1651. [CrossRef]

31. Pasqualone, A.; Piarulli, L.; Mangini, G.; Gadaleta, A.; Blanco, A.; Simeone, R. Quality characteristics of parental lines of wheat mapping populations. *Agric. Food Sci.* **2015**, *24*, 118–127. [CrossRef]

32. Mzid, N.; Todorovic, M.; Albrizio, R.; Cantore, V. *Remote Sensing Based Monitoring of Durum Wheat under Water Stress Treatments*; Spatial Analysis and GEOmatics: Rouen, France, 2017.

33. Scortichini, M.; Jianchi, C.; De Caroli, M.; Dalessandro, G.; Pucci, N.; Modesti, V.; L'Aurora, A.; Petriccione, M.; Zampella, L.; Mastrobuoni, F.; et al. A zinc, copper and citric acid biocomplex shows promise for control of *Xylella fastidiosa* subsp. *pauca* in olive trees in Apulia region (southern Italy). *Phytopathol. Mediterr.* **2018**, *57*, 48–72.

34. R Development Core Team. *R: A Language and Environment for Statistical Computing*; R Foundation for Statistical Computing: Vienna, Austria, 2013.

35. Xia, J.; Psychogios, N.; Young, N.; Wishart, D.S. MetaboAnalyst: A web server for metabolomic data analysis and interpretation. *Nucleic Acids Res.* **2009**, *37*, W652–W660. [CrossRef]

36. Xia, J.; Wishart, D.S. Using MetaboAnalyst 3.0 for comprehensive metabolomics data analysis. *Curr. Protoc. Bioinform.* **2016**, *55*. [CrossRef] [PubMed]

37. Pandey, A.; Khan, M.K.; Hakki, E.E.; Thomas, G.; Hamurcu, M.; Gezgin, S.; Gizlenci, O.; Akkaya, M.S. Assessment of genetic variability for grain nutrients from diverse regions: Potential for wheat improvement. *SpringerPlus* **2016**, *5*, 1912. [CrossRef] [PubMed]

38. Hauke, J.; Kossowski, T. Comparison of values of Pearson's and Spearman's correlation coefficients on the same sets of data. *Quaest. Geogr.* **2011**, *30*, 87–93. [CrossRef]

39. Bro, R.; Smilde, A.K. Principal component analysis. *Anal. Methods* **2014**, *6*, 2812–2831. [CrossRef]

40. Trygg, J.; Wold, S. Orthogonal projections to latent structures (O-PLS). *J. Chemom.* **2002**, *16*, 119–128. [CrossRef]

41. Triba, M.N.; Le Moyec, L.; Amathieu, R.; Goossens, C.; Bouchemal, N.; Nahon, P.; Rutledge, D.N.; Savarin, P. PLS/OPLS models in metabolomics: The impact of permutation of dataset rows on the K-fold cross-validation quality parameters. *Mol. Biosyst.* **2015**, *11*, 13–19. [CrossRef] [PubMed]

42. Aru, V.; Pisano, M.B.; Savorani, F.; Engelsen, S.B.; Cosentino, S.; Marincola, F.C. Data on the changes of the mussels' metabolic profile under different cold storage conditions. *Data Brief* **2016**, *7*, 951–957. [CrossRef] [PubMed]

43. Wheelock, Å.M.; Wheelock, C.E. Trials and tribulations of 'omics data analysis: Assessing quality of SIMCA-based multivariate models using examples from pulmonary medicine. *Mol. Biosyst.* **2013**, *9*, 2589–2596. [CrossRef] [PubMed]

44. Pettigrew, W.T. Potassium influences on yield and quality production for maize, wheat, soybean and cotton. *Physiol. Plant.* **2008**, *133*, 670–681. [CrossRef]

45. Tejera, R.L.; Luis, G.; González-Weller, D.; Caballero, J.M.; Gutiérrez, Á.J.; Rubio, C.; Hardisson, A. Metals in wheat flour; comparative study and safety control. *Nutr. Hosp.* **2013**, *28*, 506–513. [PubMed]

46. Garnett, T.P.; Graham, R.D. Distribution and remobilization of iron and copper in wheat. *Ann. Bot.* **2005**, *95*, 817–826. [CrossRef] [PubMed]

47. Peleg, Z.; Cakmak, I.; Ozturk, L.; Yazici, A.; Jun, Y.; Budak, H.; Korol, A.B.; Fahima, T.; Saranga, Y. Quantitative trait loci conferring grain mineral nutrient concentrations in durum wheat× wild emmer wheat RIL population. *Theor. Appl. Genet.* **2009**, *119*, 353–369. [CrossRef] [PubMed]

48. Hakki, E.E.; Dograr, N.; Pandey, A.; Khan, M.K.; Hamurcu, M.; Kayis, S.A.; Gezgin, S.; ÖLMEZ, F.; Akkaya, M.S. Molecular and elemental characterization of selected Turkish durum wheat varieties. *Not. Bot. Horti Agrobot. Cluj-Napoca* **2014**, *42*, 431–439. [CrossRef]

49. Witten, I.H.; Frank, E.; Hall, M.A.; Pal, C.J. *Data Mining: Practical Machine Learning Tools and Techniques*; Morgan Kaufmann: San Francisco, CA, USA, 2016.

50. Zhao, F.; Su, Y.; Dunham, S.; Rakszegi, M.; Bedo, Z.; McGrath, S.; Shewry, P. Variation in mineral micronutrient concentrations in grain of wheat lines of diverse origin. *J. Cereal Sci.* **2009**, *49*, 290–295. [CrossRef]

51. Monasterio, I.; Graham, R.D. Breeding for trace minerals in wheat. *Food Nutr. Bull.* **2000**, *21*, 392–396. [CrossRef]

52. Ficco, D.; Riefolo, C.; Nicastro, G.; De Simone, V.; Di Gesu, A.; Beleggia, R.; Platani, C.; Cattivelli, L.; De Vita, P. Phytate and mineral elements concentration in a collection of Italian durum wheat cultivars. *Field Crop. Res.* **2009**, *111*, 235–242. [CrossRef]

53. Shewry, P.R.; Piironen, V.; Lampi, A.-M.; Edelmann, M.; Kariluoto, S.; Nurmi, T.; Fernandez-Orozco, R.; Andersson, A.A.; Åman, P.; Fras, A. Effects of genotype and environment on the content and composition of phytochemicals and dietary fiber components in rye in the HEALTHGRAIN diversity screen. *J. Agric. Food Chem.* **2010**, *58*, 9372–9383. [CrossRef]

54. Dinu, M.; Whittaker, A.; Pagliai, G.; Benedettelli, S.; Sofi, F. Ancient wheat species and human health: Biochemical and clinical implications. *J. Nutr. Biochem.* **2018**, *52*, 1–9. [CrossRef]

55. Fageria, V. Nutrient interactions in crop plants. *J. Plant Nutr.* **2001**, *24*, 1269–1290. [CrossRef]

56. He, Z.L.; Yang, X.E.; Stoffella, P.J. Trace elements in agroecosystems and impacts on the environment. *J. Trace Elem. Med. Biol.* **2005**, *19*, 125–140. [CrossRef] [PubMed]

57. Suchowilska, E.; Wiwart, M.; Kandler, W.; Krska, R. A comparison of macro-and microelement concentrations in the whole grain of four Triticum species. *Plant Soil Environ.* **2012**, *58*, 141–147. [CrossRef]

58. Moreira-Ascarrunz, S.D.; Larsson, H.; Prieto-Linde, M.L.; Johansson, E. Mineral nutritional yield and nutrient density of locally adapted wheat genotypes under organic production. *Foods* **2016**, *5*, 89. [CrossRef] [PubMed]

59. Matthews, S.B.; Santra, M.; Mensack, M.M.; Wolfe, P.; Byrne, P.F.; Thompson, H.J. Metabolite profiling of a diverse collection of wheat lines using ultraperformance liquid chromatography coupled with time-of-flight mass spectrometry. *PLoS ONE* **2012**, *7*, e44179. [CrossRef] [PubMed]

60. Righetti, L.; Rubert, J.; Galaverna, G.; Folloni, S.; Ranieri, R.; Stranska-Zachariasova, M.; Hajslova, J.; Dall'Asta, C. Characterization and discrimination of ancient grains: A metabolomics approach. *Int. J. Mol. Sci.* **2016**, *17*, 1217. [CrossRef] [PubMed]

61. Garcia-Oliveira, A.L.; Chander, S.; Ortiz, R.; Menkir, A.; Gedil, M. Genetic basis and breeding perspectives of grain iron and zinc enrichment in cereals. *Front. Plant Sci.* **2018**, *9*, 937. [CrossRef] [PubMed]

62. Dimkpa, C.O.; Bindraban, P.S. Fortification of micronutrients for efficient agronomic production: A review. *Agron. Sustain. Dev.* **2016**, *36*, 7. [CrossRef]

![sustainability logo] *sustainability*

MDPI

Article

Chemical and Sensorial Characteristics of Olive Oil Produced from the Lebanese Olive Variety 'Baladi'

Milad El Riachy [1,*], Christelle Bou-Mitri [2], Amira Youssef [3], Roland Andary [4] and Wadih Skaff [5]

[1]   Department of Olive and Olive Oil, Lebanese Agricultural Research Institute (LARI), P.O. Box 287, Zahlé, Tal Amara, Lebanon

[2]   Faculty of Nursing and Health Sciences, Notre Dame University-Louaize, Zouk Mosbeh, P.O. Box 72, Zouk Mikael, Lebanon; cboumitri@ndu.edu.lb

[3]   Hasbaya station, Lebanese Agricultural Research Institute (LARI), P.O. Box 1704, Hasbaya, Lebanon; amira-youssef@lari.gov.lb

[4]   Lebanese Industrial Value Chain Development Project, US-AID, Confidence center, Dimitri Hayeck Street, P.O. Box 55463, Beirut, Horsh Tabet, Sin el Fil, Lebanon; Roland_Andary@dai.com

[5]   Ecole Supérieure d'Ingénieurs d'Agronomie Méditerranéenne, Saint Joseph University, P.O. Box 159, Zahlé, Taanayel, Lebanon; wadih.skaff@usj.edu.lb

*   Correspondence: mriachy@lari.gov.lb; Tel.: +961-8-900-037 (ext. 220)

Received: 25 October 2018; Accepted: 23 November 2018; Published: 6 December 2018

Abstract: The olive oil quality, nutritional and sensorial characteristics are associated with the chemical composition, which is the result of a complex interaction between several environmental, agronomical and technological factors. The aim of the present study is to investigate the impact of the geographical origin, harvesting time and processing system on the chemical composition and sensorial characteristics of olive oils produced from the Lebanese olive 'Baladi'. Samples ($n = 108$) were collected from North and South Lebanon, at three different harvesting times and from four processing systems. Results showed a strong effect of origin, processing system and harvest time on oil quality, fatty acid composition, total phenols and OSI. The early harvest showed higher total phenols content (220.02 mg GAE/Kg) and higher OSI (9.19 h). Moreover, samples obtained from sinolea and 3-phases recorded the lowest free acidity (0.36% and 0.64%), and the highest OSI (9.87 and 9.84 h). Consumers were not unanimous regarding the studied factors, although samples recording high ranks were mostly from South using sinolea, 3-phases and press systems at early and intermediate harvest. The overall findings suggest that the selection of the harvesting time and of the processing system could have significant influence on the characteristics of the olive oil.

Keywords: *Olea europaea* L.; olive oil; geographical origin; processing system; harvesting time; olive oil quality; fatty acid composition; sensorial evaluation; consumer preferences

1. Introduction

Olive oil is the most commonly consumed vegetable oil in the Mediterranean area owing to its sensorial quality and beneficial health effects [1–3]. The health promoting properties and overall taste of olive oils are associated in particular with their chemical composition [3]. While the product chemical and sensory characteristics determine its quality, they are the result of a complex interaction between several environmental, agronomical and technological factors. In particular, the geographical origin, the olive variety, the harvesting time and the processing system represent the most important factors influencing the olive oil composition [4–7]. Previous studies showed different fatty acid, sterol and tocopherol profiles for the same varieties cultivated in different regions [8]. Studies also reported that the phenolic and chlorophyll contents decrease along ripening with the parallel decrease of the bitter and pungent tastes and increase of the sweetness [9]. Other authors reported an increase in total and individual phenols with the increase of ripening index between 2 and 3.5 after which they decrease dramatically [10,11]. The fatty acid composition also varies during ripening where palmitic acid (C16:0) decreases, linoleic acid (C18:2) increases and oleic acid (C18:1) remains constant. This results in a decrease in monounsaturated (MUFA) to polyunsaturated fatty acids (PUFA) and saturated to unsaturated fatty acid ratio, leading to lower oil oxidative stability and loss of oil quality in general [12,13]. The selection of the processing systems could have a significant effect on the oil oxidation due to the exposure to air oxygen in the press and sinolea systems and the use of mats in the press system. These inconveniences in the aforementioned systems were completely overcome in the modern systems including the 2- and 3-phases [14,15]. However, the 3-phases system might result in a decrease in the phenolic and aromatic compounds leading to lower oxidative stability due to the use of water that could dissolve the hydrophilic phenols that will be removed with the olive mill waste water [16].

In Lebanon, olive oil production holds a very important status in the country's economy. The national production is increasing and has reached 23,000 tons in 2017. Actually, Lebanon counts as an actor in the international trade of olive oil with more than 7703 tons of exports oriented towards countries where a large Lebanese diaspora lives such as United States, Canada, and gulf countries among others [17]. Olive production mainly occurs in the North (41%) and the South-Nabatiye (36%) [17]. Among all, 'Baladi' (that means local or autochthonous in Arabic language) is the most cultivated variety for its adaptation to the local climatic conditions and for its double use value [18–22]. 'Baladi', also known in many regions as 'Soury' (according to the Lebanese town of Tyre that means in Arabic language Sour), is characterized by a medium to high oil content (around 28% expressed on fresh weight basis) and has a low pulp to pit ratio. This variety is highly productive although it has a slight alternate behavior and is highly susceptible to the olive fruit fly (*Bactorcera oleae*) and to the olive wilt disease (*Verticillium dahlia*).

In Lebanon, studies have mainly tackled the ecological characterization of some ancient olive trees in the Bshaale area and their age estimation [23]; or, the influence of the processing system and the production area on the physicochemical properties of 25 olive oil samples collected during crop season 2013/2014 [24]. Also, El Riachy et al. [25] investigated the effect of different irrigation regimes on fresh fruit weight, oil yield, quality and composition of olive oil from Baladi and Edelbi varieties. Chehade et al. [26] evaluated the impact of the cultivation area and the harvesting time on the fruit and oil characteristics of the main Lebanese olive varieties. Despite the importance of the olive oil sector in Lebanon and the increased interest in sensorial and beneficial effects of olive oil, there is a lack of data elucidating the combined effect of agro-industrial factors on the olive oil characteristics produced in this country. For this reason, the objective of the present study is to assess the effects of the geographical origin, the harvesting time and the processing system on the chemical composition of olive oil and on consumer preferences.

2. Materials and Methods

2.1. Experimental Sites

This study was implemented in 4 of the most important olive growing regions of Lebanon: Akkar and Zgharta-Koura district in the North governorate of Lebanon, Hasbaya in Nabatiye governorate (South Lebanon) and Jezzine in South Lebanon governorate (Figure 1). Akkar district is characterized by the presence of a relatively large coastal plain with high mountains to the east. Zgharta and Koura are districts that stretch from the Mediterranean Sea up to Mount Lebanon and comprise a series of foothills surrounding a low-lying plain where olive is cultivated. Hasbaya district is characterized by a long fertile valley lying at the western foot of Mount Hermon overlooking a deep amphitheater from which a brook flows to the Hasbani. Finally, the olive lots from Jezzine district originated from villages extended at altitudes from 200 to 1000 m all facing the Mediterranean Sea.

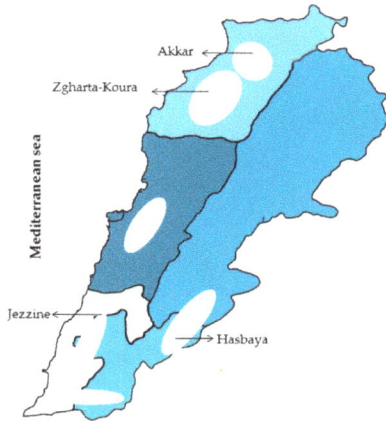

Figure 1. Map of olive groves in Lebanon indicating the sites where olives were collected. The white shapes in the map indicate the most important olive growing regions in the country [27].

2.2. Olive Fruits and Olive Oil Sampling

Samples of olive fruits and olive oils from the 'Baladi' variety were collected from the most common olive mills in the 4 olive growing regions, as follows: Press, 2- and 3-phases in Akkar and Jezzine; press, 3-phases and sinolea in Hasbaya; and, press and 3-phases in Zgharta-Koura.

The samples were collected at three harvesting times: early harvest (at the beginning of harvesting season in each region), intermediate harvest (at the middle of season) and late harvest (at the end of harvest). At each time three olive samples and three olive oil samples were collected from each region and from each processing system, except in the 2-phases system in Akkar region where 6 olive samples and 6 olive oil samples were collected (Table 1).

Table 1. Sample characteristics of olive fruits and olive oil from different geographical origins and different processing systems. The rainfall corresponds to mean values collected between 2009 and 2015 from climatic stations of the Lebanese Agricultural Research Institute (LARI).

Region	Rainfall	Altitude	Latitude	Longitude	Processing System	Olive Fruits	Olive Oil
Akkar	850 mm	300–700 m	34.5506°	36.0781°	Press	9	9
					2-phases	18	18
					3-phases	9	9
Zgharta-Koura	800 mm	200–350 m	34.2689°	35.7929°	Press	9	9
					3-phases	9	9
Hasbaya	650 mm	650–1050 m	33.3833°	35.6833°	Press	9	9
					Sinolea	9	9
					3-phases	9	9
Jezzine	750 mm	200–1000 m	33.5408°	35.5862°	Press	9	9
					2-phases	9	9
					3-phases	9	9
Total number of samples						108	108

2.2.1. Olive Fruits Samples

From each olive lot belonging to one farmer, a sample of 100 olive fruits randomly selected was taken for the determination of the ripening index (RI), as described by Frías et al. [28]. Briefly, the selected fruits were classified according to their skin color into the following 5 groups: 0 = the skin is green; 1 = the skin is yellow or yellowish-green color; 2 = the skin is a yellowish color with reddish spots; 3 = the skin is a reddish or light violet color; and, 4 = the skin is black. Then, the total number of olives in each group (n_0; n_1; .... n_4) was recorded, and the following equation was applied to determine the RI:

$$RI = \frac{[(0 \times n_0) + (1 \times n_1) + (2 \times n_2) + (3 \times n_3) + (4 \times n_4)]}{100} \tag{1}$$

2.2.2. Olive Oil Processing

The olive lots were processed to obtain the olive oil using the four following systems:

1. Press system: In this system, the olives fruits were crushed using a hummer mill that leads to a more complete breakage of olive flesh. Then, the crushed olives were grinded for around 30 min using cylindrical millstones to obtain the olive paste. The obtained olive paste was then placed on mats, stacked one above the other and pressed using a hydraulic press at a pressure up to 400 atm. The obtained oil was then pumped to a vertical centrifuge to separate the oil from vegetable water and other impurities.
2. Sinolea system (also known as cold percolation system): In this system, the olive fruits were crushed in a hammer mill. Then, the oil was separated from the olive paste using the sinolea system consisting of a series of metal discs used to mix the paste inside a perforated semi cylindrical vat including rows of metal discs or plates that dip into the paste, and the oil wets and sticks to the metal and is removed with scrapers in a continuous process. The oil dropping down the vat by gravity was collected in stainless steel recipients. To increase the efficiency of oil extraction of this system it was combined to 3-phases decanters. The oil obtained was mixed to the previously collected oil, and then separated from any remaining impurities by a vertical centrifuge.
3. 3-phases system: The olive fruits were crushed also by using a hammer mill, and then they were slowly mixed in a malaxation machine at 25 to 28 °C in order to coalesce the small oil droplets. At the end of this process, the resulting paste was homogenous with large oil spots floating on the surface and ready for separation inside the decanter. The decanter was a horizontal centrifuge rotating at around 3000 rpm to achieve the separation of the constituents of the homogenous paste into 3 different products: (i) dry pomace; (ii) vegetable water; and (iii) oil with small quantities of vegetable water that were removed by vertical centrifugation. In order to achieve better separation of the three phases in the decanter, 200–300 L of water per tons of olive paste were added.
4. 2-phases system: The 2-phases processing system was quite similar to the 3-phases system. However, the main difference is that the decanter separated the homogenous paste into only two phases: The first one is the mixture of pomace and vegetable water (wet pomace) and the second one is the oil mixed with small quantities of vegetable water. The obtained oil underwent a vertical centrifugation to clean it. In this system, no water was added to the paste.

2.3. Olive Oil Chemical Analysis

2.3.1. Free Acidity, Peroxide Values and UV Absorption

Free acidity, peroxide value, K232 and K270 (UV absorbance at 232 nm and 270 nm) were determined according to the procedures described in the Commission Regulation (EEC) No. 2568/91 [29].

2.3.2. Fatty Acids Composition

Fatty acids were determined as fatty acid methyl esters (FAMEs) according to the method described by IOC [30]. In brief, 0.1 g of oil were vigorously mixed with 2 mL of *n*-hexane and 200 µL of a methanolic solution of KOH (2 M), for 1 min. The mixture was allowed to set for 5 min,

Sustainability **2018**, *10*, 4630

and the upper phase was placed in a GC vial before injection, in duplicate, into a Shimadzu GC-2010 Plus (Tokyo, Japan) coupled to a flame ionization detector (FID) (280 °C). The used column was a fused silica capillary column (DB-wax, Agilent Technologies, Wilmington, DE) with 30 m length × 0.25 mm i.d. and 0.25 μm of film thickness. The nitrogen gas was used as carrier gas with a flow of 1.69 mL/min. Regarding the injector, the temperature was kept at 250 °C and a split ratio of 1:50 was used. A gradient oven temperature program was adopted with initial temperature set at 165 °C for 15 min, then temperature raised from 165 °C to 200 °C at a rate of 5 °C/min, and held at 200 °C for 2 min, then increased from 200 °C to 240 °C at a rate of 5 °C/min, and finally kept at 240 °C for 5 min. Authentic commercial standards were used to identify each FAME and the concentration was calculated as percentage of total peaks areas.

2.3.3. Extraction of Phenolic Compounds and Determination of Total Phenolic Content (TPC)

During prior analyses, samples were let to thaw at room temperature. The phenolic compounds in were extracted using a modification of the procedure described by Montedoro et al. [31]. An aliquot of 3 g oil was added to 2 mL of hexane in a flask and shaken for 15 s. Volumes of 1.75 mL of methanol/water (60/40, *v/v*) and 250 μL of syringic acid solution (60 ppm, internal standard) were added to the mixture and shaken for 2 min to undergo the first extraction. For the second extraction, 2 mL of methanol/water (60/40) were added and shaken for 2 min. The first and the second extracts were combined. The extracts were stored at −20 °C for further analysis.

TPC was determined spectrophotometrically using the Folin-Ciocalteu method [32]. A solution of 20 g Na_2CO_3 in 80 mL of distilled water and a solution of 100 mg of gallic acid in 100 mL methanol/water (60/40 *v/v*) were prepared. The TPC was determined precisely by introducing an aliquot of 20 μL of the methanol-water extract in an Eppendorf vial with 1.58 mL of deionized water, 300 μL of 20% Na_2CO_3 and 100 μL of Folin-Ciocalteu reagent. The mixture was manually shacked and placed in an oven at 50 °C for 5 min to accelerate the reaction, and then set to rest for 30 min at room temperature. Similarly to the samples, blank and calibration solutions of gallic acid were prepared. Then, the absorbance at 765 nm was measured using a Jenway UV/Vis spectrophotometer (Staffordshire, ST15 OSA, UK) and the TPC was expressed as mg gallic acid equivalent (GAE) per kg of oil.

2.3.4. Oxidative Stability Index (OSI)

OSI (h) was determined using a Rancimat apparatus (Model 892 Professional Rancimat, Metrohm SA, Herisau, Switzerland) according to the method described by Tura et al. [33]. This method consists of increasing the oxidation reactions by keeping 3 g of oil at 120 °C under a constant air flow of 20 L/h; and then, determining the conductivity variation of water (60 mL) due to the increase in oxidative compounds.

2.4. Consumer Preferences

2.4.1. Olive Oil Samples

The distinction of the olive oil samples determining if they are defected or without defects and the intensity of fruity, bitter and pungent sensory descriptors were evaluated only on the olive oil samples that showed chemical analyses results within the norms of extra virgin olive oil published in EEC [29]. Thus, a total of 50 olive oil samples from 108 olive oil samples (46.30%) were tested. The oil samples were geographically distributed as follows: 13 oil samples originated from Akkar, 22 from Hasbaya, 6 from Jezzine, and, 9 from Zgharta-Koura. Among the samples, 15 were processed by press system, 8 by sinolea, 5 by 2-phases and 22 by 3-phases system; from which, 22 oil samples correspond to the first harvest, 15 to the second harvest and 13 to the third one. These samples were stored for in dark, at −20 °C for a duration of 4 months and thawed to reach room temperature before tasting.

During the sessions, the samples were served in plastic cups; and a volume of approximately 10 mL was randomly served for each person with no obligation to finish the cup.

2.4.2. Sensory Characteristics

Sensory characteristics were evaluated by consumer preference test and an experienced sensorial panel.

Consumer Preference

A total of 188 consumers participated to the consumer preferences sessions. Among consumers 56.9% were females; and, 46.8% were between 19 and 30 years old and 53.2% were more than 30 years old. For each sample, consumers had to judge a sample using 'I like' or 'I don't like'. A maximum of 7 samples was given per person and session.

Experienced Sensorial Panel

The sensory analyses were also performed by a fully trained analytical sensorial panel composed of 5 trained assessors to perform olive oil sensorial analysis. Each taster first smelled the oil and judged it as defected or without defects. Then, the panel members tasted the oils without defects and marked the intensity of fruity, bitter and pungent attributes. Attributes were assessed on an oriented 10 cm line scale and quantified measuring the location of the mark from the origin according to the method of organoleptic characterization of virgin olive oil described by the IOC [34]. This method was used to classify each oil according to the intensity of the three mentioned positive attributes. An attribute was considered as delicate if the median is lower than 3; as medium if the median is between 3 and 6; and, as robust if the median is higher than 6.

2.5. Statistical Analysis

Acidity, peroxide value and UV absorbance were performed in Triplicate; and, the fatty acid composition and the OSI were done only in duplicate; however, the Folin Ciocalteu assay was performed only once. The obtained data were subjected to a multivariate analysis of variance (MANOVA) and to one-way ANOVA. Concerning the consumer preferences and the sensory profile, the collected data were analyzed using the Friedman's non-parametric test. To ensure the validity of the results, all assumptions required for the mentioned tests were checked before running them. A result is considered as statistically significant for a p-value less than 0.05. Note that, for multiple comparisons the p-value levels were adjusted using Bonferroni corrections which consist of dividing the α value (0.05) by the number of comparisons. Mean comparison (Duncan test, at $p < 0.05$) and different charts showing the interactions between regions, harvesting times and processing systems were elaborated by using the statistical package 'IBM-SPSS (version 22.0, IBM, Rochester, NY, USA).

3. Results and Discussion

3.1. Ripening Index (RI)

The difference in ripening between the olives lots processed at each harvesting time was investigated. The results showed that RI was very highly significantly different among the different harvesting times ($p < 0.001$). This difference was explained by a high eta square (70.9%). Mean comparisons showed a wide range of variation between the three harvesting times with mean values significantly increasing from early harvest (RI = 1.35) to intermediate harvest (RI = 2.34) and to late harvest (RI = 3.43). These results are in agreement with those obtained by El Riachy et al. [35] in a study on 3 varieties ('Arbequina', 'Picual' and 'Frontoio') and 12 of their segregating populations; where the RI was correlated with the harvesting time, although the evolution of ripening was different between varieties.

3.2. Oil Quality Parameters

The effects of geographical origin, processing system and harvesting time on the quality indices of olive oil including free acidity, peroxide value, K232 and K270, were investigated. Results (Table 2) showed a very highly significant effect of the interactions geographical origin * processing system * harvesting time, geographical origin * processing system and geographical origin * harvesting time on this set of variables ($p < 0.001$). Moreover, processing system * harvesting time revealed a significant effect on these quality indices ($p < 0.05$). As per each single factor, the geographical origin and the processing system showed a very highly significant effect ($p < 0.001$) and the harvesting time had a significant effect on this set of variables ($p < 0.05$).

Table 2. Results of the Multivariate Analysis of Variance (MANOVA) of the two sets of variables: quality parameters and fatty acid composition.

Parameters	Factors	Wilk's Λ	F	Partial η^2	Power
Quality parameters	Geographical origin (A)	0.35	7.83 ***	0.30	1.00
	Processing system (B)	0.55	4.01 ***	0.18	1.00
	Harvesting time (C)	0.76	2.63 *	0.13	0.92
	A * B	0.45	4.09 ***	0.18	1.00
	A * C	0.46	2.59 ***	0.18	1.00
	B * C	0.57	1.86 * [2]	0.13	0.97
	A * B * C	0.33	2.94 *** [1]	0.24	1.00
Fatty acids composition	Geographical origin (A)	0.27	8.05 ***	0.36	1.00
	Processing system (B)	0.82	1.00	0.07	0.59
	Harvesting time (C)	0.40	8.20 ***	0.37	1.00
	A * B	0.31	5.00 ***	0.25	1.00
	A * C	0.60	1.29	0.10	0.87
	B * C	0.44	2.17 **	0.15	0.99
	A * B * C	0.44	1.64 ** [3]	0.15	0.99

[1] *** $p < 0.001$; [2] * $p < 0.05$; [3] ** $p < 0.01$.

The Table 2 shows that between interactions, the three-way interaction geographical origin * processing system*harvesting time and its associated error accounted for the highest partial η^2 (24%). As per individual factors, highest partial η^2 (30% of the between subject's variance) was attributed to the geographical origin and its associated error. It is worth noting that the three studied factors and their interactions showed sufficient power to detect such effects (the power statistics > 0.80). Several previous studies showed significant effects of geographical origin, processing system and harvesting time on the olive oil quality parameters [10,36]. However, Di Giovacchino et al. [37] showed no significant differences between processing systems in free acidity, peroxide value, K232, K270.

To go further in the analysis of the effect of the geographical origin, processing system and harvesting time on each of the quality parameters, the tests of Between-Subjects effects were used (Table 3). Results showed that the observed statistical power is higher than 0.8 for all significant effects. Accordingly, the tests of Between-Subjects effects have sufficient power to detect such effects.

The interaction geographical origin * processing system was the main contributor to total variance of the free acidity accounting for 21.37%. Moreover, the interaction geographical origin * harvesting time to total variance of peroxide value (18.75%) and K232 (10.69%). In addition, the interaction geographical origin * processing system*harvesting time to total variance of K270 accounting for 22.28% (Table 3).

Table 3. Relative importance of geographical origin, processing system and harvesting time expressed as percentages of total sum of squares and significance in the ANOVA for quality parameters, fatty acid composition, total phenols (TP) and OSI. Means are in % for free acidity (FA), milliequivalent O_2/kg of oil for peroxide value (PV), % for C16:0; C18:0; C18:1; C18:2 and C18:3, mg gallic acid equivalent (GAE)/kg of oil for total phenols (TP), and hours for oxidative stability index (OSI).

Factors/Statistics	FA	PV	K232	K270	C16:0	C18:0	C18:1	C18:2	C18:3	TP	OSI
Geographical origin (A)	11.32 *[1]	18.37 *	8.21	10.36 *	25.10 *[2]	21.03 *	25.67 *	9.11 *	2.26	13.19 ***[3]	13.07 ***
Processing system (B)	10.14 *	0.37	2.46	12.44 *	0.57	0.97	0.99	1.33	1.41	15.92 ***	16.88 ***
Harvesting time (C)	0.57	4.62 *	6.14	4.15	22.59 *	2.97	4.04	6.15	2.68	0.15	5.86 **[4]
A * B	21.37 *	9.75 *	3.50	6.22	9.76 *	10.13 *	19.94 *	18.09 *	11.28 *	7.82 **	12.61 ***
A * C	2.45	18.75 *	10.69	2.59	1.02	4.40	2.62	8.06	9.59	10.23 **	1.95
B * C	5.39	5.35	4.10	7.25	8.49 *	6.04	8.50	7.75	9.17	8.36 *[5]	5.25
A * B * C	7.27	13.04 *	6.72	22.28 *	1.62	11.59	2.75	1.54	16.50 *	10.77 **	10.84 **
Error	41.50	29.75	58.18	34.72	30.86	42.86	35.49	47.97	47.11	33.55	33.53
Mean	0.94	12.99	1.60	0.14	12.56	3.55	70.47	10.88	0.60	213.09	8.13
CV[6]	69.85	34.75	21.15	31.48	8.68	17.18	2.27	11.21	13.33	29.70	39.82
S.E. Mean[7]	0.06	0.43	0.03	0.00	0.10	0.06	0.15	0.12	0.01	6.12	0.31

[1] * $p < 0.0125$ (Considering the Bonferroni correction for free acidity, peroxide value, K232 and K270; [2] * $p < 0.01$ (Considering the Bonferroni correction for C16:0; C18:0; C18:1; C18:2 and C18:3); [3] *** $p < 0.001$; [4] ** $p < 0.01$; [5] * $p < 0.05$; [6] Coefficient of variation; [7] Standard Error Mean.

As shown in Table 3, the free acidity was significantly affected by the interaction geographical origin*processing system. This revealed that whatever is the geographical origin, the processing system has a significant effect on the free acidity. In addition, the peroxide value was significantly affected by the three-way interaction geographical origin * processing system * harvesting time. This means that the interaction among the two factors (geographical origin * processing system) is different across the levels of the third factor (harvesting time).

Similarly, K270 was significantly affected by the three-way interaction geographical origin * processing system * harvesting time. The interaction among the two factors (geographical origin * processing system) is different across the three harvesting times.

The effect of each factor on the quality parameters of the olive oil was assessed. The effect of the geographical origin was significant on free acidity, peroxide value and K270 (Table 4). These results are in disagreement with those obtained by Lazzez et al. [38] that stated that the fruit ripening is the main factor influencing the olive oil qualitative parameters; and, the geographical origin has only a minor effect on these parameters.

Table 4. Effect of geographical origin on all studied parameters.

Parameters	Akkar	Hasbaya	Jezzine	Zgharta-Koura
FA [1] (%)	1.24 [a 2]	0.43 [c]	1.15 [a]	0.76 [b]
PV [3] (Meq O$_2$/kg)	14.17 [a]	13.70 [a]	13.71 [a]	8.49 [b]
K232	1.61 [a]	1.74 [a]	1.42 [b]	1.62 [a]
K270	0.15 [a]	0.14 [a]	0.11 [b]	0.15 [a]
C16:0 (%)	13.20 [a]	12.26 [c]	11.83 [d]	12.79 [b]
C18:0 (%)	3.27 [b]	3.41 [b]	3.85 [a]	3.86 [a]
C18:1 (%)	69.74 [b]	71.46 [a]	71.15 [a]	69.42 [b]
C18:2 (%)	11.22 [a]	10.37 [b]	10.63 [b]	11.34 [a]
C18:3 (%)	0.63 [a]	0.60 [a]	0.59 [a]	0.59 [a]
TP [4] (mg GAE/Kg)	208.42 [b]	235.50 [a]	193.53 [b]	217.88 [ab]
OSI [5] (h)	7.97 [b]	10.14[a]	6.42 [c]	8.01 [b]

[1] Free acidity; [2] Different letters (a, b, c) within the same row indicate significant differences ($p < 0.05$); [3] Peroxide value; [4] Total phenols; [5] Oxidative stability index.

Mean comparisons showed that Hasbaya oil recorded the lowest free acidity, Zgharta-Koura oil the lowest peroxide value and Jezzine oil the lowest K232 and K270. However, the oil from Akkar and Jezzine registered the highest free acidity exceeding 0.8% limit established by the IOC regulation for extra virgin olive oil [39], which represented an advanced level of degradation. According to Ben Temime et al. [40], the significant differences in olive oil qualitative parameters between geographical origins are not due to the cultivation area in itself, but to other factors affecting olive fruits quality such as olive fly attacks, mechanical damage during olive harvesting and transport, long delay between harvesting and processing, among others.

When considering only the two phases system that is present only in Akkar and Jezzine, the multivariate analyses of data collected from this system showed a significant effect of the geographical origin on the olive oil quality parameters ($Wilks' \Lambda = 0.53; F = 4.97, p < 0.01$). Yet, the tests of Between-Subjects effects revealed a significant effect of the region only on the K270 parameter ($F(1,25) = 15.81, p < 0.0125$); with higher values in Akkar region (0.18 vs. 0.12). However, if considering only the olive oil samples obtained through the press system, MANOVA shows a very high significant effect of the geographical origin on the olive oil quality ($Wilks' \Lambda = 0.19; F = 5.68, p < 0.001$). The Tests of Between-Subjects effects reveals a significant effect of geographical origin on free acidity and peroxide value with respectively ($F(3,32) = 13.66, p < 0.0125$) and ($F(3,32) = 11.28, p < 0.0125$). The highest acidity was recorded in Jezzine (1.66%) significantly higher than Akkar, Zgharta-Koura and Hasbaya (1.09%, 0.81% and 0.54%, respectively), with Akkar showing significantly higher acidity than Hasbaya. Regarding peroxide value, Jezzine showed significantly higher peroxide value than Hasbaya and Zgharta-Koura (15.24, 11.58, 7.70 meq O_2/kg of oil). Note that Akkar (14.04 meq O_2/kg of oil) recorded significantly higher peroxide value than Zgharta-Koura (8.49 meq O_2/kg of oil). Also, for the 3-phases processing system alone, MANOVA reveals a very high significant effect of the geographical origin on the olive oil quality ($Wilks' \Lambda = 0.30; F = 3.73, p = 0.0000$). The Tests of Between-Subjects Effects reveals a significant effect of geographical origin on free acidity ($F(3, 32) = 10.95, p < 0.0125^*$); with acidity in Jezzine (0.91%) significantly higher than that in Zgharta-Koura, Akkar and Hasbaya (0.70%, 0.54% and 0.39%, respectively). Note that oils obtained from Zgharta-Koura showed significantly higher acidity than those from Hasbaya. The results of these three comparisons on the same processing systems in different geographical origins confirm that free acidity, peroxide value and K270 are highly dependent on the geographical origins.

As per the processing system, the effect was significant on free acidity and K270 (Table 5). These results are in partial agreement with those obtained by Ben Hassine et al. [36], who indicated that the free acidity, the peroxide value, the K232 and the K270 are significantly affected by the processing system. Conversely, Salvador et al. [41] demonstrated that while oxidative stability and antioxidant content differed significantly between processing systems; free acidity, peroxide value, K232 and K270 didn't show significant differences.

Table 5. Influence of processing system on studied parameters.

Parameters	Press	2-Phases	3-Phases	Sinolea
FA [1] (%)	1.00 [b2]	1.44 [a]	0.64 [c]	0.36 [c]
PV[3] (Meq O_2/kg)	12.14 [a]	14.30 [a]	12.75 [ab]	13.46 [ab]
K232	1.58 [b]	1.57 [b]	1.57 [b]	1.88 [a]
K270	0.12 [b]	0.16 [a]	0.13 [b]	0.16 [a]
C16:0 (%)	12.49 [ab]	12.84 [a]	12.47 [ab]	12.34 [b]
C18:0 (%)	3.61 [a]	3.54 [a]	3.51 [a]	3.47 [a]
C18:1 (%)	70.35 [b]	70.14 [b]	70.65 [ab]	71.19 [a]
C18:2 (%)	11.01 [a]	10.90 [a]	10.84 [a]	10.47 [a]
C18:3 (%)	0.59 [a]	0.61 [a]	0.61 [a]	0.61 [a]
TP [4] (mg GAE/Kg)	195.86 [b]	240.91 [a]	207.02 [b]	225.89 [ab]
OSI [5] (h)	6.76 [b]	7.07 [b]	9.87 [a]	9.84 [a]

[1] Free acidity; [2] Different letters (a, b, c) within the same row indicate significant differences ($p < 0.05$); [3] Peroxide value; [4] Total phenols; [5] Oxidative stability index.

In the present study, Sinolea system recorded the lowest free acidity; and the 2-phases system, the highest one exceeding, together with the press system, the limit of 0.8% established by the IOC regulation for extra virgin olive oil [39]. These results are in disagreement with many previous studies that showed a highest free acidity in the press system and a lowest one in the centrifugation systems [36,42,43]. The high level of free acidity observed in the present study in the 2-phases system is probably related to strong infection with the olive fruit fly (*Bactrocera olea*) in both regions where the 2-phases is present: Chadra in Akkar and Bisri in Jezzine. The two regions consist of valleys with very high relative humidity and annual high infection with olive fruit fly. Indeed, several previous studies specified that the attack of the olive fly affects negatively the olive oil quality, leading to an increase in free acidity [44,45]. Previous studies stated that, when poor quality olives are industrially processed with either press or 3-phases centrifugation systems the centrifugation system, the latter gave oils with lower free acidity [46]. In the present study, the 2-phases system was unable to reduce sufficiently the free acidity maybe due to the very high infection with the olive fruit fly [44].

Press system and 3-phases decanter recorded the lowest K270 that depends on the presence of secondary oxidation products (conjugated trienes). The higher values observed in 2-phases system also may be due to the high attack of olive fruit fly indicated above; and the higher values in sinolea may be due to the observed high temperature that was used in those mills during the oil processing. These results are in agreement with those described by Gómez-Caravaca et al. [44] who reported an increase in oxidation products in olive infested by the olive fruit fly; and with those stated by Ranalli et al. [47] who also reported an increase in these products with higher malaxation temperatures.

The results (Table 6) showed that the effect of harvesting time was only significant on peroxide value. It was noticeable that the peroxide value increased significantly in the late harvesting time to 14.91 Meq O_2/kg as compared to the early and intermediate with values of 12.57 and 11.51 Meq O_2/kg, respectively. However, other studies reported an increase only in free acidity along ripening [48,49] due to the progressive activation of the lipolytic activity and to the fact that the olives are more sensitive to pathogenic infections and mechanical damage, which results in oils with higher acidity values [50]. Conversely to the results obtained in the present study, a decrease in peroxide value, K232 and K270 was observed in 'Sayali' olive oils [49] and in other monovarietal olive oils from Tunisia at late harvesting [51]. Bengana et al. [52] reported higher values of all quality indices at late harvest of olive oils from 'Chemlal' variety cultivated in Algeria.

Table 6. Evolution of studied parameters along harvesting.

Parameters	Early Harvest	Intermediate Harvest	Late Harvest
FA [1] (%)	0.89 [a][2]	0.88 [a]	1.04 [a]
PV [3] (Meq O_2/kg)	12.57 [b]	11.51 [b]	14.91 [a]
K232	1.63 [ab]	1.50 [b]	1.66 [a]
K270	0.14 [a]	0.13 [a]	0.14 [a]
C16:0 (%)	12.98 [a]	12.83 [a]	11.86 [b]
C18:0 (%)	3.43 [b]	3.54 [ab]	3.67 [a]
C18:1 (%)	70.52 [ab]	70.10 [b]	70.78 [a]
C18:2 (%)	10.50 [b]	10.94 [ab]	11.20 [a]
C18:3 (%)	0.61 [a]	0.62 [a]	0.58 [b]
TP [4] (mg GAE/Kg)	220.02 [a]	209.85 [a]	209.58 [a]
OSI [5] (h)	9.19 [a]	7.90 [b]	7.31 [b]

[1] Free acidity; [2] Different letters (a, b) within the same row indicate significant differences ($p < 0.05$); [3] Peroxide value; [4] Total phenols; [5] Oxidative stability index.

3.3. Fatty Acid Composition

MANOVA results performed on the set of the five main fatty acids of the olive oil (C16:0, C18:0, C18:1, C18:2 and C18:3) revealed a significant effect of the three-way interaction geographical origin * processing system * harvesting time on the fatty acid composition of olive oil ($p < 0.05$). On the other hand, only the geographical origin and the harvesting time revealed a very highly significant effect on the set of the main fatty acids in olive oil ($p < 0.001$) (Table 2).

The harvesting time and its associated errors accounted for high percentages of the between subject's variance expressed as partial η^2 (37%). In a previous four years study to determine the optimal harvesting period for 'Chemlali' olives, Lazzez et al. [38] also reported that the harvesting time is the factor showing the highest effect on the composition of olive oil in comparison with crop year and growing area.

However, the Tests of Between-Subjects effects revealed that the geographical origin was the main contributor to total variance of C16:0, C18:0 and C18:1. However, the interaction geographical origin * processing system * harvesting time was the main contributor to total variance of C18:3; and, the interaction geographical origin * processing system the main contributor to total variance of C18:2 (Table 3). These results are in agreement with those obtained by Bajoub et al. [53] on the 'Picholine Maroccaine' monovarietal olive oil in Morocco, who reported a significant effect of geographical origin on all fatty acids except on the minor fatty acids, heptadecenoic and myristic acids. Also, there are several studies on the use of fatty acid composition for geographical characterization of olive oils from northern countries of the Mediterranean basin [41,54,55]. On the other hand, the interaction geographical origin * processing system affected significantly all the main fatty acids in the olive oil

(Table 3). This means that whatever is the geographical origin, the main fatty acids of the olive oil are affected by the processing system.

Moreover, the interaction processing system * harvesting time affected significantly the C16:0 (Table 3). This reveals that independently of the processing system, C16:0 is affected by the harvesting time.

Regarding the interaction geographical origin * processing system * harvesting time, it was only significant for C18:3 (Table 3). The mentioned three-way interaction shows that the interaction among the two factors (geographical origin*processing system) is different across the three harvesting times.

As for the effect of each single factor, the mean comparisons showed that C16:0 and C18:2 contents were significantly higher in North Lebanon (Akkar and Zgharta-Koura) than in South Lebanon (Hasbaya and Jezzine). However, C18:1 was significantly higher in South Lebanon (Hasbaya and Jezzine) than in North Lebanon (Akkar and Zgharta-Koura). C18:0 was significantly higher in Jezzine and Zgharta-Koura than in Akkar and Hasbaya. However, the content of C18:3 was not affected by the geographical origin (Table 4). According to Beltrán et al. [56] the air temperature during oil biosynthesis could affect the amount of polyunsaturated fatty acids (linoleic and linolenic fatty acids) by means of the regulation of desaturase enzymes activities. For instance, Issaoui et al. [57] in Tunisia and Mailer et al. [8] in Australia both showed a higher content of C18:1 in cooler regions (high altitudes) and higher contents of C16:0 and C18:2 in warmer regions (low altitudes). The results obtained in the present study agree with these observations as the high C18:1 content was observed in the olive samples proceeding from Jezzine and Hasbaya where the olive fruits were harvested at altitudes up to 1000 and 1050 m, respectively; and, the high content of C16:0 and C18:2 were observed in oils from Zgharta-Koura and Akkar as the fruits were harvested from lower altitudes, up to 350 and 700 m respectively. Although, Serhan et al. [24] also previously reported strong negative correlation between altitude and C16:0, additional studies on several years and involving different regions in North and South Lebanon are essential to prove these hypotheses.

It is worth noting that, the effect of processing system on the fatty acid composition was not significant in the present study (Table 5), in concordance with the results obtained by Gimeno et al. [42] and by Serhan et al. [24] while comparing traditional and centrifugation processing systems in north Lebanon; but, in partial agreement with those obtained by Salvador et al. [41] who reported slight differences in fatty acid composition due to the processing system, although the differences were significant only in case of C16:0, C16:1 and C18:3.

However, regarding the harvesting time, the effect was only significant on C16:0 whose content decreased significantly after the intermediate harvesting time (Table 6). These results are in agreement with those reported by Cimato [5] where the delay in harvesting tended to increase the content of unsaturated fatty acids, especially linoleic, at the expense of palmitic acid. However, these results are partially in agreement with those obtained by Baccouri et al. [51] on Tunisian monovarietal olive oil and by Fuentes de Mendoza et al. [48] in a three successive years study on 'Morisca' and 'Carrasqueña' olive varieties, who reported a decrease in palmitic and linoleic acids along ripening.

3.4. Total Phenols

The effect of the studied factors and their interactions on total phenols content, determined by the Folin-Ciocalteu method, was assessed. Results showed that the interaction geographical origin * processing system * harvesting time showed a highly significant effect on total phenols ($p < 0.01$) (Table 3). This three-way interaction geographical origin * processing system * harvesting time means that whatever is the geographical origin, the processing system has a significant effect on the total phenols content for the three harvesting times. Indeed, the interaction among the two factors (geographical origin * processing system) is different across the early, intermediate and late harvest.

As per each factor alone, very highly significant effects of geographical origin and processing system were observed on the total phenols content ($p = 0.0000$), with the later showing the highest contribution (15.92%) and the former the second one (13.19%) (Table 3). Mean comparisons showed that total phenols in Hasbaya was significantly higher than in Akkar and Jezzine (235.50, 208.42 and 193.53 mg GAE/Kg of oil, respectively); while, the total phenols in Zgharta-Koura recorded an intermediate value (217.88 mg GAE/Kg of oil) (Table 4). These results don't match those shown by Baccouri et al. [51] that reported no difference in phenolic compounds according to the geographical origin in monovarietal olive oils from Tunisia; but they match those shown by Ben Temime et al. [40] and Youssef et al. [58] that reported different phenolic composition in 'Chétoui' and 'Oueslati', respectively, due to different climate and soil characteristics. Moreover, regarding the processing system, the oil from 2-phases system recorded significantly higher total phenols than 3-phases and press systems (240.91, 207.02 and 195.86 mg GAE/Kg of oil, respectively); however, sinolea system recorded an intermediate value (225.89 mg GAE/Kg of oil) (Table 5). Salvador et al. [41] previously reported that among all quality and compositional parameters of olive oil, phenolic compounds and oxidative stability stand as the main parameters affected by the processing system. In fact, it was demonstrated that the 2-phases decanter preserves more of the phenolic compounds in comparison to the 3-phases decanter where the added water causes large amounts of phenols to be eliminated with the olive mill waste water [42] (12, 16). Moreover, the high amount of O_2 dissolved in the pastes during the process of press and sinolea systems due to contact with the air result in a loss of phenolic compounds due to the activation of endogenous enzymes, polyphenoloxidase and peroxidase, that oxidize the phenolic compounds and consequently reduce their concentration in the produced oil [14].

However, regarding the harvesting time, the total phenols content decreased along ripening although the difference was not significant. This decrease in total phenols with the progress of ripening was previously well reported [35,49].

3.5. Oil Oxidative Stability (OSI)

To understand the effect of the three studied factors on the OSI, a three-way ANOVA was run. Results showed a highly significant effect ($p < 0.01$) of the three-way interaction geographical origin * processing system * harvesting time (Table 3); which means that regardless of the geographical origin, the processing system has a significant effect on the OSI for the three-harvesting time. Indeed, the interaction among the two factors (geographical origin*processing system) is different across the early, intermediate and late harvest.

As per each factor alone, the OSI was extremely highly significantly affected by the geographic origin and by the processing system ($p < 0.001$); and, highly significantly affected by the harvesting time ($p < 0.01$). It is worth to note that Hasbaya oil showed significantly higher OSI (10.14 h) followed by Zgharta-Koura, Akkar and Jezzine 8.01, 7.97, and 6.42 h, respectively) (Table 4). Interestingly, this was the same tendency observed in total phenols, in agreement with previous results showing a high positive correlation (r = 0.937) between total phenols in oils from different locations and OSI [59].

However, the 3-phases and the sinolea systems registered significantly higher OSI (9.87 and 9.84 h, respectively) in comparison with 2-phases and press systems (7.07 and 6.76 h, respectively) (Table 5). Although the 2-phases registered the highest phenolic content, the lower OSI recorded in this system could be mainly due to the higher free acidity registered in oils from this system. In previous studies, Rotondi et al. [60] have found a high positive correlation between higher free acidity and shorter shelf life of olive oil.

As per the harvesting time, the OSI decreased along ripening in parallel to the decrease of total phenols. The difference was only significant between the first and the last harvesting time (Table 6).

3.6. Sensory Characteristics

The consumer preferences and experienced sensorial panel judgment were conducted only on olive oil samples qualified chemically as extra virgin olive oil.

3.6.1. Consumer Preferences

In order to show the difference in consumer preferences among the 50 olive oil samples studied, the Friedman test was run. This test showed very highly significant differences ($\chi^2(6) = 154.85, p < 0.001$), indicating that the observed difference in the participant's choice is due to the olive oil itself and not to any other random factor. Moreover, the "Kendall's W" recorded a value of 0.53, indicating a mid-difference among the participant's choices (a Kendall's W value equal 1 indicates a complete agreement between consumers, and a Kendall's W value equal 0 indicated a complete disagreement).

The results of Friedman test (Table 7) show that the mostly preferred olive oil samples were two among three samples of Hasbaya sinolea olive oils obtained from the first harvesting time and one Jezzine press olive oil from the third harvesting time with mean ranks of 40.83. However, the secondly preferred ones were two among three of Hasbaya sinolea olive oils and Hasbaya 3-phases olive oils and 1 among three of the Hasbaya press olive oils all at the second harvesting time with mean ranks 36.7. The statement that different replicates of oils originating from the same geographical origin, processing system and harvesting time were differently judged could be due most likely to the fact that the tasters (consumers) were different or to the fact that the different replicates belong to different olive lots. It is greatly reported that the sensory characteristics of olive oil are correlated with the sample chemical composition, especially to the phenolic and fatty acid content and profile. Moreover, it is reported that the oil produced by the 2-phases has intense bitter and pungent tastes resulting from the higher phenolic content, which may be unpleasant for some consumers that are not familiar with this oil taste [16]. Therefore, it is suggested to conduct further correlation to identify the relationship between the consumer preferences and the specific chemical composition. Moreover, the quality of olive oil is not always correctly perceived by the consumer, especially since they generally appreciate what is familiar and what is strongly linked to their tradition and origin [61,62].

It was noticeable that 50% of the samples (25 samples) recorded the lowest rank. Note that the judgment of the consumers was not unanimous with regard to geographical origin, processing system and harvesting time.

Table 7. Consumer preferences and the experienced sensorial panel judgments of the samples studied.

Sample ID		Consumers Preferences				Panel Judgment		
	N	Like (%)	Don't Like (%)	Mean Rank	Classification	Fruity	Bitter	Pungent
A Press HT1R2	20	15.0	85.0	20.0	Defected	-	-	-
A Press HT3R3	31	25.8	74.2	20.0	Defected	-	-	-
A 2-Phases HT1R1	15	60.0	40.0	20.0	Missing	-	-	-
A 2-Phases HT1R2	14	7.1	92.9	20.0	EVOO	Delicate	Delicate	Medium
A 2-Phases HT1R3	16	56.3	43.8	20.0	Defected	-	-	-
A 3-Phases HT1R1	23	26.1	73.9	20.0	EVOO	Delicate	Delicate	Delicate
A 3-Phases HT1R3	24	62.5	37.5	20.0	EVOO	Delicate	Delicate	Delicate
A 3-Phases HT2R1	10	60.0	40.0	28.3	EVOO	Delicate	Delicate	Delicate
A 3-Phases HT2R2	13	61.5	38.5	24.2	EVOO	Delicate	Delicate	Delicate
A 3-Phases HT2R3	28	57.1	42.9	20.0	EVOO	Delicate	Delicate	Delicate
A 3-Phases HT3R1	14	78.6	21.4	32.5	EVOO	Delicate	Delicate	Delicate
A 3-Phases HT3R2	22	31.8	68.2	20.0	EVOO	Delicate	Delicate	Delicate
A 3-Phases HT3R3	22	63.6	36.4	20.0	EVOO	Delicate	Delicate	Delicate
H Press HT1R1	6	33.3	66.7	28.3	Defected	-	-	-
H Press HT1R2	11	45.5	54.5	20.0	Defected	-	-	-
H Press HT1R3	16	25.0	75.0	20.0	EVOO	Medium	Delicate	Medium
H Press HT2R1	11	36.4	63.6	20.0	EVOO	Delicate	Delicate	Delicate
H Press HT2R2	11	27.3	72.7	20.0	EVOO	Medium	Delicate	Medium
H Press HT2R3	9	77.8	22.2	36.7	EVOO	Medium	Delicate	Delicate
H Press HT3R2	16	18.8	81.3	20.0	Defected	-	-	-
H Press HT3R3	11	36.4	63.6	20.0	Defected	-	-	-
H 3-Phases HT1R1	7	57.1	42.9	32.5	EVOO	Medium	Delicate	Medium
H 3-Phases HT1R2	9	44.4	55.6	24.2	EVOO	Medium	Delicate	Delicate
H 3-Phases HT1R3	7	42.9	57.1	28.3	EVOO	Delicate	Delicate	Delicate
H 3-Phases HT2R1	17	64.7	35.3	20.0	EVOO	Medium	Delicate	Medium
H 3-Phases HT2R2	7	71.4	28.6	36.7	EVOO	Delicate	Delicate	Delicate
H 3-Phases HT2R3	7	71.4	28.6	36.7	EVOO	Delicate	Delicate	Medium
H Sinolea HT1R1	11	90.9	9.1	40.8	EVOO	Delicate	Delicate	Delicate
H Sinolea HT1R2	11	90.9	9.1	40.8	EVOO	Delicate	Delicate	Delicate
H Sinolea HT1R3	16	31.2	68.8	20.0	EVOO	Delicate	Delicate	Delicate
H Sinolea HT2R1	6	66.7	33.3	36.7	EVOO	Delicate	Delicate	Delicate

Table 7. *Cont.*

Sample ID	N	Consumers Preferences				Panel Judgment		
		Like (%)	Don't Like (%)	Mean Rank	Classification	Fruity	Bitter	Pungent
H Sinolea HT2R2	9	77.8	22.2	36.7	EVOO	Delicate	Delicate	Delicate
H Sinolea HT2R3	11	45.5	54.5	20.0	EVOO	Medium	Delicate	Medium
H Sinolea HT3R2	11	54.5	45.5	24.2	EVOO	Delicate	Delicate	Delicate
H Sinolea HT3R3	16	56.3	43.7	20.0	EVOO	Medium	Delicate	Delicate
ZK Press HT1R3	10	50.0	50.0	24.2	Defected	-	-	-
ZK Press HT2R1	8	50.0	50.0	28.3	Defected	-	-	-
ZK Press HT3R1	13	69.2	30.8	28.3	Defected	-	-	-
ZK Press HT3R2	8	37.5	62.5	24.2	Defected	-	-	-
ZK 3-Phases HT1R1	7	14.3	85.7	20.0	EVOO	Delicate	Delicate	Delicate
ZK 3-Phases HT1R2	13	53.8	46.2	20.0	EVOO	Delicate	Delicate	Delicate
ZK 3-Phases HT1R3	10	40.0	60.0	20.0	Defected	-	-	-
ZK 3-Phases HT2R3	10	70.0	30.0	32.5	Defected	-	-	-
ZK 3-Phases HT3R2	13	69.2	30.8	28.3	Defected	-	-	-
J Press HT3R3	7	85.7	14.3	40.8	Defected	-	-	-
J 2-Phases HT1R1	10	50.0	50.0	24.2	Defected	-	-	-
J 2-Phases HT3R1	8	62.5	37.5	32.5	Defected	-	-	-
J 3-Phases HT1R2	13	30.8	69.2	20.0	Defected	-	-	-
J 3-Phases HT1R3	10	50.0	50.0	24.2	Defected	-	-	-
J 3-Phases HT2R1	7	14.3	85.7	20.0	Defected	-	-	-

A: Akkar; H: Hasbaya; ZK: Zgharta-Koura; J: Jezzine; HT1, HT2 and HT3 represent early, intermediate and late harvest time respectively; R1, R2 and R3 represent the three repetitions (3 different olive lots)

3.6.2. Experienced Sensorial Panel

The panel judgment on olive oils without defects showed that all tasted oils fall within the delicate and medium categories. This could be due to the fact the oil samples were stored for a long period at $-20\,^{\circ}C$ before the tasting sessions. It is worth highlighting that the samples with the highest mean rank by consumers were all appreciated by the expert panelists except for the sample obtained from Jezzine using the press system in the late harvest.

It is worth to noting that the Chi square test showed a significant relation between the consumer preferences and the olive oil panel judge as defected or without defects ($\chi^2(1) = 4.87; p < 0.05$). These results show that the consumers, even naïf, were able to discriminate 70.0% of the defected samples and 62.1% of the samples without defects. Similarly, Predieri et al. [63] reported that the results from consumers and trained panelist are comparative.

The effect of geographical origin, processing system and harvesting time on the positive attributes of olive oil was also performed. The results showed no significant effect ($p > 0.05$). Similarly, Di Giovacchino et al. [37] showed no significant differences between processing systems and sensorial properties of olive oil. Although several studies reported that geographical origin, processing system and harvesting time were major factors significantly affecting the olive oil sensory characteristics, none of the studies have assessed the combined effect of a big sample size.

4. Conclusions

This study reported, for the first time, the results of the complex interaction between the geographical origin, harvesting time and processing system on the olive oil chemical and sensorial characteristics. The overall findings showed that the fatty acids composition including C16:0, C18:1 and C18:2 was mainly affected by the geographical origin. This highlights the need to conduct further studies in order to identify protected denominations of origin in Lebanon. Findings also showed a significant effect of harvesting time on the peroxide value and OSI; and, of the processing system on the free acidity, total phenols and the OSI. Moreover, this study has showed that consumer preference was not influenced by the geographical origin, harvesting time and processing system but could be affected by olive oil chemical composition. The findings of this study, therefore, may help experts and producers to draw more attention to the most adequate processing parameters and their combinations in order to produce the highest olive oil chemical and quality characteristics that suit specific customer preferences.

Author Contributions: Conceptualization, M.E.R.; Data curation, M.E.R. and W.S.; Formal analysis, W.S.; Funding acquisition, M.E.R.; Investigation, M.E.R., C.B.-M., A.Y., R.A. and W.S.; Methodology, M.E.R., C.B.-M., A.Y. and R.A.; Project administration, M.E.R.; Writing—original draft, M.E.R.; Writing—review & editing, C.B.-M. and W.S.

Funding: This project has been funded with support from the National Council for Scientific Research in Lebanon (CNRS-L). Title of the project: "Influence of agro-industrial practices on Lebanese olive oil sector". The APC has been funded by the Lebanese Agricultural Research Institute (LARI).

Acknowledgments: The Authors thanks deeply the support provided by the US-AID/LIVCD Project (Lebanese Industrial Value Chain Development). Also, the authors are thankful for the help in oil analyses provided by their colleagues in the olive and olive oil department at Tal Amara station of LARI: Ghina Jebbawi, Ghinwa Al Hawi, Joe Wakim, Maroun Jamous, Jennifer Chaanine, and, their colleagues at NDU including Jacqueline Doumit. Moreover, authors would like to thank the experts in olive oil tasting: Hussein Hoteit, Reem El Derbass, Omar Chehade, Ranim Abdulkader and Mona Keyrouz, for the help in sample sensory analysis. Finally, authors are thankfull for the help in sample collection provided by Hala Abou Trabi, Fadi Raad, Jawad Dagher, Nayef Saasouh and Wassim Syagha from LARI-Hasbaya station; and, Reem El Derbass and Issam Abou Rached from US-AID/LIVCD project.

Conflicts of Interest: The authors declare no conflict of interest. The funders had no role in the design of the study; in the collection, analyses, or interpretation of data; in the writing of the manuscript, or in the decision to publish the results.

References

1. International Olive Council (IOC). Trends in World Olive Oil Consumption-IOC Report. Available online: https://www.oliveoilmarket.eu/trends-in-world-olive-oil-consumption-ioc-report/ (accessed on 3 September 2018).
2. Servili, M.; Selvaggini, R.; Esposto, S.; Taticchi, A.; Montedoro, G.F.; Morozzi, G. Health and sensory properties of virgin olive oil hydrophilic phenols: Agronomic and technological aspect of production that affects their occurence in the oil. *J. Chromatgr.* **2004**, *1054*, 113–127. [CrossRef]
3. Bendini, A.; Cerretani, L.; Carrasco-Pancorbo, A.; Gómez-Caravaca, A.M.; Segura-Carretero, A.; Fernández-Gutiérrez, A.; Lercker, G. Phenolic molecules in virgin olive oils: A survey of their sensory properties, health effects, antioxidant activity and analytical methods. An overview of the last decade. *Molecules* **2007**, *12*, 1679–1719. [CrossRef] [PubMed]
4. Montedoro, G.F.; Garofalo, L. Caratteristiche qualitative degli oli vergini di oliva. Influenza di alcune variabili: Varietà, ambiente, conservazione, estrazione, condizionamento del prodotto finito. *Rivista Italiana Sostanze Grasse* **1984**, *LXI*, 3–11.
5. Cimato, A. Effect of agronomic factors on virgin olive oil quality. *Olivae* **1990**, *31*, 20–31.
6. Pannelli, G.; Famiani, F.; Servili, M.; Montedoro, G.F. Agro-climatic factors and characteristics of the composition of virgin olive oils. *Acta Hortic.* **1990**, *286*, 477–480. [CrossRef]
7. Inglese, P.; Famiani, F.; Galvano, F.; Servili, M.; Esposto, S.; Urbani, S. Factors affecting extra-virgin olive oil composition. *Hortic. Rev.* **2011**, *38*, 83–147. [CrossRef]
8. Mailer, R.J.; Ayton, J.; Graham, K. The Influence of growing region, cultivar and harvest timing on the diversity of Australian olive oil. *J. Am. Oil Chem. Soc.* **2010**, *87*, 877–884. [CrossRef]
9. Boustany, N. The Millenium Olives (*Olea europea* L.) in Lebanon: Rejuvenation, Micropropagation and Olive Oil Variation during Maturation and Conservation. Scuola di Dottorato in Scienze dei Sistemi Agrari e Forestali e delle Produzioni Alimentari. Ph.D. Thesis, Università degli Studi di Sassari, Sassari, Italy, 2011.
10. Salvador, M.D.; Aranda, F.; Fregapane, G. Influence of fruit ripening on Cornicabra virgin olive oil quality: A study of four crop seasons. *Food Chem.* **2001**, *73*, 45–53. [CrossRef]
11. Ben Youssef, N.; Zarrouk, W.; Carrasco-Pancorbo, A.; Ouni, Y.; Segura-Carretero, A.; Fernández-Gutiérrez, A.; Daoud, D.; Zarrouk, M. Effect of olive ripeness on chemical properties and phenolic composition of chétoui virgin olive oil. *J. Sci. Food Agric.* **2010**, *90*, 199–204. [CrossRef]
12. Ceci, L.N.; Carelli, A.A. Relation between Oxidative Stability and Composition in Argentinian Olive Oils. *Am. Oil Chem. Soc.* **2010**, *87*, 1189–1197. [CrossRef]
13. Dag, A.; Kerem, Z.; Yogev, N.; Zipori, I.; Lavee, S.; Ben David, E.A. Influence of time of harvest and maturity index on olive oil yield and quality. *Sci. Hortic.* **2011**, *127*, 358–366. [CrossRef]
14. Servili, M.; Taticchi, A.; Esposto, S.; Sordini, B.; Urbani, S. Technological Aspects of Olive Oil Production. In *Olive Germplasm—The Olive Cultivation, Table Olive and Olive Oil Industry in Italy*; IntechOpen: London, UK, 2012; pp. 151–172. [CrossRef]
15. Sciancalepore, V.; De Stefano, G.; Piacquadio, P. Effects of the cold percolation system on the quality of virgin olive oil. *Eur. J. Lipid Sci. Technol.* **2000**, *102*, 680–683. [CrossRef]
16. Angerosa, F.; Di Giovacchino, L. Natural antioxidants of virgin olive oil obtained by two and tri-phase centrifugal decanters. *Grasas Aceites* **1996**, *47*, 247–254. [CrossRef]
17. Investment Development Authority of Lebanon (IDAL). Olive Oil Industry in Lebanon: 2017 Factsheet. p. 12. Available online: http://investinlebanon.gov.lb/Content/uploads/SideBlock/171011013554317~{}Olive%20Oil%20Factsheet%202017.pdf (accessed on 3 September 2018).
18. Kassab, S. *Monographie des Variétés D'olive Libanaise*; Institut de Recherches Agronomiques Libanais: Abdeh, Liban, 1973.
19. Chamoun, R.; Baalbaki, R.; Kalaitzis, P.; Talhouk, S. Molecular characterization of Lebanese olive germplasm. *Tree Genet. Genomes* **2009**, *5*, 109–115. [CrossRef]
20. Chalak, L.; Chehade, A.; Elbitar, A.; Hamadeh, B.; Youssef, H.; Nabbout, R.; Maha, S.; Haj, A.K.; Awada, A.; Bouaram, G.; et al. Morphological characterization of cultivated olive trees in Lebanon. In Proceedings of the 4th International Conference on Olive Culture, Biotechnology and Quality of Olive Tree Products (OLIVEBIOTEQ), Chania, Greece, 31 October–4 November 2011; Volume 1, pp. 51–56.

21. Chehade, A.; El Bittar, A.; Choueiri, E.; Kadri, A.; Nabbout, R.; Youssef, H.; Smeha, M.; Awada, A. Characterization of the Main Lebanese Olive Germplasm. Available online: https://www.researchgate.net/publication/286186212_Characterization_of_the_main_Lebanese_olive_germplasm (accessed on 27 November 2018).

22. Chalak, L.; Haouane, H.; Essalouh, L.; Santoni, S.; Besnard, G.; Khadari, B. Extent of the genetic diversity in Lebanese olive (*Olea europaea* L.) trees: A mixture of an ancient germplasm with recently introduced varieties. *Genet. Resour. Crop. Evol.* **2015**, *62*, 621–633. [CrossRef]

23. Bou Yazbeck, E.; Abi Rizk, G.; Hassoun, G.; El-Khoury, R.; Geagea, L. Ecological characterization of ancient olive trees in Lebanon-Bshaaleh area and their age estimation. *IOSR J. Agric. Vet. Sci.* **2018**, *11*, 35–44. [CrossRef]

24. Serhan, M.; Younes, H.; Chami, J. Physicochemical changes in Baladi olive oil as a function of production area and extraction system in North Lebanon. *J. Food Technol. Nutr. Sci.* **2016**, *1*, 1–9.

25. El Riachy, M.; Haber, A.; Abou Daya, S.; Jebbawi, G.; Al Hawi, G.; Talej, V.; Houssein, M.; El Hajj, A. Influence of irrigation regimes on quality attributes of olive oils from two varieties growing in Lebanon. *Int. J. Environ. Agric. Biotechnol.* **2017**, *2*, 895–905. [CrossRef]

26. Chehade, A.; El Bitar, A.; Kadri, A.; Choueiri, E.; Nabbout, R.; Youssef, H.; Smeha, M.; Awada, A.; Al Chami, Z.; Dubla, E.; et al. In situ evaluation of the fruit and oil characteristics of the main Lebanese olive germplasm. *J. Sci. Food Agric.* **2015**, *96*, 2532–2538. [CrossRef]

27. Investment Development Authority in Lebanon (IDAL). Map Source: Ministry of Agriculture. 2010. Available online: http://investinlebanon.gov.lb/Content/uploads/SideBlock/180606110102652North%20Lebanon%20Presentation%202018.pdf (accessed on 10 September 2018).

28. Frías, L.; García-Ortiz, A.; Hermoso, M.; Jiménez, A.; Llavero Del Pozo, M.P.; Morales, J.; Ruano, T.; Uceda, M. *Analistas de Laboratorio de Almazara*, 3rd ed.; Junta de Andalucía: Sevilla, Spain, 1999; ISBN 84-89802-61-0.

29. European Union Commission Regulation (EUC). No. 2568/91: Characteristics of olive and olive pomace oils and their analytical methods. *Off. J. Eur. Commun.* **1991**, *248*, 1–82.

30. International Olive Council (IOC). Preparation of the Fatty Acid Methyl Esters from Olive Oil and Olive-Pomace Oil. Available online: www.internationaloliveoil.org/documents/viewfile/3892-testing8eng (accessed on 10 September 2018).

31. Montedoro, G.; Servili, M.; Baldioli, M.; Miniati, E. Simple and hydrolysable phenolic compounds in virgin olive oil. 2. Initial characterization of the hydrolysable fraction. *J. Agric. Food Chem.* **1992**, *40*, 1571–1576. [CrossRef]

32. Singleton, V.L.; Rossi, J.A. Colorimetry of total phenolics with phosphomolybdic-phosphotungstic acid reagents. *Am. J. Enol Viticult.* **1965**, *16*, 144–158.

33. Tura, D.; Gigliotti, C.; Pedo, S.; Failla, O.; Bassi, D.; Serraiocco, A. Influence of cultivar and site of cultivation on levels of lipophilic and hydrophilic antioxidants in virgin olive oils (*Olea europeae*) and correlation with oxidative stability. *Sci. Hortic.* **2007**, *112*, 108–109. [CrossRef]

34. International Olive Council (IOC). Sensory Analysis of Olive Oil. Method for the Organoleptic Assessment of Virgin Olive Oil. COI/T.20/Doc. No. 15/Rev. 10. 2018. Available online: http://www.internationaloliveoil.org/estaticos/view/224-testing-methods (accessed on 10 September 2018).

35. El Riachy, M.; Priego-capote, F.; Rallo, L.; Luque de Castro, M.D.; León, L. Phenolic profile of virgin olive oil from advanced breeding selections. *Span. J. Agric. Res.* **2012**, *10*, 443–453. [CrossRef]

36. Ben Hassine, K.; El Riachy, M.; Taamalli, A.; Malouche, D.; Ayadi, M.; Talmoudi, K.; Aouini, M.; Jlassi, Y.; Benincasa, C.; Romano, E.; et al. Consumer discrimination Of Chemlali and Arbequina Olive Oil Cultivars according To Their Cultivar, Geographical Origin, And Processing System. *Eur. J. Lipid Sci. Technol.* **2014**, *116*, 812–824. [CrossRef]

37. Di Giovacchino, L.; Sestili, S.; Di Vincenzo, D. Influence of olive processing on virgin olive oil quality. *Eur. J. Lipid Sci. Technol.* **2002**, *104*, 587–601. [CrossRef]

38. Lazzez, A.; Vichi, S.; Grati-Kammoun, N.; Néji-Arous, M.; Khlif, M.; Romero, A.; Cossentini, M. A four year study to determine the optimal harvesting period for Tunisian Chemlali olives. *Eur. J. Lipid Sci. Technol.* **2011**, *113*, 796–807. [CrossRef]

39. International Olive Council (IOC). International Trade Standard Applying to Olive Oils and Olive-Pomace Oils International Trade Standard Applying to Olive Oils and Olive-Pomace Oils. Available online: www.internationaloliveoil.org/documents/viewfile/3615-normaen1 (accessed on 12 September 2018).

40. Ben Temime, S.; Baccouri, B.; Taamalli, W.; Abaza, L.; Daoud, D.; Zarrouk, M. Location effects on Chétoui virgin olive oil stability. *J. Food Biochem.* **2006**, *30*, 659–670. [CrossRef]

41. Salvador, M.D.; Aranda, F.; Gómez-Alonso, S.; Fregapane, G. Influence of extraction system, production year and area on Cornicabra virgin olive oil: A study of five crop seasons. *Food Chem.* **2003**, *80*, 359–366. [CrossRef]

42. Gimeno, E.; Castellote, A.I.; Lamuela-Raventós, R.M.; De la Torre, M.C.; López-Sabater, M.C. The effects of harvest and extraction methods on the antioxidant content (phenolics, a-tocopherol, and b-carotene) in virgin olive oil. *Food Chem.* **2002**, *78*, 207–211. [CrossRef]

43. Torres, M.M.; Maestri, D.M. The effects of genotype and extraction methods on chemical composition of virgin olive oils from traslasierra valley (Córdoba, Argentina). *Food Chem.* **2006**, *96*, 507–511. [CrossRef]

44. Gómez-Caravaca, A.M.; Cerretani, L.; Bendini, A.; Segura-Carretero, A.; Fernández-Gutiérrez, A.; Del Carlo, M.; Compagnone, D.; Cichelli, A. Effects of fly attack (Bactrocera oleae) on the phenolic profile and selected chemical parameters of Olive Oil. *J. Agric. Food Chem.* **2008**, *56*, 4577–4583. [CrossRef] [PubMed]

45. Mraicha, F.; Ksantini, M.; Zouch, O.; Ayadi, M.; Sayadi, S.; Bouaziz, M. Effect of olive fruit fly infestation on the quality of olive oil from Chemlali cultivar during ripening. *Food Chem. Toxicol.* **2010**, *48*, 3235–3241. [CrossRef] [PubMed]

46. Di Giovacchino, L.; Solinas, M.; Miccoli, M. Effect of extraction systems on the quality of virgin olive oil. *J. Am. Oil Chem. Soc.* **1994**, *71*, 1189–1194. [CrossRef]

47. Ranalli, A.; Contento, S.; Schiavone, C.; Simone, N. Malaxing temperature affects volatile and phenol composition as well as other analytical features of virgin olive oil. *Eur. J. Lipid Sci. Technol.* **2001**, *103*, 228–238. [CrossRef]

48. Fuentes de Mendoza, M.; De Miguel Gordillo, C.; Expóxito, J.M.; Casas, J.S.; Cano, M.M.; Vertedor, D.M.; Baltasar, M.F. Chemical composition of virgin olive oils according to the ripening in olives. *Food Chem.* **2013**, *141*, 2575–2581. [CrossRef] [PubMed]

49. Nsir, H.; Taamalli, A.; Valli, E.; Bendini, A.; Gallina-Toschi, T.; Zarrouk, M. Chemical composition and sensory quality of Tunisian 'Sayali' virgin olive oils as affected by fruit ripening: Toward an appropriate harvesting time. *J. Am. Oil Chem. Soc.* **2017**, *94*, 913–922. [CrossRef]

50. Gutiérrez, F.; Jiménez, B.; Ruiz, A.; Albi, M.A. Effect of olive ripeness on the oxidative stability of virgin olive oil extracted from the varieties Picual and Hojiblanca and on the different components involved. *J. Agric. Food Chem.* **1999**, *47*, 121–127. [CrossRef]

51. Baccouri, O.; Guerfel, M.; Baccouri, B.; Cerretani, L.; Bendini, A.; Lercker, G.; Daoud Ben Miled, D. Chemical composition and oxidative stability of Tunisian monovarietal virgin olive oils with regard to fruit ripening. *Food Chem.* **2008**, *109*, 743–754. [CrossRef]

52. Bengana, M.; Bakhouche, A.; Lozano-Sánchez, J.; Amir, Y.; Youyou, A.; Segura-Carretero, A.; Fernández-Gutiérrez, A. Influence of olive ripeness on chemical properties and phenolic composition of Chemlal extra-virgin olive oil. *Food Res. Int.* **2013**, *54*, 1868–1875. [CrossRef]

53. Bajoub, A.; Hurtado-Fernández, E.; Ajal, E.A.; Fernández-Gutiérrez, A.; Carrasco-Pancorbo, A.; Ouazzani, N. Quality and chemical profiles of monovarietal north Moroccan olive oils from "Picholine Marocaine" cultivar: Registration database development and geographical discrimination. *Food Chem.* **2015**, *179*, 127–136. [CrossRef] [PubMed]

54. Dugo Mo, G.; Alfa, M.; La Pera, L.; Mavrogeni, E.; Pallicino, D.; Pizzimenti, G.; Maisano, R. Characterization of Sicilian virgin olive oils, note X: A comparasion between Cerasuola and Nocellera del Bellice varieties. *Grasas Aceites* **2004**, *55*, 415–422.

55. Gürdeniz, G.; Ozen, B.; Tokatli, F. Classification of Turkish olive oils with respect to cultivar, geographic origin and harvest year, using fatty acid profile and mid-IR spectroscopy. *Eur. Food Res. Technol.* **2008**, *227*, 1275–1281. [CrossRef]

56. Beltrán, G.; del Río, C.; Sánchez, S.; Martínez, L. Seasonal changes in olive fruit characteristics and oil accumulation during ripening process. *J. Sci. Food Agric.* **2004**, *84*, 1783–1790. [CrossRef]

57. Issaoui, M.; Flamini, G.; Brahmi, F.; Dabbou, S.; Ben Hassine, K.; Taamali, A.; Chehab, H.; Ellouz, M.; Zarrouk, M.; Hammami, H. Effect of the growing area conditions on differentiation between Chemlali and Chétoui olive oils. *Food Chem.* **2010**, *119*, 220–225. [CrossRef]

58. Youssef, O.; Guido, F.; Manel, I.; Ben Youssef, N.; Pier Luigi, C.; Mohamed, H.; Daoud, D.; Zarrouk, M. Volatile compounds and compositional quality of virgin olive oil from Oueslati variety: Influence of geographical origin. *Food Chem.* **2011**, *124*, 1770–1776. [CrossRef]
59. Guerfel, M.; Ouni, Y.; Taamalli, A.; Boujnah, D.; Stefanoudaki, E.; Zarrouk, M. Effect of location on virgin olive oils of the two main Tunisian olive cultivars. *Eur. J. Lipid Sci. Technol.* **2009**, *111*, 926–932. [CrossRef]
60. Rotondi, A.; Bendini, A.; Cerretani, L.; Mari, M.; Lercker, G.; Gallina Toschi, T. Effect of olive ripening degree on the oxidative stability and organoleptic properties of cv. Nostrana di Brisighella extra virgin olive oil. *J. Agric. Food Chem.* **2004**, *52*, 3649–3654. [CrossRef] [PubMed]
61. Bendini, A.; Cerretani, L.; Salvador, M.D.; Fregapane, G.; Lercker, G. Stability of the Sensory Quality of Virgin Olive Oil during Storage: An Overview. *Ital. J. Food Sci.* **2009**, *21*, 389–406.
62. Lozano-Sánchez, J.; Cerretani, L.; Bendini, A.; Segura-Carretero, A.; Fernández-Gutierrez, A. Filtration Process of Extra Virgin Olive Oil: Effect on Minor Components, Oxidative Stability and Sensorial and Physicochemical Characteristics. *Trends Food Sci. Technol.* **2010**, *21*, 201–211. [CrossRef]
63. Predieri, S.; Medoro, C.; Magli, M.; Gatti, E.; Rotondi, A. Virgin olive oil sensory properties: Comparing trained panel evaluation and consumer preferences. *Food Res. Int.* **2013**, *54*, 2091–2094. [CrossRef]

sustainability

MDPI

Article

Fatty Acid Compositions of Selected Polish Pork Hams and Sausages as Influenced by Their Traditionality

Michał Halagarda *, Władysław Kędzior, Ewa Pyrzyńska and Wanda Kudełka

Department of Food Commodity Science, Cracow University of Economics, ul. Sienkiewicza 5,
30-033 Kraków, Poland; wladyslaw.kedzior@uek.krakow.pl (W.K.); ewa.pyrzynska@uek.krakow.pl (E.P.);
wanda.kudelka@uek.krakow.pl (W.K.)
* Correspondence: michal.halagarda@uek.krakow.pl; Tel.: +48-12-293-78-46

Received: 24 September 2018; Accepted: 24 October 2018; Published: 25 October 2018

Abstract: Sausages and hams are perceived as important components of culinary heritage for many regions all over the world. Consumers believe that traditional foods are characterized by unique sensory properties and high quality. However, the fats found in all pork meat products are generally not associated with favorable dietary patterns. The aim of this study was to verify the possible differences regarding the composition of fatty acids between traditional Polish pork hams and *wiejska* sausages, and their conventional equivalents. For this purpose, the fat content and fatty acid profiles were determined. The research material consisted of 2 varieties of traditional hams and 5 varieties of sausages, as well as 4 varieties of both conventional hams and sausages. The results of this study demonstrated that traditional hams contained significantly higher percentage of C 20:3 (*cis*-11,14,17) acid than their conventional equivalents. Traditional sausages were characterized by lower shares of C 18:2 (*cis*-9,12) and Polyunsaturated Fatty Acids (PUFA), whereas higher content of C 18:1 (*cis*-9), C 18:3 (*cis*-9,12,15), C 20:0 and Monounsaturated Fatty Acids (MUFA). This resulted in significantly higher amounts of *n*-3 and lower of *n*-6 acids than in conventional sausages. All of the tested meat products were also characterized by an unfavorable *n*-6/*n*-3 ratio.

Keywords: traditional sausages; conventional sausages; traditional hams; conventional hams; traditional meat products; pork; fatty acids

1. Introduction

Consumers today look for foods characterized not only by health safety and proper nutritional value but also by unique sensory properties, high quality and natural composition. Therefore, an increased interest in traditional and regional food products can be noticed. Such foods, thanks to the usage of characteristic methods of growing plants, breeding animals and essentially traditional processing technologies, are characterized by a unique appearance, smell and taste [1–3]. Moreover, they are associated with the local culture and identity of people in various parts of the world, thus contributing to the folklore and traditions of specific communities and, at the same time, becoming an emblem or a flagship product of a certain region. Traditional foods have a representational value and can contribute to the development and sustainability of the countryside. They also ensure greater diversity of food choice for customers [4]. Consumers are generally convinced that traditional foods have a positive impact on health characteristics [5] that has been proved over time, and unique sensory properties [6].

The Polish market of traditional foods has been developing dynamically. Since 2005, the food producers have a possibility to register their products on the List of Traditional Products of Polish Ministry of Agriculture and Rural Development. Only highest quality products whose uniqueness results from traditional method of production (successfully used for over 25 years) can be listed.

In addition, they tend to be a part of the identity of the local community and an element of the cultural heritage of the region of origin. The production methods and product features do not have to be inevitably linked to the specific place. The List is aimed to promote the unique products and facilitate the acquisition of the well-known European Union mark Traditional Specialty Guaranteed [7].

Sausages and hams are an important aspect of culinary heritage of many regions all over the world and have been consumed for centuries [8]. In Poland, smoked hams and sausages of world-renowned quality have been produced for centuries [9]. They are characterized by their incomparable sensory features. The manufacturers use only natural ingredients such as meat and natural spices in the production process. The meat comes from animals that are reared using traditional feeds (e.g., potatoes and green fodder) what positively affects its sensory characteristics and nutritional value. Substances such as artificial additives, fillers, improvers or preservatives are not permitted, except for a mixture of salt and nitrite. To extend the shelf life, hams and sausages are dried and/or smoked with the use of carefully chosen wood [10].

Beside traditional meat products, a large selection of high-performance products is available on the market. These products have lower prices but are produced by means of the technology that involves the use of various food additives which substitute high-priced raw materials [11]. As a consequence, such meat products notably differ in the sensory characteristics and partly in nutritional value [12–14].

Meat-based products contain a lot of fat whose physiological role is not only to provide energy but also to carry various substances such as hormones or vitamins. It also positively influences the sensory characteristics of meat products, mostly the juiciness and smell [15,16]. However, due to the presence of cholesterol and relatively large amounts of saturated fatty acids, as well as very low levels of n-3 polyunsaturated fatty acids (PUFA), the fats found in meat products are not regarded as a positive content from a nutritional point of view [17,18]. Since the current human diet is usually already rich in fats, especially saturated and n-6 polyunsaturated, and there is also a need to balance the n-6/n-3 fatty acid intake for prevention of chronic diseases, it is particularly essential to evaluate the fatty acid profile in daily eaten foods, such as meat products [17,19]. Yet, the composition of meat fat depends on various factors, such as the individual characteristics of the animal (age, sex, species, breed) and its diet [17,20,21]. Proper selection of raw materials can therefore have a positive effect on the fatty acid profile. The aim of this study was then to investigate the possible differences regarding the composition of various fatty acids between traditional Polish smoked hams and *wiejska* sausages, and their conventional (mass produced) equivalents.

2. Materials and Methods

2.1. Product Samples

The research material can be divided into two groups: Meat-based products that are registered on the List of Traditional Products of Polish Ministry of Agriculture and Rural Development and conventional ones. Among those listed, there were 2 varieties of smoked hams (letter codes A-B) and 5 varieties of traditional *wiejska* sausages (G-K). The conventional products tested consisted of 4 varieties of smoked hams (letter codes C-F) and 4 varieties of sausages (L-O).

The sausage samples were uniform in terms of their diameter, they were all made of pork and were semi-coarsely ground. In the initial phase of the manufacturing process of all varieties of *wiejska* sausages, firstly the raw materials are prepared and then the meat is comminuted. Subsequently, the ingredients (meat and seasoning) are mixed, natural casings are filled and left to settle (which usually takes a few hours). In the next step, the sausages are wood hot-smoked and then they are baked. The production processes mostly differ in terms of the raw materials or the type of wood used, as well as in terms of smoking or baking parameters applied. Traditional recipes allow the use of curing salts.

The hams examined were also made of pork meat. The production process of traditional hams involves the use of herbs and spices in which the meat is marinated. Then the product is tied with string or put in a special mesh and wood hot-smoked. Afterwards, the ham is scalded in hot water.

As in the case of sausages, the differences in the manufacturing process include the use of different raw materials and wood, as well as smoking parameters.

The production of conventional meat products is aimed at receiving high yields in a relatively short time. Such mass-products are manufactured with the use of various types of food additives and modern-day production technologies. Detailed ingredient lists are presented in Table 1.

The main selection criterion for conventional products was their similar appearance to the traditional counterparts. The tests included six products of each variety but produced by different manufacturers and were repeated three times. The meat products were bought in various marketplaces: The manufacturer owned shops, delicatessens and supermarkets, between 3 and 7 days after their production. All of the analyses were performed and completed within 4 days after the acquisition.

Table 1. The ingredient lists of the product varieties tested (compiled on the basis of manufacturers' declarations).

Product	Ingredient List
A [a]	pork meat, salt, natural spices, preservative: sodium nitrite
B [a]	pork meat, salt, natural spices: allspice, bay leaf, pickling salt
C [b]	pork meat, salt, milk protein, pork collagen protein, lactose, glucose, stabilizers: triphosphates E451, sodium citrate E331, antioxidants: sodium ascorbate E301, sodium isoascorbate E316, flavor enhancer: monosodium glutamate E621, natural spices and their extracts, yeast extracts, aromatic preparations, acidity regulator: citric acid E330, preservative: sodium nitrite E250.
D [b]	pork meat, water, soy protein, stabilizers: E451, E452, E262, thickener E407, animal protein, sugars, antioxidants: E316, E301, flavor enhancer: E621, salt, vegetable fiber, aromas, spice extracts, acidity regulators: E331, preservative: E250.
E [b]	pork meat, water, stabilizer: E451, thickener E407, soy protein, wheat fiber, maltodextrins, animal protein, glucose, E621, E316, salt, E261, E326, natural spice extracts, aromas, E250.
F [b]	pork meat, water, salt, soy protein, stabilizers: E451, E508, E331, gelling agents: E407, E415, E425i, flavor enhancer: E621, antioxidant: E316, glucose, soy protein hydrolyzate, spice extracts, aromas, smoke flavor, preservative: E250.
G [c]	pork meat, salt, natural spices, sugar
H [c]	pork meat, salt, black pepper, natural garlic
I [c]	pork meat, salt, spices (garlic, pepper), sugar
J [c]	pork meat, garlic, salt, pickling salt
K [c]	pork meat, pork fat, salt, garlic, spices in various proportions
L [d]	pork meat, water, mechanically separated pork, potato starch, pork fat, stabilizers: E450, E451, soy protein, gelling agent (E407), pork collagen protein, antioxidant: sodium ascorbate, salt, flavor enhancer: monosodium glutamate, seasoning, aromas, plant protein hydrolysate, glucose, antioxidant: ascorbic acid, paprika extract, acidity regulator: sodium acetate, sodium citrate, hemoglobin, preservative: sodium nitrite
M [d]	pork meat, water, pork connective tissue, pork fat, potato starch, salt, aroma, dextrose, glucose syrup, taste enhancer (E621), antioxidants: E300, E301, E316, E315, stabilizers: E262, E331, modified cellulose (E461), plasma protein, preservative: E250
N [d]	pork meat, potato starch, salt, soy protein, stabilizer (E450), E451, E262, flavor enhancer (E621)
O [d]	pork meat, water, pork fat, potato starch, wheat fiber, salt, soy protein, dextrose, stabilizer: triphosphate (E407a), thickeners: xanthan gum, tare gum, acidity regulator: potassium chloride, antioxidants: isoascorbic acid, sodium isoascorbate, pork protein, grape sugar, flavor enhancers: disodium inosinate, monosodium glutamate, seasonings and extracts of spices, aroma, aroma of smoke, maltodextrin, dye: cochineal, preservative: sodium nitrite

Note: [a]—traditional hams; [b]—conventional hams; [c]—traditional sausages; [d]—conventional sausages.

2.2. Fat

The soxhlet method was used to determine the fat content. The procedure was as follows: Minced samples (5 g each) were dried (oven-drying method), then transferred to thimbles and extracted for three hours with the use of petroleum ether in the Büchi Extraction System B-811. After extraction, the thimbles were dried at $103 \pm 2\,°C$ for one hour, then they were cooled down to room temperature in a desiccator and afterwards they were weighted. Drying, cooling and weighing was repeated until the results of two successive weightings did not vary by more than 0.1% by weight of the sample [22].

2.3. Fatty Acids

In order to determine the fatty acid profile, samples of hams and sausages were subjected to extraction according to Soxhlet method. The extracts were esterified according to ISO 12966-2:2017 standard [23]. The analysis of the esterified samples was carried out in accordance with ISO 12966-1:2014 standard [24] on the SRI 8610C gas chromatograph with Restek RTX-2330 column length l = 105 m, Ø = 0.25 mm with FID detector, using hydrogen as a carrier gas. The Food Industry FAME Mix from Restek was used as a reference material.

2.4. Statistical Analysis

The measurement of the fat concentration and fatty acid profiles were analyzed with the use of statistical methods and the R 3.5.0 software. The required value for statistical significance was set at 0.05. The Shapiro-Wilk test was used to verify the normality of variable distribution. The test showed that the distributions were not normal. That is why, the differences between product varieties were analyzed using the Kruskal-Wallis test with post-hoc Dunn's test. The differences between product groups (traditional and conventional) were analyzed with the use of linear mixed models with product group as fixed effect and product source as random effect, mostly in order to account for the variance resulting from the purchase sites.

As a result of dimension reduction, performed by the means of principal component analysis (PCA), a 2-dimensional sample map was developed. It was used to identify the most similar and dissimilar samples.

3. Results and Discussion

3.1. Fat Contents and Fatty Acid Profiles of Tested Traditional and Conventional Hams and Sausages

The results of the fat content and fatty acid profiles for the tested hams are presented in Tables 2 and 3, whereas the results for the tested sausages are presented in Tables 4 and 5.

Table 2. Fat content (g/100 g) and saturated fatty acid profiles (% of total fatty acids) of selected Polish pork hams.

Parameter	Product Group	x̄ (sd)	p
Fat	I (N = 2) II (N = 4)	8.56 (5.53) 5.81 (3.68)	0.329
C 12:0	I (N = 2) II (N = 4)	0.01 (0.03) 0.43 (1.04)	0.332
C 13:0	I (N = 2) II(N = 4)	0.00 (0.00) 0.05 (0.18)	0.505
C 14:0	I (N = 2) II (N = 4)	1.60 (0.24) 1.76 (0.25)	0.182
C 15:0	I (N = 2) II (N = 4)	0.00 (0.00) 0.03 (0.09)	0.505

Table 2. *Cont.*

Parameter	Product Group	x̄ (sd)	*p*
C 16:0	I (N = 2)	25.9 (1.79)	0.664
	II (N = 4)	26.35 (1.60)	
C 17:0	I (N = 2)	0.27 (0.24)	0.298
	II (N = 4)	0.70 (0.99)	
C 18:0	I (N = 2)	12.30 (1.50)	0.364
	II (N = 4)	11.76 (1.02)	
C 20:0	I (N = 2)	0.39 (0.23)	0.881
	II (N = 4)	0.41 (0.22)	
C 21:0	I (N = 2)	0.33 (0.22)	0.346
	II (N = 4)	0.23 (0.18)	
SFA	I (N = 2)	40.80 (1.98)	0.581
	II (N = 4)	41.71 (2.80)	

Group I—traditional hams; Group II—conventional hams; *p*—indicates significant differences between groups of products; sd—standard deviation; N—number of varieties tested.

Table 3. Unsaturated fatty acid profiles (% of total fatty acids) of selected Polish pork hams.

Parameter	Product Group	x̄ (sd)	*p*
C 14:1 (*cis*-9)	I (N = 2)	0.18 (0.44)	0.885
	II (N = 4)	0.22 (0.52)	
C 16:1 (*cis*-9)	I (N = 2)	3.96 (0.47)	0.466
	II (N = 4)	4.26 (0.85)	
C 17:1 (*cis*-10)	I (N = 2)	0.50 (0.28)	0.503
	II (N = 4)	0.35 (0.39)	
C 18:1 (trans-9)	I (N = 2)	0.14 (0.19)	0.063
	II (N = 4)	0.50 (0.45)	
C 18:1 (*cis*-9)	I (N = 2)	44.61 (1.73)	0.645
	II (N = 4)	43.96 (3.23)	
C 18:2 (*cis*-9,12)	I (N = 2)	8.37 (2.24)	0.887
	II (N = 4)	8.10 (2.50)	
C 18:3 (*cis*-6,9,12)	I (N = 2)	0.12 (0.14)	0.720
	II (N = 4)	0.16 (0.15)	
C 18:3 (*cis*-9,12,15)	I (N = 2)	0.66 (0.24)	0.203
	II (N = 4)	0.55 (0.14)	
C 20:3 (*cis*-11,14,17)	I (N = 2)	0.55 (0.23)	0.003
	II (N = 4)	0.19 (0.16)	
MUFA	I (N = 2)	49.39 (1.78)	0.929
	II (N = 4)	49.28 (2.84)	
PUFA	I (N = 2)	9.70 (2.52)	0.721
	II (N = 4)	9.00 (2.53)	
UFA	I (N = 2)	59.10 (1.94)	0.617
	II (N = 4)	58.28 (2.81)	
n-6	I (N = 2)	9.04 (2.35)	0.768
	II (N = 4)	8.45 (2.57)	
n-3	I (N = 2)	0.66 (0.24)	0.203
	II (N = 4)	0.55 (0.14)	
n-6/*n*-3	I (N = 2)	14.37 (3.65)	0.679
	II (N = 4)	16.69 (7.76)	

Group I—traditional hams; Group II—conventional hams; *p*—indicates significant differences between groups of products; sd—standard deviation; N—number of varieties tested.

Table 4. Fat content (g/100 g) and saturated fatty acid profiles (% of total fatty acids) of selected Polish pork *wiejska* sausages.

Parameter	Product Group	x̄ (sd)	p
Fat	I (N = 5)	(26.39) 3.47	0.010
	II (N = 4)	(21.80) 2.90	
C 12:0	I (N = 5)	0.24 (0.45)	0.520
	II (N = 4)	0.34 (0.24)	
C 13:0	I (N = 5)	0.19 (0.53)	0.317
	II(N = 4)	0.43 (0.74)	
C 14:0	I (N = 5)	1.93 (0.51)	0.980
	II (N = 4)	1.93 (0.26)	
C 15:0	I (N = 5)	0.00 (0.00)	0.261
	II (N = 4)	0.03 (0.10)	
C 16:0	I (N = 5)	27.07 (2.01)	0.922
	II (N = 4)	27.00 (1.11)	
C 17:0	I (N = 5)	0.62 (0.43)	0.121
	II (N = 4)	0.33 (0.14)	
C 18:0	I (N = 5)	12.64 (0.82)	0.083
	II (N = 4)	11.67 (0.98)	
C 20:0	I (N = 5)	0.38 (0.17)	<0.001
	II (N = 4)	0.24 (0.73)	
C 21:0	I (N = 5)	0.34 (0.13)	0.744
	II (N = 4)	0.32 (0.12)	
SFA	I (N = 5)	43.40 (2.42)	0.386
	II (N = 4)	42.74 (1.24)	

Group I—traditional sausages; Group II—conventional sausages; *p*—indicates significant differences between groups of products; sd—standard deviation; N—number of varieties tested.

Table 5. Unsaturated fatty acid profiles (% of total fatty acids) of selected Polish pork *wiejska* sausages.

Parameter	Product Group	x̄ (sd)	p
C 14:1 (*cis*-9)	I (N = 5)	0.05 (0.18)	0.535
	II (N = 4)	0.12 (0.41)	
C 16:1 (*cis*-9)	I (N = 5)	3.92 (0.59)	0.964
	II (N = 4)	3.93 (0.54)	
C 17:1 (*cis*-10)	I (N = 5)	0.48 (0.44)	0.418
	II (N = 4)	0.75 (0.93)	
C 18:1 (trans-9)	I (N = 5)	0.25 (0.20)	0.539
	II (N = 4)	0.30 (0.20)	
C 18:1 (*cis*-9)	I (N = 5)	43.54 (2.10)	0.001
	II (N = 4)	40.27 (1.33)	
C 18:2 (*cis*-9,12)	I (N = 5)	7.18 (2.13)	<0.001
	II (N = 4)	10.90 (1.34)	
C 18:3 (*cis*-6,9,12)	I (N = 5)	0.17 (0.13)	0.931
	II (N = 4)	0.16 (0.11)	
C 18:3 (*cis*-9,12,15)	I (N = 5)	0.65 (0.09)	0.004
	II (N = 4)	0.54 (0.11)	
C 20:3 (*cis*-11,14,17)	I (N = 5)	0.15 (0.26)	0.672
	II (N = 4)	0.11 (0.26)	
MUFA	I (N = 5)	48.23 (1.97)	0.001
	II (N = 4)	45.38 (1.06)	

Table 5. *Cont.*

Parameter	Product Group	x̄ (sd)	p
PUFA	I (N = 5) II (N = 4)	8.23 (2.14) 11.81 (1.37)	<0.001
UFA	I (N = 5) II (N = 4)	56.46 (2.21) 57.19 (1.25)	0.310
n-6	I (N = 5) II (N = 4)	7.58 (2.13) 11.27 (1.40)	<0.001
n-3	I (N = 5) II (N = 4)	0.65 (0.09) 0.54 (0.11)	0.004
n-6/n-3	I (N = 5) II (N = 4)	11.96 (3.89) 21.98 (5.83)	<0.001

Group I—traditional sausages; Group II—conventional sausages; *p*—indicates significant differences between groups of products; sd—standard deviation; N—number of varieties tested.

3.1.1. Hams

It was shown that the average fat concentration among all tested hams did not differ significantly between the analyzed groups of products. Nevertheless, conventional product F was characterized by the lowest (2.01 g/100 g) content of this nutrient, statistically significantly lower ($p < 0.001$) than for products B (10.07 g/100 g), D (8.04 g/100 g) and E (9.57 g/100 g).

The analyzed ham varieties contained more unsaturated (with predominant share of C 18:1 (*cis*-9) acid) than saturated fatty acids, what is in accordance with the results received by Kasprzyk, Tyra and Babicz [21]. They were characterized by high concentrations of C 16:0 and C 18:1 (*cis*-9), as well as a moderate content of C 16:1 (*cis*-9), C 18:0 and C 18:2 (*cis*-9,12). However, although stearic acid (C 18:0) belongs to the group of saturated acids, it was shown that its consumption does not affect the HDL or LDL cholesterol or the total cholesterol/HDL ratio, which is considered the risk factor for cardiovascular diseases [25]. Nevertheless, C:14 and C:16 acids exhibit artherogenic effects [26]. Their average concentration in hams accounted for 27.50% of total fatty acids in traditional products and 28.11% in conventional ones. Similar results were obtained by Pietrzak-Fiećko and Modzelewska-Kapituła [27]—27.05%, whereas in traditional Spanish meat products they reached 25% of the total fatty acids [28]. The slight differences between Polish and Spanish products may result from the use of different breeds of pigs that are fed differently [29,30].

All of the hams, especially variety D (27.46), were characterized by an unfavorable *n*-6/*n*-3 ratio, which should preferably vary from 1/1 to 4/1 [19]. Similar results were obtained by Garbowska, Pietrzak-Fiećko and Radzymińska [31] for Polish local, traditional and conventional pork hams. The average ratio for these groups of products ranged from 14.87 to 15.33. Pietrzak-Fiećko and Modzelewska-Kapituła [27] also received the average ratio of 15.52 in Polish smoked ham. This maybe the result of similar production methods and the use of breeds characteristic for this geographical area that are fed with the use of similar forage, since these factors mainly affect the fatty acid profile [29,30]. According to Grześkowiak et al. [32], the *n*-6/*n*-3 ratio typical for raw pork is close to 14.9. It should, however, be noted that some differences in the unsaturated fatty acids content might be attributed to the amounts of spices with anti-oxidative potential (e.g., allspice, black pepper, garlic) [33] used in the traditional products as well as the levels of artificial antioxidants added to conventional hams.

The traditional ham varieties contained significantly higher amounts of C 20:3 (*cis*-11,14,17) acid (0.55%) than their conventional equivalents (0.19%) in relation to total fatty acids. Moreover, variety B (0.61%) contained significantly more of this compound than variety F (0.00%). Considering the fatty acid profiles, no other statistically significant differences both between the ham varieties tested and between traditional and conventional products were detected. Correspondingly, Garbowska, Pietrzak-Fiećko and Radzymińska [31] showed that the fatty acid profile in pork products does not depend either on the origin of raw materials, or on the production methods. In their research local meat products showed similar fatty acid profile to the mass produced counterparts.

In the discussed study it was also shown that the fatty acid profiles of Polish hams are typical of processed pork products both from Poland [31] and from other parts of the world [17,29]. As in the case of Iberian hams examined by Fernández et al. [17], dry cured hams analyzed by Žlender et al. [29] and Polish local, traditional and conventional pork meat products tested by Garbowska, Pietrzak-Fiećko and Radzymińska [31], a high MUFA concentration indicates that these products can be an element of a healthy diet. Consuming MUFAs has a favorable effect on the cardiovascular system [34]. Moreover, besides the D variety (27.46), the other samples tested in this research had a similar *n*-6/*n*-3 ratio (12.05 to 15.93) to hams examined by Garbowska, Pietrzak-Fiećko and Radzymińska [31] (from 8.26 to 16.80), Fernández et al. [17] (from 9.36 to 13.75) and Žlender et al. [29] (from 10.4 to 16.1).

3.1.2. Sausages

Sausages are produced from meat of various fat concentrations and connective tissue of different proportions that depend on the recipe. The fat content is therefore higher than in hams and may reach even 35% [27]. In the case of the sausages tested, traditional products were characterized by statistically significantly higher content of fat than their conventional equivalents (26.39 g/100 g and 21.8 g/100 g, respectively). Sausage H had the highest content of fat (28.83 g/100 g), significantly higher than for varieties L (21.23 g/100 g), M (20.9 g/100 g) and N (19.34 g/100 g), whereas sausage N—the lowest (19.34 g/100 g) and significantly lower than for varieties J (27.44 g/100 g), K (26.87 g/100 g) and O (26.54 g/100 g) ($p < 0.001$).

Considering the fatty acid profiles, the analyzed sausages contained more unsaturated than saturated fatty acids. Similarly to the hams tested, they were characterized by high concentrations of C 16:0 and C 18:1 (*cis*-9), as well as moderate content of C 16:1 (*cis*-9), C 18:0 and C 18:2 (*cis*-9,12) in total fatty acids, what seems to be typical for Polish pork sausages [31]. The average concentrations of the sum of C:14 and C:16 acids were a bit higher than those noted for hams and equaled 29.00% for traditional products and 28.93% for the conventional ones. Pietrzak-Fiećko and Modzelewska-Kapituła [27] received a bit lower average concentration in Polish pork sausages—27.5% of total fatty acids. The significant differences between the analyzed meat product varieties concerned the concentrations of C 17:0, C 18:0, C 18:1 (*cis*-9), C 18:2 (*cis*-9,12) acids as well as monounsaturated fatty acids (MUFAs) and *n*-6 acids. Statistically significant differences were found in the samples with the highest and the lowest shares of these compounds. In the case of C 17:0 acid, they were associated with samples I (1.18%) and J (0.24%), respectively ($p = 0.025$). In relation to C 18:0 acid, J (13.75%), as well as N (10.65%) and O (11.1%) ($p = 0.008$); whereas, in the case of C 18:1 (*cis*-9), H (46.53%) as well as N (39.35%) and O (39.6%), respectively ($p = 0.005$). In the case of C 18:2 (*cis*-9,12), the differences were mostly associated with samples N (12.14%) and H (5.02%) ($p = 0.008$).The sausage H contained the highest percentages of MUFAs (51.07%), which were significantly higher than in the samples with the lowest concentration of these compounds—L (45.12%) and N (44.97%) ($p = 0.01$). It also had the lowest content of *n*-6 acids (5.56%), which was significantly lower than in the sample with the highest share of these fatty acids—N (12.37%) ($p = 0.015$).

Contrary to the results of research by Garbowska, Pietrzak-Fiećko and Radzymińska [31], in this research some significant differences between the analyzed groups of products were detected. Traditional products, in comparison to their conventional equivalents, were characterized by statistically significantly C 18:2 (*cis*-9,12) acid (7.18% and 10.9%, respectively), whereas a higher share of C 18:1 (*cis*-9) acid (43.54% and 40.27%, respectively), C 18:3 (*cis*-9,12,15) acid (0.65% and 0.54%, respectively) and C 20:0 acid (0.38% and 0.24%, respectively). The differences in C 18:2 (*cis*-9,12) and C 18:1 (*cis*-9) acids may result from the pork used for production. Thus, the proportion of C 18:2 (*cis*-9,12) is higher [35] and C 18:1 (*cis*-9) acid lower [36] in fat of the lean pigs. Traditional food producers, most probably use breeds that are richer in fat as a source of pork for sausages. This is also reflected in fat content differences among two groups of the analyzed products (Table 4). Moreover, in the discussed research traditional products also contained significantly more monounsaturated fatty acids (MUFAs) (48.23% and 45.38%, respectively) but lower amounts of polyunsaturated fatty acids (PUFAs) (8.23% and 11.81%, respectively) in the total fatty acid content. These differences can also be attributed to the fatter pigs used for production of traditional sausages [36]. As previously mentioned, a regular intake of MUFAs has a positive effect on the cardiovascular system [34]. On the other hand, PUFAs have anti-atherosclerosis, anti-inflammatory and anti-aggregation properties [37]. Similarly as in the case of hams, it should be indicated that some differences in the unsaturated fatty acids content may result from the amounts of spices with antioxidative potential (e.g., allspice, black pepper, garlic) [33] used in traditional products as well as from the supplementation of conventional sausages with artificial antioxidants.

For comparison purposes, Polish sausages examined by Pietrzak-Fiećko and Modzelewska-Kapituła [27] contained on average 49.45% of MUFAs and 9.20% of PUFAs in the total fatty acid content, whereas Swiss cooked pork sausages had 46.42% of MUFAs and 8.50% of PUFAs [38]. In the fermented dry Sremska sausages made of pork meat, examined by Parunović et al. [39], the share of MUFAs ranged from 42.50% to 52.79% and of PUFAs from 8.70% to 16.79% of total fatty acids. Amaral et al. [40] showed that pork Frankfurter type sausages contained from 45.29% to 51.53% of MUFAs and from 12.21% to 16.54% of PUFAs. Similarly, the average concentration of MUFAs was 44.89% and 11.79% of PUFAs in Chorizo ahumado (smoked sausage) [41].

Nevertheless, far more important is the content of essential unsaturated fatty acids which includes two series of compounds: *n*-3 and *n*-6 [42]. Traditional sausages had significantly higher shares of *n*-3 acids (0.65% and 0.54%, respectively) and, especially due to C 18:2 (*cis*-9,12) acid, lower concentration of *n*-6 acids (7.58% and 11.27%, respectively) than conventional products. Therefore, they were characterized by better *n*-6/*n*-3 ratio (11.96 and 21.98, accordingly), yet still quite unfavorable [19]. These differences might be attributed to the source of raw meat that comes from local breeds which are reared using traditional feeds. Petrón et al. [30] showed that the genotype and feed have a strong effect on the fatty acids profile of meat. The diet is of crucial meaning as in monogastric animals, fats are absorbed unmodified [43].

Similarly, Garbowska, Pietrzak-Fiećko and Radzymińska [31] received more favorable ratio for traditional sausages (8.26) in comparison to local (16.80) and conventional (13.75) products. Pietrzak-Fiećko and Modzelewska-Kapituła [27] obtained an average ratio of 13.87 for Polish sausages, whereas Amaral et al. [40] a ratio of 9 to 13 for pork Frankfurter type sausages, and Romero et al. [41] 10.55 for Chorizo ahumado (smoked sausage). On the other hand, Parunović et al. [39] received much less favorable ratios for fermented dry pork Sremska sausages (from 17.33 to 38.94) depending on the pig breed used as a source of meat.

3.2. The Verification of the Differences between Traditional and Conventional Hams and Sausages with the Use of Principal Component Analysis (PCA)

With the intention of illustrating the differences between the traditional and conventional meat products tested, the PCA method was employed. The first two main components obtained explained the 45% of the total variance; the first component explained 26.23% and the second one 18.77%.

The first component indicates a high content of: C 13:0, C 17:1 (*cis*-10), C 18:2 (*cis*-9,12), C 20:0, and low concentration of C 18:0, C 18:1 (*cis*-9) fatty acids in the product.

The second component indicates a high concentration of C 12:0, C 16:1 (*cis*-9), C 18:1 (trans-9) and low content of C 20:3 (*cis*-11,14,17) fatty acids in the product.

The variables are presented as a factor map (Figure 1a). The individuals factor map (Figure 1b) shows that traditional products are mostly gathered in the bottom left corner. Although the product H has a positive value of the second component, it is very low and close to 0. The exception is, however, sausage K, which has positive values of both components and differs much from other traditional meat products tested. Therefore, on the basis of PCA, it can be stated that Polish traditional smoked hams and sausages are generally characterized by relatively higher shares of C 18:0, C 18:1 (*cis*-9), C 20:3 (*cis*-11,14,17) and lower of C 13:0, C 17:1 (*cis*-10), C 18:2 (*cis*-9,12), C 20:0, C 12:0, C 16:1 (*cis*-9), C 18:1 (trans-9) than conventional meat products.

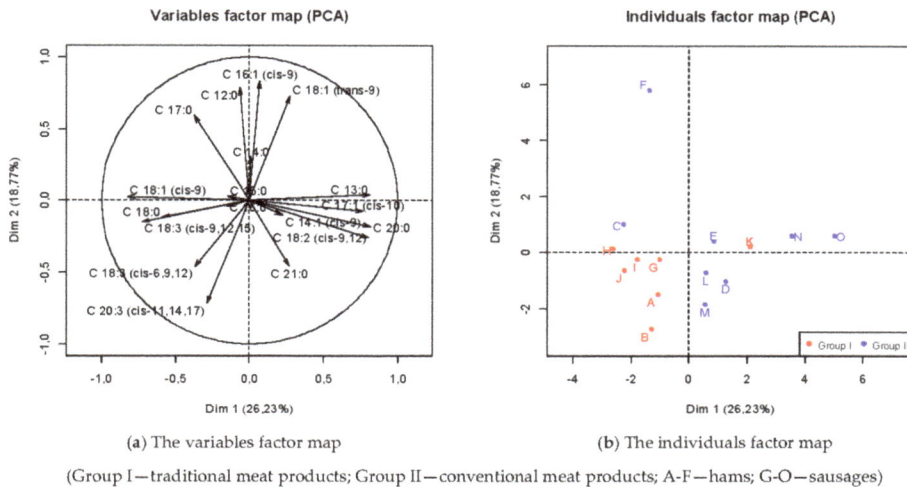

(a) The variables factor map   (b) The individuals factor map

(Group I—traditional meat products; Group II—conventional meat products; A-F—hams; G-O—sausages)

Figure 1. The results of the principal component analysis (PCA).

4. Conclusions

Traditional food products are believed to be of higher quality than their conventional counterparts. The results of this study demonstrated that the average fat content in the ham samples tested did not differ significantly between the analyzed groups of products, whereas traditional sausages had statistically significantly higher fat concentration than their conventional equivalents. Therefore, when considering fatty acid profiles traditional and conventional hams are similar. It means that the breads of pigs (and the systems of their feeding) used as raw materials were alike. Moreover, traditional production methods did not have an influence on the content of fatty acids.

It was also shown that the fatty acid profiles of the Polish pork hams and sausages tested are typical for processed pork products both from Poland and other parts of the world. In the case of hams, traditional products contained significantly higher amounts of C 20:3 (*cis*-11,14,17) acid than their conventional equivalents. More significant differences were found in sausages. Traditional products were characterized by lower concentrations of C 18:2 (*cis*-9,12) acids, whereas higher content of C 18:1 (*cis*-9), C 18:3 (*cis*-9,12,15) and C 20:0 acids. What is more, they contained significantly more MUFAs and less PUFAs. All of the tested meat products were characterized by unfavorable n-6/n-3 ratio. However, in the case of sausages, traditional ones contained significantly higher amounts of n-3 and lower of n-6 acids than their conventional equivalents. Consequently, they were characterized by a better n-6/n-3 ratio that is more favorable from a nutritional point of view. The differences between two groups of sausages tested are most probably the result of using fatter pig breeds for production of traditional products. The producers choose this type of pork because it has a positive influence on the flavor as fat is its carrier. Therefore, it can be stated that this is one of the reasons why traditional sausages represent unique sensory characteristics.

On the basis of PCA it was determined that Polish traditional smoked hams and sausages are generally characterized by relatively higher percentages of C 18:0, C 18:1 (*cis*-9), C 20:3 (*cis*-11,14,17), whereas lower shares of C 13:0, C 17:1 (*cis*-10), C 18:2 (*cis*-9,12), C 20:0, C 12:0, C 16:1 (*cis*-9), C 18:1 (trans-9) than conventional meat products. This proves that the use of traditional raw materials, and perhaps to some extent the processing methods, has an influence on the fatty acid profiles of pork meat products such as hams or sausages.

Author Contributions: Conceptualization and research methodology, M.H. and W.K. (Władysław Kędzior); Formal analysis, M.H.; Investigation, M.H. and E.P.; Resources, M.H.; Data Curation, M.H.; Writing, M.H.; Supervision, W.K. (Władysław Kędzior), W.K. (Wanda Kudełka).

Funding: The research was financed by the Ministry of Science and Higher Education—grant for the maintenance of the research potential, awarded to the Faculty of Commodity Science and Product Management of the Cracow University of Economics.

Conflicts of Interest: The authors declare no conflict of interest.

References

1. Lücke, F.K.; Vogeley, I. Traditional 'air-dried' fermented sausages from Central Germany. *Food Microbiol.* **2012**, *29*, 242–246. [CrossRef] [PubMed]
2. Cayot, N. Sensory quality of traditional foods. *Food Chem.* **2007**, *101*, 154–162. [CrossRef]
3. Talon, R.; Lebert, I.; Lebert, A.; Leroy, S.; Garriga, M.; Aymerich, T.; Drosinos, E.H.; Zanardi, E.; Ianieri, A.; Fraqueza, M.J.; et al. Traditional dry fermented sausages produced in small-scale processing units in Mediterranean countries and Slovakia. 1: Microbial ecosystems of processing environments. *Meat Sci.* **2007**, *77*, 570–579. [CrossRef] [PubMed]
4. Guerrero, L.; Guàrdia, M.D.; Xicola, J.; Verbeke, W.; Vanhonacker, F.; Zakowska-Biemans, S.; Sajdakowska, M.; Sulmont-Rossé, C.; Issanchou, S.; Contel, M.; et al. Consumer-driven definition of traditional food products and innovation in traditional foods. A qualitative cross-cultural study. *Appetite* **2009**, *52*, 345–354. [CrossRef] [PubMed]
5. Trichopoulou, A.; Soukara, S.; Vasilopoulou, E. Traditional foods: A science and society perspective. *Trends Food Sci. Technol.* **2007**, *18*, 420–427. [CrossRef]
6. Lengard Almli, V.; Verbeke, W.; Vanhonacker, F.; Naes, T.; Hersleth, M. General image and attribute perception of traditional food. *Food Qual. Prefer.* **2011**, *22*, 129–138. [CrossRef]
7. Rudawska, E.D. Customer loyalty towards traditional products—Polish market experience. *Br. Food J.* **2014**, *116*, 1710–1725. [CrossRef]
8. Campos, S.D.; Alves, R.C.; Mendes, E.; Costa, A.S.G.; Casal, S.; Oliveira, M.B.P.P. Nutritional value and influence of the thermal processing on a traditional Portuguese fermented sausage (alheira). *Meat Sci.* **2013**, *93*, 914–918. [CrossRef] [PubMed]
9. Marianski, S.; Marianski, A. *Home Production of Quality Meats and Sausages*; Bookmagic LLC: Seminole, FL, USA, 2010.
10. The List of Traditional Products of Polish Ministry of Agriculture and Rural Development. Available online: http://www.minrol.gov.pl/pol/Jakosc-zywnosci/Produkty-regionalne-i-tradycyjne/Lista-produktow-tradycyjnych/ (accessed on 29 June 2018).
11. Feiner, G. *Meat Products Handbook—Practical Science and Technology*; Woodhead Publishing Limited: Cambridge, UK, 2006; ISBN 9781845690502.
12. Murphy-Gutekunst, L.; Uribarri, J. Hidden phosphorus-enhanced meats: Part 3. *J. Ren. Nutr.* **2005**, *15*, e1–e4. [CrossRef]
13. Halagarda, M.; Kędzior, W.; Pyrzyńska, E. Nutritional Value and Potential Chemical Food Safety Hazards of Selected Traditional and Conventional Pork Hams from Poland. *J. Food Qual.* **2017**, *2017*, 1–10. [CrossRef]
14. Halagarda, M.; Kędzior, W.; Pyrzyńska, E. Nutritional value and potential chemical food safety hazards of selected Polish sausages as influenced by their traditionality. *Meat Sci.* **2018**, *139*, 25–34. [CrossRef] [PubMed]
15. Ruiz, J.; Pèrez-Palacios, T. Ingredients. In *Handbook of Fermented Meat and Poultry*; Toldra, F., Hui, Y.H., Astiasarán, I., Sebranek, J.G., Talon, R., Eds.; Wiley: New York, NY, USA, 2014; pp. 55–67. ISBN 978111852269.
16. Biesalski, H.K. Meat as a component of a healthy diet—Are there any risks or benefits if meat is avoided in the diet? *Meat Sci.* **2005**, *70*, 509–524. [CrossRef] [PubMed]
17. Fernández, M.; Ordóñez, J.A.; Cambero, I.; Santos, C.; Pin, C.; de la Hoz, L. Fatty acid compositions of selected varieties of Spanish dry ham related to their nutritional implications. *Food Chem.* **2007**, *101*, 107–112. [CrossRef]

18. Muguerza, E.; Gimeno, O.; Ansorena, D.; Bloukas, J.G.; Astiasarán, I. Effect of replacing pork backfat with pre-emulsified olive oil on lipid fraction and sensory quality of Chorizo de Pamplona—A traditional Spanish fermented sausage. *Meat Sci.* **2001**, *59*, 251–258. [CrossRef]

19. Simopoulos, A.P. Omega-6/omega-3 essential fatty acid ratio and chronic diseases. *Food Rev. Int.* **2004**, *20*, 77–90. [CrossRef]

20. Nuernberg, K.; Nuernberg, G.; Priepke, A.; Dannenberger, D. Sea buckthorn pomace supplementation in the finishing diets of pigs—Are there effects on meat quality and muscle fatty acids? *Arch. Anim. Breed* **2015**, *58*, 107–113. [CrossRef]

21. Kasprzyk, A.; Tyra, M.; Babicz, A. Fatty acid profile of pork from a local and a commercial breed. *Arch. Anim. Breed* **2015**, *58*, 379–385. [CrossRef]

22. ISO 1444. *Meat and Meat Products. Determination of Free Fat Content*; ISO: Geneva, Switzerland, 1996.

23. ISO 12966-2. *Animal and Vegetable Fats and Oils. Gas Chromatography of Fatty Acid Methyl Esters. Part 2: Preparation of Methyl Esters of Fatty Acids*; ISO: Geneva, Switzerland, 2017.

24. ISO 12966-1. *Animal and Vegetable Fats and Oils. Gas Chromatography of Fatty Acid Methyl Esters. Part 1: Guidelines on Modern Gas Chromatography of Fatty Acid Methyl Esters*; ISO: Geneva, Switzerland, 2014.

25. Siri-Tarino, P.W.; Sun, Q.S.; Frank, B.H. Saturated Fatty Acids and Risk of Coronary Heart Disease: Modulation by Replacement Nutrients. *Curr. Atheroscler. Rep.* **2010**, *12*, 384–390. [CrossRef] [PubMed]

26. Williamson, C.S.; Foster, R.K.; Stanner, S.A.; Buttriss, J.L. Red meat in the diet. *Nutr. Bull.* **2005**, *30*, 323–355. [CrossRef]

27. Pietrzak-Fiećko, R.; Modzelewska-Kapituła, M. Fatty Acids Profile of Polish Meat Products. *Ital. J. Food Sci.* **2014**, *26*, 363–369.

28. Jimenez-Colmenero, F.; Pintado, T.; Cofrades, S.; Ruiz-Capillas, C.; Bastida, S. Production variations of nutritional composition of commercial meat products. *Food Res. Int.* **2010**, *43*, 2378–2384. [CrossRef]

29. Žlender, B.; Polak, T.; Špacapan, D.; Andronikov, D.; Gašperlin, L. Influence of Raw Matter Origin and Production Period on Fatty acid Composition of Dry-Cured Hams. *Acta Agric. Slov.* **2008**, *92*, 53–60.

30. Petrón, M.J.; Muriel, E.; Timón, M.L.; Martín, L.; Antequera, T. Fatty acids and triacylglycerols profiles from different types of Iberian dry-cured hams. *Meat Sci.* **2004**, *68*, 71–77. [CrossRef] [PubMed]

31. Garbowska, B.; Pietrzak-Fiećko, R.; Radzymińska, M. Fatty acid composition of local, traditional and conventional pork meat products. In *Current Trends in Commodity Science: New Trends in Food Quality, Packaging and Consumer Behavior*; Juś, K., Jasnowska-Małecka, J., Bińczak, O., Eds.; Poznań University of Economics and Business: Poznań, Poland, 2015; pp. 35–46. ISBN 978-83-938018-8-6.

32. Grześkowiak, E.; Borzuta, K.; Lisiak, D.; Strzelecki, J.; Janiszewski, P. Właściwości fizykochemiczne i sensoryczne oraz skład kwasów tłuszczowych mięśnia longissimus dorsi mieszańców PBZ X WBP oraz PBZ X (D X P). *Zywn-Nauk Technol. Jakość* **2010**, *6*, 189–198.

33. Paur, I.; Carlsen, M.H.; Halvorsen, B.L.; Blomhoff, R. Antioxidants in Herbs and Spices. In *Herbal Medicine: Biomolecular and Clinical Aspects*, 2nd ed.; Benzie, I.F.F., Wachtel-Galor, S., Eds.; CRC Press/Taylor & Francis: Boca Raton, FL, USA, 2011; pp. 11–35.

34. Garcia Rebollo, A.J.; Macia Botejara, E.; Ortiz Cansado, A.; Morales Blanco, P.J.; Martin Bellido, M.; Fallola Sanchez, A.; Mena Arias, P.; Campillo Alvarez, J.E. Effects of Consumption of Meat Product Rich in Monounsaturated Fatty Acids (the Ham from the Iberian Pig) on Plasma Lipids. *Nutr. Res.* **1998**, *18*, 743–750. [CrossRef]

35. Wood, J.D.; Nute, G.R.; Richardson, R.I.; Whittington, F.M.; Southwood, O.; Plastow, G.; Mansbridge, R.; Da Costa, N.; Chang, K.C. Effects of breed, diet and muscle on fat deposition and eating quality in pigs. *Meat Sci.* **2004**, *67*, 651–667. [CrossRef] [PubMed]

36. Wood, J.D.; Enser, M.; Fisher, A.V.; Nute, G.R.; Sheard, P.R.; Richardson, R.I.; Hughes, S.I.; Whittingtonn, F.M. Fat deposition, fatty acid composition and meat quality: A review. *Meat Sci.* **2008**, *78*, 343–358. [CrossRef] [PubMed]

37. Gerhard, G.T.; Ahmann, A.; Meeuws, K. Effect of a Low-fat Diet Compared with those of a High-monounsaturated Fat Diet on Body Weight, Plasma Lipids and Lipoproteins, and Glycemic Control in Type 2 Diabetes. *Am. J. Clin. Nutr.* **2004**, *80*, 668–673. [CrossRef] [PubMed]

38. Schmid, A.; Collomb, M.; Hadorn, R. Fatty acid composition of Swiss cooked sausages. *Fleischwirtschaft* **2009**, *24*, 56–59.

39. Parunović, N.; Petrović, M.; Đorđević, V.; Marković, B.; Petrović, Z.; Radović, Č.; Savić, R. Fatty Acids Profiles, Cholesterol Content and Sensory Properties of Fermented Dry "Sremska" Sausages Made of Pork Meat from Various Breeds. In Proceedings of the 61st International Congress of Meat Science and Technology, Clermont-Ferrand, France, 23–28 August 2015.

40. Amaral, J.S.; Soares, S.; Mafra, I.; Oliveira, M.B.P.P. Assessing the variability of the fatty acid profile and cholesterol content of meat sausages. *Riv. Ital. Sostanze Grasse* **2014**, *91*, 261–272.

41. Romero, M.C.; Romero, A.M.; Doval, M.M.; Judis, M.A. Nutritional value and fatty acid composition of some traditional Argentinean meat sausages. *Food Sci. Technol. (Campinas)* **2013**, *33*, 161–166. [CrossRef]

42. Block, R.; Pearson, T. The Cardiovascular Implications of Omega-3 Fatty Acids. *Folia Cardiol.* **2006**, *13*, 557–569.

43. Gläser, K.R.; Wenk, C.; Scheeder, M.R. Effects of feeding pigs increasing levels of C 18:1 trans fatty acids on fatty acid composition of backfat and intramuscular fat as well as backfat firmness. *Arch. Tierernahrung* **2002**, *56*, 117–130. [CrossRef]

sustainability

MDPI

Perspective

Upland Italian Potato Quality—A Perspective

Daniela Pacifico

Council for Agricultural Research and Economics Analysis (CREA)—Research Centre for Cereal and Industrial Crops (CI)—Via Corticella, 133, 40128 Bologna, Italy; daniela.pacifico@crea.gov.it; Tel.: +39-0516316811

Received: 18 September 2018; Accepted: 26 October 2018; Published: 30 October 2018

Abstract: Upland potatoes satisfies consumer demand for high quality foods linked to traditional areas of origin and for new specialties and niche products endowed with added nutritional value, as it is commonly thought that the crop and environment synergy improves the potential beneficial properties of the tuber and gives it a special taste and a renowned quality. Herein, we report considerations on Italian germplasm and the effect of altitude on the sensorial and nutritional value of potato tubers, and investigate the possibility of addressing the nutritional challenge through mountain, eco-friendly, and social agriculture. Finally, we discuss the molecular and biochemical results concerning the impact of altitude on the compositional quality of the tuber, in order to justify promotional claims.

Keywords: mountain; sustainability; altitude; food; health

1. From Andean Potato to Italian Local Ecotypes

1.1. A Taste of History

The potato finds its center of origin in South America, around Lake Titicaca (partly in Peru and partly in Bolivia), where the indigenous people (Quechua) have been consuming this traditional food for about 8000 years. The Inca Empire would dry tubers (chuño) to store or to exchange for goods, which would forever affect the Andean culture and religion [1]. Even today, Peru and Bolivia have an annual consumption of 70–80 kg per person (fresh and processed) and host the largest biodiversity in the world [2]. From the 17th century, the potato spread to the European continent, where, without any doubt, it played an important role in mountain economy [3]. From the Alps to the Southern Apennines, the potato became a major staple food for more than three centuries, during which agronomic practice and selection created many locally-adapted genotypes. When urbanization and the Industrial Revolution caused mass emigration from mountain communities toward the city, the intensive plain agriculture, based on monoculture, replaced upland potato production, marking the decline of its fragile economies. [3]. Fortunately, however, many old local varieties and traditional knowhow did not disappear forever and are still present in the Italian territory, as showed in Figure 1 [4]. Currently, the modern varieties show a reduction in structure genetics caused by the domestication of selective breeding [5]. Thus, the interest of the scientific community towards future global germplasm enhancement, along with changing food lifestyles increasingly sensitive to the geographical area of origin, have stimulated a return to the great biological diversity of the potato, which has a precious source of unexplored quality traits such as biotic resistances and adaptability to a wide spectrum of environments [6].

Figure 1. Italian local varieties—(**a**) "Ricciona di Napoli" field in flower; (**b**) "Fiocco di Neve" plant in flower; (**c**) "Ricciona di Napoli" tubers; (**d**) "Patata 'E Moru" tuber; (**e**) "Rossa di Cetica" tubers; (**f**) "Viola Calabrese" tubers. The different colors of the list indicate the regions of origin. Source: Mandolino et al., 2015 [4].

1.2. Traditional Italian Varieties

Over the last fifty years, the cultivation of traditional ecotypes and well-known, popular cultivars have overlapped within the same areas and, despite these traditional local varieties having different names, they are sometimes in fact genetically identical to the cultivars. A surprising comparative study was recently carried out on 28 ecotypes, with a total of 54 accessions recovered in eight different Italian regions, and between over 2000 varieties belonging to EU Common Catalogue and to the SASA variety collection [4]. The outcomes of such fingerprinting carried out by twelve SSR (microsatellites or single sequence repeat) markers, allowed for highlighting the genetic similarity, for example, of "Castagno d'Andrea" (from Tuscany) to cv. Kennebec, an American white-fleshed, mid-late variety, and similarly

"Rossa del Dottore" (from Piemonte) to cv. Desiree, a Dutch red-skinned and yellow-fleshed variety. Moreover, an example is "Malva Arnaldi" (from Piemonte), which resulted in being genetically identical to cv. Vitelotte Noir, a traditional purple-fleshed potato whose color is conferred from the anthocyanins—phenols appreciated for their well-known health properties [7]. A careful fingerprinting is thus necessary to guarantee the genetic distinctiveness of traditional local potatoes. Recently, "Ricciona di Napoli" (from Campania) has represented an excellent example of an ecotype that did not show a genetic profile similar to any other genotype [4]. Its well-established historical and geographical identity, its traditional use by farmers, cultural fondness of consumers, and good agronomic and organoleptic characteristics were the main reasons for its rediscovery (project «Pat.Ri.Na.»—"Progetto di valorizzazione della patata Ricciona di Napoli", funded by the Campania Region and coordinated by OP Campania Patate; Figure 2) [8]. Nevertheless, had it not shown a unique genetic profile, it would not have been reintroduced to the market properly. Many local genotypes find their typical and traditional areas of cultivation in the Italian mountains, but it is necessary to distinguish them unambiguously from the most commonly used varieties, such as cv. Kennebec, cv. Vitellotte, and cv. Desiree, in order to promote them properly.

Figure 2. Packaging of cv. "Riciona".

1.3. Today's Mountain Potato

Beyond the genetic features, the traditional, local, and rural farming, often typical of mountain families [3], perfectly matches a renewed consumer interest in territoriality, generally perceived as a guarantee of healthier and safer food for humans and the environment, and in contrast with the world-wide, urban, fast, and cheap identity, typical of the extensive plain cultivation. Of all of the major crops, the potato is arguably one of the most important species sustaining the mountain family farming [3]. Nevertheless, many factors penalize small-hold farmers in mountain areas who struggle to compete with large-scale producers from the lowlands, and potato production in Italy remains concentrated in flat areas. The main reasons are the productive performance of upland ecotypes, which are decidedly lower than plain cultivars, and the negative influence of the mountain environment on the tuber yield [9]. It is also true that plain and mountain potatoes do not share the same harvest time, which makes the late summer–autumn harvest of the latter potentially profitable. But, it is for the same reason, Italian mountain potatoes compete with the cheaper ones from North Europe. The result is an elite product whose target consumer is a high-end spender. However, its price, albeit higher, reflects tradition and territory, and many elements have to be considered to contribute to its full potential and economic value, namely: quality; awareness of sustainable and organic farming;

and the combination of economic, cultural, and social factors. Its promotion should also be seen to provide local smallholders an opportunity to increase profits, and should represent a source of employment and mountain depopulation alleviation. Low-income family farming is poorly integrated in the commodity market, and specific policy interventions could represent an opportunity to receive greater support. Along with the debate as to whether upland potato farming should be sustained, in 2018, European politics have moved towards the labelling of mountain products, as the Italian denomination "Prodotto di Montagna" [10], which should guarantee premium prices. Similarly to the protected denomination of origin (PDO) ("Patata di Bologna") and the protected geographical indication (PGI) ("Patata della Sila"), but with fewer procedures and costs, this new provenience certification scheme was created to sustain the traditional linkage between the environment and quality, and communicating to consumers that the geographical area influences the final composition of the product [11]. Is this however true? The environment may greatly influence the overall biochemical and nutritional status of tubers, with consequent effects on human diet [12,13], but little or no information is available concerning tuber response to local provenience and its putative surplus value for justifying claims to mountain denominations, and protecting consumers from the possible exploitation of their trust in upland potato quality.

2. Influence of Upland Farming on Potato Quality

"Mountain potato" is a catchy term, and its increasing popularity with the consumer is influencing its true market power. Overall, European consumers recognize high quality in mountain products [14]. Here, we give an overview of how altitude impacts the organoleptic characteristics and the content of components with potentially health-promoting effects, especially the antioxidant contents generated as a response to reactive oxygen species in plants.

2.1. Nutritional Properties

The nutritional contribution of potato tubers to the human diet is mainly due to carbohydrates (c.ca 20%); proteins of fairly high quality, with an amino-acid pattern well matched to human requirements (2%); lipids (0.1%); and organic acids (0.4%–1%) [15]. A single medium-sized potato contains about half the recommended daily intake of vitamin C (20 mg/g d.w) [16] and a fifth of the recommended daily value of potassium [17]. However, in the past, interest has been focused greatly on secondary plant metabolites, such as phenols (0.2 to 219 mg/g DW) [18], flavonoids (0 to 45 mg/g DW) [17], and carotenoids (1.10 to 12.2 mg/kg DW) [19], due to their antioxidant activity conferring protection against degenerative and age-related diseases [18]. Anthocyanins, the major plants flavonoids, are the main pigments responsible for the red-blue color of many crops [7], such as potatoes, which are only present in colored tubers. Red potatoes have acylated glucosides of pelargonidin, whereas purple potatoes, such as cv. "Vitelotte", also have acylated glucosides of malvidin. Anthocyanins confer health benefits in the human diet as antimutagen, anticarcinogen, and antiglycaemic agents, through a hydrophilic antioxidant activity comparable to those provided by Brussel sprouts or spinach [7]. Their physiological role in the epidermal tissues of plant organs is to protect the underlying tissues from UV radiation damage, strongly affecting the photo-biological metabolism related to protection and their repair mechanisms [20]. Therefore, flavonoids and phenolics are named "UV-absorbing compounds" (UVACs) [21,22]. As many other factors contribute to the complexity of the mountain ecosystem, solar UV-B radiation depends, to a large extent, on altitude [23]. As elevation increases, the air mass decreases, which links to greater atmospheric transparency, especially with regards to shorter wavelength radiation as in the UV-B range (280–315 nm) [24]. UV-B enhances the accumulation of flavonoids and related phenolics [20–22], probably triggered by genes that encode phenylalanine ammonia lyase (PAL) and chalcone synthase (CHS) [25,26], which has a strict light requirement to be expressed in *Arabidopsis thaliana* [27]. CHS is induced in response to UV-B, by means of a signaling pathway model that passes through the UVR8 regulation (UV-B receptor UV resistance locus 8a) [28]. Anthocyanin synthesis is controlled by three loci, *D*, *P*, and *R*, which have been localized to chromosomes 11, 2, and 10, respectively [26]. *P* and *R* were found to be the genes encoding the biosynthetic enzymes involved in purple anthocyanin synthesis flavonoid 3′,5′-hydroxylase (F3′5′H) and dihydroflavonol 4-reductase (DFR), whereas *D* encodes an R2R3 MYB named StAN1, which is similar to petunia AN2 [29–32]. StAN1 is the key regulator of anthocyanin biosynthesis in the coloration of potatoes and is modulated by sucrose [33], whereas its involvement in the response to environmental factors such as UV-B has not yet been shown. Unfortunately, most studies are focused on the above-ground plant parts, such as the leaves [34]. What is less clear, but more interesting, is the effect of light on the subterranean organ, as the dietary value of the potato is a tuber. In the roots of *Gynura bicolor*, for example, anthocyanins only accumulate when directly exposed to light [35], whilst an indirect effect of light on the tuber was suggested. Indeed, increasing the phenolic content was observed in the tubers grown at increased elevation sites (1141 m and 2010 m a.s.l) [9]. Even though the long transport of anthocyanins from source to sink has never been demonstrated, it has been convincingly shown that their precursors (e.g., naringenin) can be transported from shoots to roots, where all of the enzymes required to complete their biosynthesis may be present [36]. It is therefore possible that the anthocyanin precursors, rather than anthocyanins per se, are synthesized in the shoot, and then transported to the organs that grow in the dark. Additionally, it has recently been hypothesized that some GSTs are also involved in this long-distance transport system [37]. Nevertheless, the mechanisms of anthocyanin transport remain poorly understood, as does its response to UV radiations, and they warrant further investigation. However, UV-B radiation is not the only highland factor influencing the biochemical changes in tuber composition. In accordance, imposing UV-B treatment at +20% compared to the UV-B level observed at the upper experimental site, the increase of phenolics in the artificial experiment was −38% than the natural field [9]. Other factors, therefore, must be influential on the phenolic content. Besides protecting against UV radiation, anthocyanins also respond to cold temperatures, probably in association with DFR gene activation [38,39]. Indeed, purple- and red-fleshed potatoes grown at northern latitudes showed higher anthocyanins and total phenolic content (2.5 and 1.4 times, respectively) [40], probably in response to an induction of expression of the hydroxycinnamoyl-CoA quinate hydroxycinnamoyl transferase (HQT) and cinnamate-4-hydroxylase (C4H), key enzymes for phenol biosynthesis in potato tubers [33].

Based on the above discussion, can differences in the secondary metabolites be considered sufficient to affect the nutritional quality of the upland potato? In contrast to common belief, the potato is one of the richest sources of phytonutrients in the world in diets. Indeed, even though other crops have a higher content of beneficial bio-compounds [41], the potato is one of the most widely consumed [42]. Therefore, even modest changes in tuber composition can cause big changes in human nutritional intake [43], and the significance of altitude on tuber composition can be considered a useful approach to achieving the global goal of increasing the availability of nutrients to the consumer. In addition, often mountain farming is organic farming that could itself be considered relevant to diet and health, even though a great debate is currently ongoing, and an objective and exhaustive assessment has not been reached [44,45]. Considering that upland potato cultivation is part of a system that goes beyond the effects of altitude on tubers and embraces a global concept of environmental agroecosystem, we shall consider that in European mountain regions, organic family farms are often the most viable forms of agriculture. A reduction in the environmental impact of agriculture preserves the naturalness of the products. At higher altitudes, fewer phytochemicals are required, because the potato is less challenged by nematodes; insects, such as the potato tuber moth and potato beetle; aphides; and soil-borne viruses. In addition, a more sustainable management of water resources and anti-sprouting agents is possible. Traditional mountain knowhow often leads small farmers to adopt organic practices that have been used locally for decades. Sometimes, for example during the winter, tubers can be stored on Italian slopes in fern-covered holes in the soil, avoiding cold storage chambers or anti-sprouting agents (Figure 3).

Figure 3. August/September tuber storage on the slopes of Etna (Randazzo, CT, 765 m a.s.l; Italy).

2.2. Organoleptic Properties

Moreover, the sensory evaluation of the upland potato is necessary in order to assess its global quality. Even though potato producers usually carry out the panel test to select the most suitable genotypes for cultivation in upland areas, little scientific evidence supports the hypothesis of an influence of highland cultivation on the organoleptic characteristics of the potato. An increase in the amylose/amylopectin ratio was observed in the tubers grown at higher elevations, where high values of amylose impact positively on texture [46]. This observation is consistent with its significant improvement in boiled potato grown at a higher altitude, but studies are still few and fragmentary [47]. A possible explanation is provided by evidence of alterations in the glucan chain length distribution of the starch structure, which is observed in the tubers grown under low temperature conditions [48]. This response could be triggered by the activation of the plastidial phosphorylase 1 and the inhibition of the starch synthase [48]. However, it is currently impossible to draw unequivocal conclusions and further studies should be implemented.

3. Future Perspectives

Biofortification programs aimed at optimizing tuber phytonutrients for disease prevention are ongoing, and not only include molecular breeding for the selection of new genotypes, but also a suitable choice of environment to improve the natural beneficial composition of the plant. This is particularly relevant to the potato. New biofortified genotypes are hard to obtain through conventional selective breeding with wild potato species, often described as particularly rich in beneficial phytonutrients [49]. For example, group Phureja has contributed different traits to modern yellow-fleshed potato cultivars [6]. The presence of incompatibility barriers caused by the genetic complexity of the *Solanum* genera, characterized by a tetrasomic inheritance and high heterozygosis, could hamper the backcrosses necessary to obtain hybrids [6], unless new plant breeding techniques are adopted (NPBTs) [44].

Mechanisms Underlining Natural Adaptation to Upland Environments

A future perspective could be to investigate the environmentally triggered epigenetic traits that could control the quality features to further identify the stable and hereditable marks for potato quality improvement. Indeed, even though the effects of environmental stimuli on potato transcriptome can explain a wide range of transient changes, the emerging new sources of phenotypic variations could explain the stable changes that persist after stimulus, and underlie the adaptation of potato to different habitats, such as mountain areas. Unfortunately, new technologies have so far been applied to model species such as *Arabidopsis thaliana* [50] and *Zea mays* [51], with evident limitations in transfering knowledge to non-model species, due to specific genomic features and ploidy levels. Nevertheless, Liu et al. [52] have successfully shown that lycopene accumulation during tomato fruit ripening is under epigenetic control, as follows: carotenoid biosynthesis depends on the DEMETER-LIKE DNA Demethylase 2 (DML2) responsible for the DNA methylation level. This example illustrates how epigenetic marks can influence the accumulation of a metabolite that contributes to the quality and the nutritional value of tomatoes. We could speculate on a complex genetic and epigenetic network that could emerge as a source in the explanation of the impact on quality traits of a transgenerational adaptation of potato to upland areas. Indeed, plant adaptation to an environment depends on a complex interaction between genetic variation and phenotypic plasticity. Growing evidence shows that the genetic sequence alone cannot explain the full spectrum of phenotypic diversity and its heritability in plants. The so called "plant memory" could play a role in the fitness of the plants through the heritable and stable modifications not mediated by the changes in the underlying DNA sequence. Epigenomes are remodeled during the plant development and in the response to environmental stimuli, such as drought, salt, temperature, and pathogen/herbivory attacks, generating a hereditable phenotypic variation, classified as cis-acting meQTL and trans-acting meQTL [53]. The former is due to SNPs tagging structural variants such as transposable elements (TE), causing siRNA silencing or spreading DNA methylation into flanking regions of the coding genes. The latter affects the methylation levels at the genome-wide scale, causing changes inside the chromatin control genes, with effects that can be adaptive [54]. Numerous studies have addressed the silencing of TE by DNA methylation as a mechanism of trans-generational memory. The functional relevance that epigenetic mechanisms have on evolutionary adaptation to a specific habitat was shown by Dubin et al. [55]. Using genome wide association studies (GWAS), they revealed that CHH methylation increases with temperature and CpG gene body (GBM) methylation with the latitude of origin. The consequence is that chromatin could mark local adaptation, contributing to plant phenotypic plasticity. Chromomethylase 2 (CMT-2) is supposed to be an adaptive locus, whose role is the genetic controller of plant temperature tolerance through the regulation of epigenetic modifications in natural adaptation [56,57]. Similarly, Rius et al. [51] identified in maize a crosslink between a quality trait, such as the content of flavonoids, and an epigenetic regulation mediated by the altitude and the level of UV-B. The DNA methylation in the promotor, intron 1 and intron 2 of P1, a R2R3-MYB transcription factor that regulates the accumulation of flavonoids, is lower in the landaraces of maize adapted to a high altitude than in a low altitude inbred line, probably caused directly by an adaptation to UV-B radiations. Indeed, UV-B radiations can change the chromatin structure of plants [58], showing how plant adaptation to contrasting environments could evolve in different epigenomes. Beside the importance of the epigenetic mechanism underlining the plant adaptation to environment, a great debate on a possible restoration of epigenome changes through reverse epimutation put at risk the basis upon which transgenerational stress memory is built, because an environmental induced epigenetic status could be reset [50].

To date, few studies have focused on the epigenetics of the potato. Recently the miniature inverted-repeat transposable elements (MITEs) involvement in the fine regulation of the potato tuber skin color (Class II, MITE F3′5′H) [59] has been identified. Their insertion within the first exon of the gene encoding for F3′5′H, causes its inactivation. However, the possibility of a "transgenerational stress memory" to explain long-term upland potato adaptation has not so far been addressed. Interestingly, Ibanez [9] reported that natural populations of wild potato species *S. kurtianum* that have already adapted to different altitudes showed a different increase in UVAC when exposed to +UV-B treatments, and epigenetic regulation of altitude mediated by the effect of UV-B radiation cannot be excluded, as previously shown in maize [51]. A future direction could be the effective exploitation of epigenetic-based phenotypic diversity of potatoes traditionally grown and produced in mountain areas in shaping quality traits, and the identification of adaptive loci associated with different altitudes in genotypes of *Solanum tuberosum* grown and adapted to different habitats and their potential use for plant improvement.

4. Conclusions

In conclusion, the paper aims to stimulate fresh thinking on the close link between the Italian mountain areas and the traditional local varieties of potato, and investigate the effects of altitude on the nutritional quality of potato tubers, together with the environmental and social implications. The molecular features underlining the potato adaptation to upland environments and the corresponding biological processes involved are also discussed.

It is important to point out the predominance of genetics over environmental effects when explaining the variations of biochemical tuber composition, in order to support the importance of the genetic distinctiveness of the ecotypes. However, our view is to treat this crop as part of a system that extends beyond the genetic features, to create a connection with the environmental ecosystem. Whether and in which ways altitude really affects the organoleptic and nutritional value of tubers has not been clarified exhaustively, and supplementary information to both consumers and producers is necessary. Nonetheless, the variation in metabolites observed in response to collective environmental influences raises questions about the extent to which these pathways can be stimulated by environmental inputs in a manner that enriches nutrients for health improvement while, at the same time, protecting yields [33]. Many studies are focused on the response to UV, documenting its importance above the other environmental factors that compose the complex upland system. However, it is necessary to extent these studies to what actually happens in the real-world, and future efforts shall approach omic techniques applied to potato grown in experimental gardens at different altitudes. In our context, the magnitude of environmental variations between study sites and the difficulties in showing how altitude can improve tuber health promoting compounds suggest caution, primarily because the nutraceutical value of a food product needs to be accurately demonstrated before adopting promotional health claims, as stated by the European Commission concerted Action on Functional Food Science in Europe (FU.FO.S.E.). The promotion of upland potato farming should also comply with consumer preferences and increasing community resilience in the face of mountain depopulation. Our perspective is to encourage studies that compare plains and mountains, as well as to investigate the molecular mechanisms underlining the natural adaptation of potatoes to upland environments.

Funding: This study, as well as the potato germoplasm collection located at CREA-CI (Bologna; Italy), were funded by the Project of Conservation of Plant Genetic Resources (RGV/FAO; Mipaaf).

Acknowledgments: The author would like to thank Giuseppe Mandolino, Bruno Parisi, and Vincenzo Rossi for their precious suggestions.

Conflicts of Interest: The author declares no conflict of interest.

References

1. Peñarrieta, J.M.; Alvarado, J.A.; José, K.; Bravo, A.; Bergenståhl, B. Chuño and Tunta; the Traditional Andean sun-dried potatoes. In *Potatoes: Production, Consumption and Health Benefits*; Caprara, C., Ed.; Nova Science Publishers: Hauppauge, NY, USA, 2011; pp. 1–12, ISBN 978-1-62100-703-6.

2. PotatoPRO. Available online: https://www.potatopro.com/south-america/potato-statistics (accessed on 22 October 2018).

3. Gentilcore, D. *Italy and the Potato: A History, 1550–2000*; Bloomsbury Academic: London, UK, 2013; ISBN1 1472526317, ISBN2 9781472526311.

4. Mandolino, G.; Parisi, B.; Andrenelli, L.; Ferrari, A.; Ventisei, H.; Reid, A. Molecular fingerprinting of traditional Italian potato varieties. In Proceedings of the Joint Congress SIBV-SIGA, Milano, Italy, 8–11 September 2015; ISBN 978-88-904570-5-0.

5. Navarre, D.A.; Shakya, R.; Hellmann, H. Vitamins, Phytonutrients and Minerals in potato. In *Advances in Potato Chemistry and Technology*; Singh, J., Kaur, L., Eds.; Academic Press: Cambridge, MA, USA, 5 February 2016; ISBN 9780128005767.

6. Bethke, P.C.; Halterman, D.A.; Jansky, S. Are We Getting Better at Using Wild Potato Species in Light of New Tools? *Crop. Sci.* **2017**, *57*, 1241–1258. [CrossRef]

7. Brown, C.R. Antioxidants in potato. *Am. J. Potato Res.* **2005**, *82*, 163–172. [CrossRef]

8. Pentangelo, A.; Raimo, F.; Mandolino, G.; Parisi, B. La patata Ricciona di Napoli torna nei campi. *L'Informatore Agrario* **2014**, *70*, 53–55.

9. Ibañez, V.N.; Berli, F.J.; Masuelli, R.W.; Bottini, R.A.; Marfil, C.F. Influence of altitude and enhanced ultraviolet-B radiation on tuber production, seed viability, leaf pigments and morphology in the wild potato species Solanum kurtzianum Bitter & Wittm collected from an elevational gradient. *Plant Sci.* **2017**, *261*, 60–68. [CrossRef] [PubMed]

10. Ministero Delle Politiche Agricole, Forestali e del Turismo. Available online: https://www.politicheagricole.it/flex/cm/pages/ServeBLOB.php/L/IT/IDPagina/12257 (accessed on 18 September 2018).

11. Santini, F.; Guri, F.; Gomez y Paloma, S. Labelling of agricultural and food products of mountain farming. In *JRC [Joint Research Centre] Scientific and Policy Report*; European Commission, JRC: Seville, Spain, 2013.

12. Pacifico, D.; Casciani, L.; Ritota, M.; Mandolino, G.; Onofri, C.; Moschella, A.; Parisi, B.; Cafiero, C.; Valentini, M. NMR-Based Metabolomics for Organic Farming Traceability of Early Potatoes. *J. Agric. Food Chem.* **2013**, *61*, 11201–11211. [CrossRef] [PubMed]

13. Herencia, J.F.; García-Galavís, P.A.; Dorado, J.A.R.; Maqueda, C. Comparison of nutritional quality of the crops grown in an organic and conventional fertilized soil. *Sci. Hortic.* **2011**, *129*, 882–888. [CrossRef]

14. Project EURO MARC Euromontana. 2007–2009 EuroMarc—Mountain Agrofood Products in Europe, Their Consumers, Retailers and Local Initiatives. Available online: https://www.euromontana.org/en/?s=euromarc (accessed on 18 September 2018).

15. Lisinska, G.; Leszczynski, W. *Potato Science and Technology*; Springer: Berlin, Germany, 1989; ISBN 185166307X.

16. Love, S.L.; Pavek, J.J. Positioning the Potato as a Primary Food Source of Vitamin C. *Am. J. Potato Res.* **2008**, *85*, 277–285. [CrossRef]

17. Freedman, M.R.; Keast, D.R. Potatoes, including French fries, contribute key nutrients to diets of U.S. adults: NHANES 2003–2006. *J. Nutr. Ther.* **2012**, *1*, 1–11. [CrossRef]

18. Akyol, H.; Riciputi, Y.; Capanoglu, E.; Caboni, M.F.; Verardo, V. Phenolic Compounds in the Potato and Its Byproducts: An Overview. *Int. J. Mol. Sci.* **2016**, *17*, 835. [CrossRef] [PubMed]

19. Evers, D.; Deußer, H. Potato Antioxidant Compounds: Impact of Cultivation Methods and Relevance for Diet and Health. In *Nutrition, Well-Being and Health*; Bouayed, J., Ed.; IntechOpen: Rijeka, Croatia, 2012; ISBN 978-953-51-0125-3.

20. Agati, G.; Tattini, M. Multiple functional roles of flavonoids in photoprotection. *New Phytol.* **2010**, *186*, 786–793. [CrossRef] [PubMed]

21. Jaakola, L.; Hohtola, A. Effect of latitude on flavonoid biosynthesis in plants. *Plant Cell Environ.* **2010**, *33*, 1239–1247. [CrossRef] [PubMed]

22. Berli, F.J.; Alonso, R.; Bressan-Smith, R.; Bottini, R. UV-B impairs growth and gas exchange in grapevines grown in high altitude. *Physiol. Plant.* **2013**, *149*, 127–140. [CrossRef] [PubMed]

23. Chen, L.; Niu, K.; Wu, Y.; Geng, Y.; Mi, Z.; Flynn, D.F.; He, J.S. UV radiation is the primary factor driving the variation in leaf phenolics across Chinese grasslands. *Ecol. Evol.* **2013**, *3*, 4696–4710. [CrossRef] [PubMed]

24. Ziska, L.H.; Teramura, A.H.; Sullivan, J.H. Physiological sensitivity of plants along an elevational gradient to UV-B radiation. *Am. J. Bot.* **1992**, *86*, 863–871. [CrossRef]

25. Casati, P.; Walbot, V. Gene Expression Profiling in Response to Ultraviolet Radiation in Maize Genotypes with Varying Flavonoid Content. *Plant Phys.* **2003**, *132*, 1739–1754. [CrossRef]

26. Holton, T.A.; Cornish, E.C. Genetics and biochemistry of anthocyanin biosynthesis. *Plant Cell* **1995**, *7*, 1071–1083. [CrossRef] [PubMed]

27. Saslowsky, D.; Winkel-Shirley, B. Localization of flavonoid enzymes in Arabidopsis roots. *Plant J.* **2001**, *27*, 37–48. [CrossRef] [PubMed]

28. Rizzini, L.; Favory, J.L.; Cloix, C.; Faggionato, D.; O'Hara, A.; Kaiserli, E.; Baumeister, R.; Schäfer, E.; Nagy, F.; Jenkins, G.I.; et al. Perception of UV-B by the Arabidopsis UVR8 Protein. *Science* **2011**, *332*, 103–106. [CrossRef] [PubMed]

29. Liu, Y.; Lin-Wang, K.; Espley, R.V.; Wang, L.; Yang, H.; Yu, B.; Dare, A.; Varkonyi-Gasic, E.; Wang, J.; Zhang, J.; et al. Functional diversification of the potato R2R3 MYB anthocyanin activators AN1, MYBA1, and MYB113 and their interaction with basic helix-loop-helix cofactors. *J. Exp. Bot.* **2016**, *67*, 2159–2176. [CrossRef] [PubMed]

30. Jung, C.S.; Griffiths, H.M.; De Jong, D.M.; Cheng, S.; Bodis, M.; Kim, T.S.; De Jong, W.S. The potato developer (D) locus encodes an R2R3 MYB transcription factor that regulates expression of multiple anthocyanin structural genes in tuber skin. *Theor. Appl. Genet.* **2009**, *120*, 45–57. [CrossRef] [PubMed]

31. Zhang, B.; Hu, Z.; Zhang, Y.; Li, Y.; Zhou, S.; Chen, G. A putative functional MYB transcription factor induced by low temperature regulates anthocyanin biosynthesis in purple kale (Brassica oleracea var. acephala f. tricolor). *Plant Cell Rep.* **2012**, *31*, 281–289. [CrossRef] [PubMed]

32. D'Amelia, V.; Aversano, R.; Ruggiero, A.; Batelli, G.; Appelhagen, I.; Dinacci, C.; Hill, L.; Martin, C.; Carputo, D. Subfunctionalization of duplicate MYB genes in Solanum commersonii generated the cold-induced ScAN2 and the anthocyanin regulator ScAN1. *Plant Cell Environ.* **2018**, *41*, 1038–1051. [CrossRef] [PubMed]

33. Payyavula, R.S.; Navarre, D.A.; Kuhl, J.C.; Pantoja, A.; Pillai, S.S. Differential effects of environment on potato phenylpropanoid and carotenoid expression. *BMC Plant Boil.* **2012**, *12*, 39. [CrossRef] [PubMed]

34. Santos, I.; Fidalgo, F.; Almeida, J.M.; Salema, R. Biochemical and ultrastructural changes in leaves of potato plants grown under supplementary UV-B radiation. *Plant Sci.* **2004**, *167*, 925–935. [CrossRef]

35. Shimizu, Y.; Maeda, K.; Kato, M.; Shimomura, K. Co-expression of GbMYB1 and GbMYC1 induces anthocyanin accumulation in roots of cultured Gynura bicolor DC. plantlet on methyl jasmonate treatment. *Plant Phys. Biochem.* **2011**, *49*, 159–167. [CrossRef] [PubMed]

36. Buer, C.S.; Muday, G.K.; Djordjevic, M.A. Flavonoids are differentially taken up and transported long distances in Arabidopsis. *Plant Physiol.* **2007**, *145*, 478–490. [CrossRef] [PubMed]

37. Zhao, J.; Dixon, R.A. The 'ins' and 'outs' of flavonoid transport. *Trends Plant Sci.* **2010**, *15*, 72–80. [CrossRef] [PubMed]

38. Christie, P.J.; Alfenito, M.R.; Walbot, V. Impact of low-temperature stress on general phenylpropanoid and anthocyanin pathways: Enhancement of transcript abundance and anthocyanin pigmentation in maize seedlings. *Planta* **1994**, *194*, 541–549. [CrossRef]

39. Ahmed, N.U.; Park, J.I.; Jung, H.J.; Yang, T.J.; Hur, Y.; Nou, I.S. Characterization of dihydroflavonol 4-reductase (DFR) genes and their association with cold and freezing stress in Brassica rapa. *Gene* **2014**, *550*, 46–55. [CrossRef] [PubMed]

40. Reyes, L.; Miller, J.; Cisneros-Zevallos, L. Environmental conditions influence the content and yield of anthocyanins and total phenolics in purple- and red-flesh potatoes during tuber development. *Am. J. Potato Res.* **2004**, *81*, 187–193. [CrossRef]

41. Ezekiel, R.; Singh, N.; Sharma, S.; Kaur, A. Beneficial phytochemicals in potato—A review. *Food Res. Int.* **2013**, *50*, 487–496. [CrossRef]

42. United States Department of Agriculture. Economic Research Service. Available online: https://www.ers.usda.gov/data-products/ag-and-food-statistics-charting-the-essentials/food-availability-and-consumption/ (accessed on 18 September 2018).

43. Chun, O.K.; Kim, D.; Smith, N.; Schroeder, D.; Han, J.T.; Lee, C.Y. Daily consumption of phenolics and total antioxidant capacity from fruit and vegetables in the American diet. *J. Sci. Food Agric.* **2005**, *85*, 1715–1724. [CrossRef]

44. Pacifico, D.; Paris, R. Effect of Organic Potato Farming on Human and Environmental Health and Benefits from New Plant Breeding Techniques. Is It Only a Matter of Public Acceptance? *Sustainability* **2016**, *8*, 1054. [CrossRef]

45. Huber, M.; Rembiałkowska, E.; Średnicka, D.; Bügel, S.; van de Vijver, L.P.L. Organic food and impact on human health: Assessing the status quo and prospects of research. *NJAS Wagening. J. Life Sci.* **2011**, *58*, 103–109. [CrossRef]

46. Simkova, D.; Lachman, J.; Hamouz, K.; Vok, B. Effect of cultivar, location and year on total starch, amylose, phosphorus content and starch grain size of high starch potato cultivars for food and industrial processing. *Food Chem.* **2013**, *141*, 3872–3880. [CrossRef] [PubMed]

47. Pevicharova, G. Correlations between sensory traits of boiled potatoes. *Bulg. J. Agric. Sci.* **2015**, *21*, 877–881.

48. Orawetz, T.; Malinova, I.; Orzechowski, S.; Fettke, J. Reduction of the plastidial phosphorylase in potato (Solanum tuberosum L.) reveals impact on storage starch structure during growth at low temperature. *Plant Phys. Biochem.* **2016**, *100*, 141–149. [CrossRef] [PubMed]

49. Andre, C.M.; Guislain, M.; Bertin, P.; Outir, P.; del Rosario, M.; Herrera, M.; Hoffmann, L. Andean potato cultivars (*Solanum tuberosum* L.) as a source of antioxidant and mineral micronutrients. *J. Agric. Food Chem.* **2007**, *55*, 366–378. [CrossRef] [PubMed]

50. Becker, C.; Hagmann, J.; Muller, J.; Koenig, D.; Stegle, O.; Borgwardt, K.; Weigel, D. Spontaneous epigenetic variation in the *Arabidopsis thaliana* methylome. *Nature* **2011**, *480*, 245–249. [CrossRef] [PubMed]

51. Rius, S.P.; Emiliani, J.; Casati, P. P1 Epigenetic Regulation in Leaves of High Altitude Maize Landraces: Effect of UV-B Radiation. *Front. Plant Sci.* **2016**, *7*, 523. [CrossRef] [PubMed]

52. Liu, R.; How-Kit, A.; Stammitti, L.; Teyssier, E.; Rolin, D.; Mortain-Bertrand, A.; Halle, S.; Liu, M.; Kong, J.; Wu, C.; et al. A DEMETER-like DNA demethylase governs tomato fruit ripening. *Proc. Natl. Acad. Sci. USA* **2015**, *112*, 10804–10809. [CrossRef] [PubMed]

53. Lauria, M.; Rossi, V. Origin of Epigenetic Variation in Plants: Relationship with Genetic Variation and Potential Contribution to Plant Memory. In *Memory and Learning in Plants Signaling and Communication in Plants*; Baluska, F., Gagliano, M., Eds.; Springer: Cham, Switzerland, 2018.

54. Taudt, A.; Colomé-Tatché, M.; Joannes, F. Genetic sources of population epigenomic variation. *Nat. Rev. Genet.* **2016**, *17*, 319–332. [CrossRef] [PubMed]

55. Dubin, M.; Zhang, P.; Meng, D.; Remigereau, M.S.; Osborne, E.; Casale, F.P.; Drewe, P.; Kahles, A.; Jean, G.; Vilhjálmsson, B. DNA methylation in Arabidopsis has a genetic basis and shows evidence of local adaptation. *Elife* **2015**, *4*, e05255. [CrossRef] [PubMed]

56. Qüesta, J.I.; Walbot, V.; Casati, P. Mutator transposon activation after UV-B involves chromatin remodeling. *Epigenetics Off. J. DNA Methylation Soc.* **2010**, *5*, 352–353. [CrossRef]

57. Shen, X.; De Jonge, J.; Forsberg, S.K.G.; Pettersson, M.E.; Sheng, Z.; Hennig, L.; Carlborg, Ö. Natural CMT2 Variation Is Associated with Genome-Wide Methylation Changes and Temperature Seasonality. *PLoS Genet.* **2014**, *10*, e1004842. [CrossRef] [PubMed]

58. Müller-Xing, R.; Xing, Q.; Goodrich, J. Footprints of the sun: Memory of UV and light stress in plant. *Front. Plant Sci.* **2014**, *5*, 474. [PubMed]

59. Momose, M.; Abe, Y.; Ozeki, Y. Miniature Inverted-Repeat Transposable Elements of *Stoneway* are active in potato. *Genetics* **2010**, *186*, 59–66. [CrossRef] [PubMed]

![sustainability logo] *sustainability*

MDPI

Article

Influence of Ripening on Chemical Characteristics of a Traditional Italian Cheese: Provolone del Monaco

Nadia Manzo [1], Antonello Santini [2,*], Fabiana Pizzolongo [1], Alessandra Aiello [1], Andrea Marrazzo [1], Giuseppe Meca [3], Alessandra Durazzo [4], Massimo Lucarini [4] and Raffaele Romano [1]

[1]   Department of Agriculture, University of Napoli Federico II, Via Università, 100-80055 Portici (Napoli), Italy; nadia.manzo@unina.it (N.M.); fabpizzo@unina.it (F.P.); alessandra.aiello@unina.it (A.A.); draven88@hotmail.it (A.M.); rafroman@unina.it (R.R.)

[2]   Department of Pharmacy, University of Napoli Federico II, Via D. Montesano, 49-80131 Napoli, Italy

[3]   Laboratory of Food Chemistry and Toxicology, Faculty of Pharmacy, University of Valencia, Av. Vicent Andrés Estellés s/n, 46100 Burjassot, Spain; giuseppe.meca@uv.es

[4]   CREA—Research Centre for Food and Nutrition, Via Ardeatina 546, 00178 Roma, Italy; alessandra.durazzo@crea.gov.it (A.D.); massimo.lucarini@crea.gov.it (M.L.)

*   Correspondence: asantini@unina.it; Tel.: +39-81-253-9317

Received: 5 March 2019; Accepted: 20 April 2019; Published: 30 April 2019

Abstract: The envisaged promotion of local products contributes to environmental protection and is a valid tool for the promotion of socioeconomic development, enhancement of territories, and biodiversity preservation and sustainability. Provolone del Monaco is a semi hard pasta filata cheese granted PDO (Protected Designation of Origin) designation by the European Union. Provolone del Monaco is obtained from raw cow's milk, produced in the specific areas of the Lattari Mountains and Sorrento Peninsula (Naples, Italy), and ripened for at least six months. To the best of our knowledge, no studies concerning the complete chemical characterization of Provolone del Monaco cheese are available. In the present study; the chemical characterization (moisture; pH; titratable acidity; nitrogen; and fat content), fatty acid composition determined by using gas-chromatography-flame-ionization-detector (GC-FID); volatile organic compounds by solid-phase microextraction followed by gas chromatography-mass spectrometry (SPME-GC/MS), and maturation indices were evaluated during ripening. Two different average typical cheese sizes (3 kg and 5 kg) and two different internal portions were studied. After 6 months of ripening, the most important changes recorded were the loss of water, the increase in acidity, the nitrogen (as ammonia) release, and the production of volatile organic compounds. The cheese size did not affect the chemical composition of Provolone del Monaco.

Keywords: Provolone del Monaco; traditional foods; biodiversity; sustainability

---

1. Introduction

The great progress in technological processes, agricultural practices, and changes in life style has led toward paying greater attention to local products as principal elements for food product quality improvement, and in the meantime, supporting local agro-biodiversity. The role of biodiversity for food quality and healthy status is well recognized: traditional, local, and season foods are produced by techniques based on historic and cultural traditions of a specific territory and which occur only in a local place. The promotion of local foods is well addressed towards a sustainable and environmentally friendly production system, as well as to validated traceability systems. An appropriate use of farmland, protection of animal health and welfare, and environmental conservation are linked to climate knowledge, soil quality, and landscaping, leading to substantial improvements in the product quality. The envisaged promotion of local products contributes to environmental protection and it is a valid tool for the promotion of socioeconomic development, enhancement of territories, and biodiversity preservation and sustainability. There is the need to assess the extent to which it will become clear that local/traditional foods play an important role in the food pattern of many population groups in European countries. Components of traditional diets can be reinforced and promoted for their nutritional quality, health properties, and safety. From this perspective, nutritional science should support sustainable ecosystems, ecological resources, and healthy environments, because nutrition and environmental sustainability are strictly linked through the food system.

Provolone del Monaco is a semi-hard pasta filata cheese produced from raw cow's milk in the Lattari Mountains and in Sorrento Peninsula areas within the Campania Region (Italy). This cheese, shown in Figure 1, is ripened for at least six months before being considered a PDO (Protected Designation of Origin) cheese, according to the Italian Regulation set in 2010 [1].

The PDO designation indicates the particular characteristics of quality of this cheese depending on local factors, such as: (i) milk origin (the milk used for cheese making is produced by cattle living exclusively in the manufacturing area), (ii) traditional cheese making process, and (iii) microclimate conditions of the production area. This cheese is cylindrical shape; it is characterized by having a creamy white pasta filata, and a smooth and yellowish rind. The smell of Provolone del Monaco is strong and penetrant, while the taste can be sweet or spicy, depending on the ripening time.

According to the PDO regulation for the production of this cheese, an amount ≥ 20% of the milk used for making the Provolone del Monaco cheese must be obtained from cows belonging to the "Agerolese" breed, which originates from the Sorrento Peninsula area (located in the Campania Region, South of Italy). This breed produces milk with high butterfat content. The remaining 80% of the milk used for making the Provolone del Monaco cheese comes from different local cattle breeds (e.g., Frisona, Jersey, Brunalpina, and Podolica). According to the regulation for producing the Provolone del Monaco cheese, at least 40% of the dry matter feed for cows must come from fresh fodder.

Only a few studies regarding Provolone del Monaco PDO cheese are available in the scientific literature. Aponte et al. [2] focused on lactic acid bacteria occurring during cheese manufacture and ripening. Romano et al. [3] studied how ripening can influence the amounts of cheese fatty acids, $\omega-3$ and conjugated linoleic acids (CLA). Also, diet can influence the composition of milk fat [4].

The results of these studies indicated that for Provolone del Monaco, the main fatty acids present were medium molecular weight fatty acids (43%), ranging from C11:0 to C16:1 cis-9, during the first six months of ripening, and after this period of time, high molecular weight fatty acids (41%) and low molecular weight fatty acids (8%) were present. The minor acidic component, represented by the c9t11 conjugated linoleic acid (CLA), did not show significant differences during ripening. Di Monaco et al. [5] observed that Provolone del Monaco cheese sensorial attributes recognized as typical by consumers are vinegar odor, ripened flavor, and appearance of peculiarities, such as the number of holes (eyelets) of different forms and dimensions inside the cheese.

The total fat, pH, protein, and moisture contents of Provolone del Monaco cheese have a significant effect on the degree of release of volatile compounds, including several free fatty acids (FFAs), and can affect lipase activity [6,7]. Short chain fatty acids directly contribute to flavor, but fatty acids can also act as precursors for the production of a wide range of other volatile flavor compounds [8].

The principal pathways for the formation of flavor compounds in cheese during ripening are glycolysis, lipolysis, and proteolysis [8–10]. Glycolysis refers to the metabolism of residual lactose, lactate, and citrate. Lactate contributes to the flavor of acid-curd cheeses and probably also contributes to the flavor of ripened cheese varieties, particularly early in maturation. Acidification of the cheese has a major indirect effect on flavor since it determines the buffering capacity of the cheese and the growth of various microorganisms during ripening other than the activity of the enzymes involved in cheese ripening. Lactate can be oxidized to acetate and CO_2 by the non-starter lactic acid bacteria present in cheeses, and the availability of O_2 can be influenced by the size of the cheese and by the oxygen permeability of the rind or packaging material [11]. The lipid fraction of the cheese can undergo oxidative or enzymatic hydrolytic degradation with the production of FFAs (lipolysis). The agents responsible for lipolysis in cheese made from raw milk, such as Provolone del Monaco, are indigenous milk lipase, raw milk microflora, and non-starter lactic acid bacteria. The impact of FFA on the flavor can also be influenced by the pH since carboxylic acids and their salts are perceived differently by the consumers, and a higher pH value lowers the FFA perception. Free fatty acids also act as precursor molecules for a cascade of catabolic reactions that lead to the production of other flavor compounds, such as methyl ketones, lactones, esters, alkanes, and secondary alcohols [11].

Proteolysis is the most complex biochemical reaction that occurs during cheese ripening and consists of catabolic reactions and degradation of the casein matrix to a range of peptides and free amino acids (FAAa), and of subsequent reactions involved in the catabolism of FAA. Proteolysis plays a vital role in the development of textural changes and flavor. The contribution to flavor is due to the formation of peptides and free amino acids, the liberation of substrates (amino acids) for secondary catabolic changes, and the change of the cheese matrix, which facilitate the release of sapid compounds during mastication [11].

There is a lack of studies regarding the complete chemical characterization of Provolone del Monaco cheese correlating the chemical characterization, fatty acid composition, volatile organic compounds (VOCs), and maturation index during ripening. The aim of the present paper is to fully characterize the nutritional properties of Provolone del Monaco, an example of traditional Italian cheese. Moreover, the effect of ripening, which also includes comparisons between two different cheese sizes and between two different internal parts of the cheese are evaluated.

2. Materials and Methods

2.1. Cheese-Making

The typical protocol for Provolone del Monaco production is quite long and starts by mixing together the milk collected in the morning and in the evening, after a preliminary cleaning step by centrifugation. The following steps are: (i) the coagulation (40–60 min) by adding lamb rennet paste without starter addition; (ii) curd cutting to hazelnut size first and to corn bean later; (iii) cooking at T in the range from 48 to 52 °C for 30 min; (iv) whey draining on flaxen cloths; (v) curd acidification on a wood table at room temperature for t = 12–14 h; (vi) stretching in water at T in the range 85–95 °C; (vii) molding in pieces; (viii) salting in brine (t = 8–12 h/kg); (ix) ripening at T in the range 8–15 °C, and 85% relative humidity for a minimum of 6 months. After ripening, the cheese pieces must have a weight in the range from 2.5 to 8 kg, and the ratio between fat and dry matter must be >40.5%.

2.2. Sampling

Three batches of Provolone del Monaco were produced in June 2016 in the factory Perrusio S.r.L. located in Meta di Sorrento (Napoli, Italy) according to the PDO regulation. For each batch, two cheese sizes, namely, of 3 and 5 kg of weight, were studied.

Three samples of both size cheeses for each batch were collected every 90 days for analysis (time 0, 90, 180, and 270 days) in a ripening chamber with controlled temperature and humidity, equipped with a data logger (Testo mod. 174H) for temperature and humidity data acquisition. The collection was repeated two times. Sample at time 0 was considered immediately after salting the cheese in brine.

In each sample, two different areas of the cheese were collected: the core (C), at 4–6 cm from the rind, and the portion just below the rind (S), at 0.3–0.5 cm from the rind. The samples were vacuum-sealed and kept in a freezer at −20 °C until the analysis.

2.3. pH and Moisture Content

The pH of the cheeses was determined by direct insertion of a suitable pH meter (sensION+, Hach Lange, CO, USA) in three different parts of the cheese. Moisture content was determined by drying 5 g of sample at 102.0 ± 2.0 °C until constant weight [12]. Results were expressed as weight percentage (%, w/w).

2.4. Titratable Acidity

Ten grams of grated cheese was added with 50 mL of deionized water and put in a bath at 40 °C for 5 min. The mixture was homogenized for 5 min with an Ultra-turrax T25 homogenizer (Janke & Kunkel, GmbH & CO, Staufen, Germany). The suspension was transferred in a 100 mL volumetric task and made up to the mark using deionized water. The suspension was filtrated, and 25 mL was titrated using 0.1 NaOH and phenolphthalein as indicator (D.M. 1986, FIL-IDF 5A:1969). The results were expressed in percent by weight of lactic acid (% w/w).

2.5. Nitrogen Content

Cheese samples were analyzed using the official method Kjeldahl method [13] to determine nitrogen fractions: the total nitrogen (TN) and the pH 4.6-soluble nitrogen (SN). To determine TN, 0.6 g of grated cheese was inserted into a Kjeldahl tube of 20 mL sulfuric acid 98% and 2 g of catalyst (cupric sulfate and potassium sulfate in the ratio 1:24) was added. In order to determine SN, 20 g of grated cheese was put into a 300 mL beaker, and 80 mL of 0.5 M trisodium citrate at pH 7 was added and then kept at 40 °C for 1 h. After this time, 3 drops of formaldehyde were added and the mixture was homogenized at 24.000 rpm for 1 min. After this step, the pH was adjusted to 4.6 using HCl 37% and the mixture was brought to a volume of 500 mL using deionized water. After 1 h resting, the solution was filtrated through Whatman n°42 filter paper. Sulfuric acid and catalyst were added to 25 mL of the filtrated solution. The results were expressed in percent on dry weight (%, w/dw).

2.6. Fat Content

The fat content was gravimetrically determined according to the method described in D.M. 1986 [13], which is based on the Schimith–Bondzynski–Ratzlaff traditional method of extracting lipids, with some modifications. Ten grams of grated cheese were hydrolyzed using 10 mL of 37% hydrochloric acid (d = 1.125) and 7 mL of 95% (v/v) ethyl alcohol. The cheese suspension was homogenized with an Ultra-turrax T25 homogenizer (Janke & Kunkel, GmbH & CO, Staufen, Germany) in a glass beaker for 30 min at 50 °C with constant magnetic stirring. After cooling, the fatty matter was extracted using 20 mL of ethyl ether-petroleum ether 1:1 solution under constant stirring for 15 min. The suspension was then rested for 10 min to allow phase separation. This extraction protocol was repeated 3 times. Three organic extracts were pooled, dried over anhydrous sodium sulfate, filtered with a cellulose filter, evaporated under reduced pressure in a rotary evaporator and weighed. The results were expressed in percent on dry weight (%, w/dw). The fatty extracts were analyzed for fatty acids.

2.7. Gas Chromatographic Analysis of Fatty Acids

Fatty acid methyl esters (FAMEs) were prepared using 2 N potassium hydroxide in a methanol solution as described by Nota et al. [14], with some modifications. Ten milligrams of extracted fat were weighed in a 2 mL vial and dissolved in 1 mL of hexane. Three hundred microliters of 2N methanolic potassium-hydroxide were added, the mixture was shaken vigorously for 30 s and allowed to react for a total of 6 min at room temperature (about 20 °C). The hexane phase (1 μL), containing the FAMEs, was analyzed by high-resolution gas chromatography (HRGC). A DANI Master gas-chromatograph (DANI Instruments SpA, Milan, Italy), equipped with a PTV (programmed temperature vaporizer), a split injector, and a flame ionization detector (FID) was used. Analysis was performed with a Quadrex 007-23 column for FAME (60 m, 0.25 mm i.d., 0.25 mm film thickness; Quadrex Corp., Belthany, CT, USA). The carrier gas was high-purity helium with a flow rate of 1.2 mL/min. The injector and detector temperatures were kept at 240 °C. The column oven temperature was programmed at 80 °C for 5 min, from 80 to 165 °C at 5 °C/min for 5 min, from 165 to 230 °C at 3 °C/min for 0.5 min and from 230 to 260 °C at 50 °C/min for 2 min. The FID conditions were a 10:1 ratio of air:hydrogen and a temperature of 260 °C. The identification and the quantification of separated peaks were performed using the Supelco 37 Component FAME MIX (Supelco Bellofonte, PA, USA) as external standards. The fatty acids were determined using the external standard method and were expressed as percentage of total fatty acids.

2.8. Volatile Organic Compounds Analysis

The extraction and analysis of volatile organic compounds (VOC) was performed using SPME-GC/MS, according to Lee et al. [15], with some modifications. The solid-phase microextraction (SPME) device (Supelco Co., Bellefonte, PA, USA) equipped with a 50/30-μm thickness divinyl-benzene/carboxen/polydimethylsiloxane (DVB/CAR/PDMS) fiber coated with 2-cm length stationary phase was used. Twenty-five grams of frozen grated cheese was transferred into a 100 mL bottle, previously added with 25 mL of deionized water, 50 μL of 2-methyl-3-heptanone as internal standard (408 mg L^{-1}), and 12.5 g of sodium phosphate (Sigma, St. Louis, MO, USA). Samples were homogenized and heated on a heating magnetic stirrer. Then the SPME device was hermetically put in the bottles containing the samples and left for 30 min at 40 °C. The SPME was introduced directly into the GC injector where the thermal desorption of the analytes was performed at 250 °C for 10 min. A GC system 6890N equipped with a mass detector 5973 (Agilent Technologies, Palo Alto, CA, USA) was used. The VOCs were separated on a 30 m × 0.250 mm capillary column coated with a 0.25 μm film of 5% diphenyl l95% dimethylpolysiloxane (HP5MS J&W Scientific, Folson, CA, USA). Splitless injection was used for the samples. The column oven temperature was programmed at 10 °C/min from an initial temperature of 50 (held for 2 min) to 150 °C, then at 15 °C/min to 300 °C, which was held for 10 min. The injection and ion source temperatures were 250 and 230 °C, respectively. Helium (99.999%) was used as carrier gas at a flow rate of 1 mL/min. The ionizing electron energy was 70 eV and the mass range scanned was 40–450 amu in full-scan acquisition mode. The compounds were identified using the NIST Atomic Spectra Database version1.6 and verified by the retention indices. The VOCs were calculated by the internal standard method and were expressed as μg/kg of cheese.

2.9. Statistical Analysis

All analyses were performed in triplicate. Significant differences among the different samples were determined by one-way ANOVA statistical analysis and PCA (Principal Component Analysis). Tukey's test was used to discriminate among the means of the variables. Differences with $p < 0.05$ were considered significant. The data elaboration was carried out using XLStat version 2009.3.02 (Addinsoft Corp., Paris, France).

3. Results

3.1. Moisture, pH, and Titratable Acidity

Provolone del Monaco's typical shape is shown in Figure 1; the chemical composition at different ripening times and sampled at different portions is reported in Table 1.

Figure 1. Provolone del Monaco.

Table 1. Chemical composition of cheese samples (mean values ± SD) during 270 days of ripening.

Samples	Ripening Days	pH	Moisture %	Titratable Acidity (Lact. ac. %)	Total N (g/100 g dw)	Soluble N (g/100 g dw)	SN/TN*100	Fat (g/100 g dw)
C3	0	5.03 [b] ± 0.25	46.20 [a] ± 1.34	0.17 ± 0.02	7.17 ± 0.15	0.49 [c] ± 0.01	6.87 [c] ± 0.05	42.96 ± 2.81
	90	5.63 [a,b] ± 0.02	40.03 [b] ± 0.85	0.14 ± 0.01	6.82 ± 0.13	1.00 [b] ± 0.01	14.74 [b] ± 0.15	41.65 ± 1.13
	180	5.77 [a] ± 0.18	40.61 [b] ± 0.54	0.22 ± 0.03	7.02 ± 0.03	1.08 [b] ± 0.03	15.36 [b] ± 0.42	43.45 ± 0.13
	270	5.68 [a,b] ± 0.10	38.35 [b] ± 0.13	0.22 ± 0.03	7.06 ± 0.18	1.70 [a] ± 0.08	24.02 [a] ± 0.59	39.89 ± 2.09
C5	0	5.30 [c] ± 0.04	47.05 [a] ± 0.92	0.16 [a,b] ± 0.01	7.10 ± 0.23	0.50 [d] ± 0.02	6.99 [c] ± 0.58	42.84 ± 1.83
	90	5.79 [b] ± 0.03	42.93 [b] ± 0.64	0.13 [b] ± 0.01	7.11 ± 0.08	1.12 [c] ± 0.06	15.70 [b] ± 0.61	42.65 ± 1.26
	180	5.97 [a,b] ± 0.01	42.11 [b,c] ± 1.05	0.21 [a] ± 0.01	7.12 ± 0.24	1.30 [b] ± 0.04	18.33 [b] ± 1.14	43.51 ± 1.17
	270	6.02 [a] ± 0.08	39.51 [c] ± 0.27	0.22 [a] ± 0.03	7.17 ± 0.03	2.04 [a] ± 0.06	28.46 [a] ± 0.88	41.88 ± 1.24
S3	0	5.23 [b] ± 0.06	43.80 [a] ± 1.89	0.18 [b] ± 0.01	7.01 ± 0.56	0.40 [c] ± 0.01	5.73 [c] ± 0.27	41.67 ± 3.28
	90	5.49 [a,b] ± 0.03	31.63 [b] ± 1.18	0.13 [c] ± 0.01	6.62 ± 0.12	0.80 [b] ± 0.01	12.11 [b] ± 0.40	39.40 ± 2.39
	180	5.70 [a] ± 0.11	31.92 [b] ± 3.03	0.24 [a] ± 0.01	7.15 ± 0.25	1.03 [a] ± 0.04	14.44 [a] ± 0.06	42.10 ± 3.35
	270	5.69 [a] ± 0.06	29.35 [b] ± 0.61	0.21 [a,b] ± 0.01	7.03 ± 0.04	1.14 [a] ± 0.06	16.16 [a] ± 0.84	39.84 ± 0.17
S5	0	5.29 [c] ± 0.02	45.76 [a] ± 0.02	0.18 [b] ± 0.01	7.53 [a] ± 0.04	0.47 [b] ± 0.01	6.30 [b] ± 0.12	42.91 ± 3.22
	90	5.46 [b] ± 0.01	36.89 [b] ± 2.69	0.13 [c] ± 0.01	6.74 [c] ± 0.04	1.29 [a] ± 0.09	19.12 [a] ± 1.28	41.39 ± 1.86
	180	5.73 [a] ± 0.02	34.81 [b] ± 2.33	0.22 [a] ± 0.01	7.36 [b,c] ± 0.21	1.18 [a] ± 0.04	16.06 [a] ± 0.11	42.18 ± 4.72
	270	5.77 [a] ± 0.08	31.69 [b] ± 0.31	0.21 [a,b] ± 0.01	7.02 [b,c] ± 0.12	1.21 [a] ± 0.11	17.25 [a] ± 1.34	37.18 ± 0.53

a–c: Different letters in the same column correspond to significant differences ($p < 0.05$); C3 core portion of 3 kg size; C5 core portion of 5 kg size; S3 under rind portion of 3 kg size; S5 under rind portion of 5 kg size.

The cheese at time 0 showed a water content ranging from 43.80 to 47.05%. In particular, the moisture percentage was 46.20% and 47.05% in the core (C) portions and 43.80 and 45.76% in the portion under the rind (S) portions, in the 3-kg and 5-kg sized cheeses, respectively. At 90 days of ripening, the highest loss of moisture was observed. In fact, the values decreased to 40.03% and 42.93% in C portions and to 31.63% and 36.89% in S portions in the 3-kg and 5 kg-sized cheeses, respectively. After 90 days of ripening, the loss of water was less marked, probably due to the definitive rind formation. The S portions of both sizes had lower moisture contents than the C portions because the S portion was located closer to the external surface of the cheese in comparison with the C portions, producing a faster loss of water.

The pH values detected in the analyzed samples ranged from 5.03 to 6.02 in the C portions and from 5.23 to 5.77 in the S portions in the 3-kg and 5 kg-sized cheeses, respectively. The increase of pH in cheese during ripening has been associated with the proteolytic process that released large amounts of nitrogenated alkaline compounds [16]. After 270 days of ripening, the 5-kg sized cheese showed slightly higher pH values than the 3-kg sized cheese, probably because of the more abundant water content in the 5-kg sized cheese [17].

With reference to the titratable acidity, the data obtained showed values ranging from 0.17 to 0.22 of lactic acid/100 g of cheese in the C portions and from 0.19 to 0.21 g of lactic acid/100 g of cheese in the S portions during 270 days of ripening in the 3-kg and 5-kg sized cheeses, respectively. In particular, during the first 90 days of storage, all the samples presented a decrease of titratable acidity in comparison with the data observed in the first stage of the maturation period. A significant increase in this parameter was observed at 270 days, possibly due to lactic acid formation from residual lactose still present in the cheese [18].

3.2. Nitrogen Content

The total nitrogen (TN) content values measured in cheese samples ranged from 6.82 to 7.17% dw in the C portions and from 6.62 to 7.53% dw in the S portions in the 3-kg and 5-kg sized cheeses, respectively (Table 1). No significant increase was observed during the storage process in all examined samples. The S samples showed higher TN contents in comparison with the C samples, probably due to the loss of water and to the dry matter (DM) concentration [3].

Regarding soluble nitrogen (SN) at pH 4.6, at time 0, the data analyzed ranged from 0.49 to 0.50% in the C portions and from 0.40 to 0.47% in the S portions in the 3-kg and 5-kg sized cheeses, respectively. A significant increase was observed in all samples at 90 days of ripening, with values ranging from 1.00 to 1.12% in the C portions and from 0.80 to 1.29% in the S portions in both cheese sizes (3 kg and 5 kg, respectively). This increment of the parameter studied can be explained by the fact that proteolysis is directly correlated with the increase in pH during the ripening phase [19]. After 270 days of ripening, the C portions showed SN values higher than the S portions (1.70% and 2.04% with respect to 1.14% and 1.21%, in 3-kg and 5-kg sized cheeses, respectively), according to Gobbetti et al. [20]. This difference is probably due to their water contents; in fact, the moisture content of the substrate can determine a minor proteolytic activity [13]. Considering that proteolysis is one of the principal biochemical transformations produced during the ripening of cheese, the ratio between soluble nitrogen at pH 4.6 and total nitrogen (SN/TN) was calculated. These data, called the maturation index, represent a good indicator of cheese ripening and of its protolithic activity.

In particular, at the beginning of the ripening period, the SN/TN data detected ranged from 6.87 to 6.99 for the C samples and from 5.73 to 6.30 for the S samples, in the 3 kg and 5 kg cheeses, respectively. These data drastically increased during the storage period, particularly at 90 and 180 days of ripening, with data ranging from 15.36 to 18.33 for the C samples and from 14.44 to 16.06 for the S samples in the 3-kg and 5-kg sized cheeses, respectively. After 270 days of ripening, the C portions of both 3 kg and 5 kg samples showed SN/TN values higher than those of the S portions because of the more abundant water content in the C portions [17]. In particular, the SN/TN data detected in the C portions were 24.02 and 28.46 for 3-kg and 5-kg sized cheeses, respectively, whereas in the S portion, they were 16.16

and 17.25 for 3-kg and 5-kg sized cheeses, respectively, indicating that Provolone del Monaco PDO cheese is a firm/hard pasta filata cheese [21].

3.3. Fat Content

One of the most important analytical parameters that allow Provolone del Monaco cheese to obtain the PDO brand is the fat content on dry matter (F/DM). No significant differences were observed compared with the time zero data, even if after 6 months of ripening, the fat content (g/100 g dw) detected was higher than 42% and ranged from 42.10 to 43.51%. At 270 days ripening, cheese samples showed fat content (g/100 g dw) mean values of 39.86% and 39.53% for 3 kg and 5 kg cheese, respectively.

3.4. Fatty Acid Profile

Twenty-two fatty acids in concentration >0.1% were identified in Provolone del Monaco cheese samples, from C4:0 (butyric acid) to C20:3n3 (eicosatrienoic acid), including c9t11-conjugated linoleic acid isomer (CLAc9t11). The identification of the fatty acids is reported in Table 2a,b.

The fatty acid profile of cheese largely reflected that of the raw milk from which cheese was made [22–25]. Medium chain fatty acids (MCFAs, from C10:0 to C16:1) were the most abundant in all the samples, with values reaching 49%. Among MCFAs, palmitic acid (C16:0) was the most representative, with a percentage of approximately 29%. Long chain fatty acids (LCFAs, from C17:0 to C20:3n3) reached approximately 44%, with oleic acid (C18:1n9c) as the most abundant (21%). Short chain fatty acids (SCFAs, from C4:0 to C8:0) reached values of approximately 6%. Concentrations of approximately 0.65% of c9t11 CLA were detected. In recent years, attention has been paid towards the potential benefits of short chain fatty acids and CLA [26,27]. For instance, the roles and importance of butyric acid has been expanding steadily: immunomodulatory activity of butyric acid in the gastrointestinal tract, the positive effects on irritable bowel syndrome, in non-specific inflammatory bowel diseases, in cancer prevention and treatment [28,29].

The analysis revealed that saturated fatty acids (ΣSFAss) represented approximately 69%, with C16:0 as the most abundant ΣSFA. Among unsaturated fatty acids (ΣUFAs), monounsaturated (ΣMUFA) represented approximately 24% of ΣUFA, and the most abundantly detected ΣMUFA was oleic acid (C18:1n9c). Finally, polyunsaturated fatty acids (ΣPUFA) represented just 6%, with linoleic acid (C18:2n6c) as the most representative ΣPUFA (3.6%). All these results are in agreement with Romano et al. [3]. A few significant differences in fatty acid concentrations were detected during ripening. Regarding the 3 kg cheese samples, capric acid (C10:0) showed a significant increase (from 2.49% to 2.83%) in the C portions; the linolelaidic acid value (C18:2n6t), however, decreased significantly (from 0.57% to 0.53%) in the S portions. In the 5 kg samples, caproic acid (C6:0) showed a significant increase (from 1.90% to 2.21%) in the S portions; oleic acid (C18:1n9c), on the opposite, showed a significant decrease (from 21.19% to 20.77%) in the S portions.

The reduction of the LCFA and the increase of the SCFA observed could be due to the action of indigenous milk lipases [6] or to the action of gastric lipases present in the goat rennet used to produce Provolone del Monaco cheese [3]. These lipases preferentially hydrolyze long chain fatty acids to produce short chain fatty acids. Depending on their concentration and perception threshold, volatile fatty acids can either positively or negatively contribute to the aroma of the cheese or to a rancidity defect [8].

Lipolysis is classified as spontaneous or induced [30]. Free fatty acid flavors are produced by lipase activity and originate from short-chain FFA, with described flavors that are reminiscent of vinegar, cheese, sweat, and soap [31]. These FFA are generated by the enzymatic hydrolysis of ester bonds of triglycerides. Oxidized flavors are due to the autoxidation of fatty acids and are characterized by cardboard, metallic, and mushroom flavors [32,33].

Table 2. (a) Composition of fatty acids (% ± SD) in 3-kg size samples during 270 days of ripening. (b) Composition of fatty acids (% ± SD) in 5-kg size samples during 270 days of ripening.

(a)

	C				S			
	0	90	180	270	0	90	180	270
C4:0	3.04 ± 0.05	3.07 ± 0.00	3.13 ± 0.02	3.10 ± 0.04	2.81 ± 0.18	3.00 ± 0.18	3.01 ± 0.28	3.22 ± 0.22
C6:0	1.93 [a,b] ± 0.05	1.84 [b] ± 0.05	2.03 [a,b] ± 0.05	2.13 [a,b] ± 0.07	1.93 [a,b] ± 0.07	1.97 [a,b] ± 0.09	2.05 [a,b] ± 0.11	2.19 [a] ± 0.13
C8:0	1.18 ± 0.10	1.14 ± 0.00	1.23 ± 0.02	1.42 ± 0.24	1.18 ± 0.08	1.27 ± 0.08	1.20 ± 0.12	1.26 ± 0.13
C10:0	2.49 [c] ± 0.04	2.51 [b,c] ± 0.01	2.78 [a,b] ± 0.05	2.83 [a] ± 0.07	2.65 [a,b,c] ± 0.08	2.63 [a,b] ± 0.02	2.55 [a,b,c] ± 0.11	2.53 [b,c] ± 0.12
C12:0	2.94 ± 0.01	2.88 ± 0.01	2.97 ± 0.08	3.00 ± 0.05	3.00 ± 0.06	2.98 ± 0.03	2.93 ± 0.07	2.88 ± 0.02
C14:0	10.90 [b] ± 0.16	10.94 [a,b] ± 0.01	11.23 [a] ± 0.06	10.77 [b] ± 0.04	11.04 [a,b] ± 0.08	11.01 [a,b] ± 0.09	10.97 [a,b] ± 0.02	10.75 [b] ± 0.04
C14:1	0.95 ± 0.00	0.94 ± 0.00	0.96 ± 0.01	0.96 ± 0.01	0.97 ± 0.02	0.97 ± 0.00	0.96 ± 0.01	0.95 ± 0.01
C15:0	1.07 [b] ± 0.02	1.10 [a,b] ± 0.01	1.08 [a,b] ± 0.00	1.09 [a,b] ± 0.01	1.12 [a] ± 0.00	1.12 [a,b] ± 0.00	1.11 [a,b] ± 0.01	1.11 [a,b] ± 0.02
C16:0	29.52 ± 0.10	29.63 ± 0.03	29.28 ± 0.07	28.93 ± 0.05	29.34 ± 0.29	29.31 ± 0.05	29.31 ± 0.39	29.21 ± 0.36
C16:1	1.49 [b,c] ± 0.02	1.48 [c] ± 0.03	1.48 [c] ± 0.01	1.53 [a,b] ± 0.01	1.55 [a,b] ± 0.02	1.51 [a,b,c] ± 0.00	1.52 [a,b,c] ± 0.01	1.56 [a] ± 0.00
C17:0	0.89 [b] ± 0.01	0.94 [a,b] ± 0.01	0.92 [a,b] ± 0.02	0.95 [a,b] ± 0.04	0.97 [a,b] ± 0.04	0.98 [a] ± 0.00	0.99 [a] ± 0.03	1.01 [a] ± 0.02
C17:1	0.22 [d] ± 0.00	0.23 [c,d] ± 0.00	0.23 [b,c,d] ± 0.01	0.25 [a,b,c] ± 0.01	0.25 [a,b,c] ± 0.01	0.25 [a,b] ± 0.00	0.25 [a] ± 0.00	0.25 [a,b] ± 0.00
C18:0	15.26 ± 0.07	15.26 ± 0.04	14.96 ± 0.10	15.05 ± 0.10	15.07 ± 0.17	15.13 ± 0.01	15.08 ± 0.17	15.16 ± 0.22
C18:1n9t	0.45 ± 0.03	0.48 ± 0.04	0.42 ± 0.01	0.47 ± 0.04	0.45 ± 0.01	0.48 ± 0.01	0.45 ± 0.04	0.51 ± 0.02
C18:1n9c	21.24 [a] ± 0.11	21.18 [a,b] ± 0.01	20.96 [b,c] ± 0.02	21.03 [a,b,c] ± 0.09	20.99 [b,c] ± 0.05	20.82 [c] ± 0.01	20.96 [b,c] ± 0.08	20.86 [c] ± 0.01
C18:2n6t	0.52 [b] ± 0.01	0.51 [b] ± 0.00	0.51 [b] ± 0.01	0.52 [b] ± 0.00	0.57 [a] ± 0.02	0.54 [a,b] ± 0.00	0.54 [a,b] ± 0.00	0.53 [b] ± 0.01
C18:2n6c	3.62 [a] ± 0.03	3.55 [a,b] ± 0.01	3.52 [b] ± 0.04	3.59 [a,b] ± 0.02	3.60 [a] ± 0.02	3.61 [a] ± 0.00	3.61 [a] ± 0.00	3.58 [a,b] ± 0.01
C20:0	0.21 ± 0.01	0.21 ± 0.01	0.20 ± 0.01	0.22 ± 0.01	0.22 ± 0.00	0.13 ± 0.00	0.23 ± 0.00	0.23 ± 0.00
C18:3n3	0.63 ± 0.02	0.62 ± 0.00	0.63 ± 0.02	0.65 ± 0.02	0.66 ± 0.01	0.67 ± 0.01	0.67 ± 0.01	0.66 ± 0.00
C9t11	0.64 [a,b] ± 0.00	0.63 [a,b] ± 0.00	0.61 [b] ± 0.01	0.65 [a] ± 0.02	0.65 [a] ± 0.00	0.65 [a] ± 0.00	0.65 [a] ± 0.00	0.66 [a] ± 0.01
C20:3n6	0.17 [c] ± 0.00	0.17 [c] ± 0.00	0.17 [b,c] ± 0.00	0.17 [b,c] ± 0.00	0.19 [a,b] ± 0.01	0.19 [a] ± 0.00	0.19 [a] ± 0.00	0.18 [a,b,c] ± 0.00
C20:3n3	0.10 [c] ± 0.00	0.10 [b,c] ± 0.00	0.10 [b,c] ± 0.01	0.11 [a,b,c] ± 0.00	0.11 [a,b,c] ± 0.00	0.11 [a,b,c] ± 0.00	0.11 [a,b] ± 0.00	0.12 [a] ± 0.00
SFAs	69.76 ± 0.14	69.94 ± 0.07	70.21 ± 0.14	69.87 ± 0.21	69.77 ± 0.06	69.94 ± 0.03	69.84 ± 0.10	69.91 ± 0.02
MUFAs	24.45 ± 0.17	24.38 ± 0.05	24.14 ± 0.06	24.33 ± 0.15	24.32 ± 0.06	24.15 ± 0.01	24.23 ± 0.11	24.23 ± 0.01
PUFAs	5.77 [a,b,c] ± 0.04	5.66 [b,c] ± 0.02	5.63 [c] ± 0.09	5.79 [a,b,c] ± 0.06	5.88 [a] ± 0.01	5.87 [a] ± 0.01	5.86 [a] ± 0.01	5.82 [a,b] ± 0.01
UFAs	30.21 ± 0.15	30.04 ± 0.07	29.77 ± 0.14	30.12 ± 0.21	30.20 ± 0.05	30.02 ± 0.02	30.12 ± 0.10	30.05 ± 0.02
n3	0.75 ± 0.02	0.74 ± 0.01	0.74 ± 0.03	0.78 ± 0.03	0.78 ± 0.01	0.80 ± 0.01	0.80 ± 0.00	0.79 ± 0.00
n6	4.38 [a,b,c] ± 0.02	4.30 [c,d] ± 0.01	4.27 [d] ± 0.05	4.35 [b,c,d] ± 0.02	4.45 [a] ± 0.00	4.42 [a,b] ± 0.00	4.42 [a,b] ± 0.01	4.37 [a,b,c] ± 0.00
n3/n6	0.17 ± 0.00	0.17 ± 0.01	0.17 ± 0.01	0.18 ± 0.01	0.18 ± 0.00	0.18 ± 0.00	0.18 ± 0.00	0.18 ± 0.00
SCFAs (4–8)	6.15 ± 0.20	6.06 ± 0.04	6.39 ± 0.01	6.65 ± 0.35	5.92 ± 0.32	6.24 ± 0.35	6.25 ± 0.51	6.67 ± 0.49
MCFs (10–16)	49.49 [a,b] ± 0.24	49.58 [a,b] ± 0.05	49.90 [a] ± 0.26	49.23 [a,b] ± 0.02	49.81 [a,b] ± 0.06	49.66 [a,b] ± 0.18	49.48 [a,b] ± 0.19	49.11 [b] ± 0.22
LCFAs (17–24)	44.36 ± 0.03	44.36 ± 0.00	43.71 ± 0.25	44.12 ± 0.37	44.27 ± 0.26	44.10 ± 0.16	44.27 ± 0.32	44.23 ± 0.26

Table 2. *Cont.*

(b)

	C				S			
	0	90	180	270	0	90	180	270
C4:0	2.81 [a,b] ± 0.08	2.72 [a,b] ± 0.24	2.75 [a,b] ± 0.25	2.77 [a,b] ± 0.08	2.84 [a,b] ± 0.12	2.62 [b] ± 0.13	2.99 [a,b] ± 0.21	3.32 [a] ± 0.19
C6:0	1.84 [b] ± 0.04	1.90 [a,b] ± 0.05	1.87 [b] ± 0.07	1.85 [b] ± 0.06	1.90 [b] ± 0.04	1.76 [b] ± 0.06	2.00 [a,b] ± 0.16	2.21 [a] ± 0.05
C8:0	1.13 ± 0.02	1.15 ± 0.09	1.18 ± 0.04	1.15 ± 0.02	1.16 ± 0.00	1.13 ± 0.02	1.16 ± 0.04	1.30 ± 0.08
C10:0	2.52 ± 0.03	2.46 ± 0.07	2.58 ± 0.01	2.51 ± 0.01	2.49 ± 0.00	2.50 ± 0.01	2.54 ± 0.10	2.52 ± 0.02
C12:0	2.89 ± 0.04	2.85 ± 0.02	2.91 ± 0.03	2.89 ± 0.02	2.86 ± 0.00	2.89 ± 0.01	2.94 ± 0.12	3.08 ± 0.13
C14:0	10.93 ± 0.01	10.86 ± 0.00	10.93 ± 0.08	10.92 ± 0.06	10.89 ± 0.02	10.96 ± 0.04	10.99 ± 0.15	10.94 ± 0.04
C14:1	0.95 ± 0.00	0.95 ± 0.01	0.97 ± 0.00	0.97 ± 0.00	0.95 ± 0.00	0.97 ± 0.02	0.97 ± 0.02	0.98 ± 0.02
C15:0	1.10 ± 0.01	1.12 ± 0.00	1.11 ± 0.00	1.12 ± 0.00	1.11 ± 0.00	1.13 ± 0.01	1.12 ± 0.01	1.10 ± 0.03
C16:0	29.40 ± 0.13	29.33 ± 0.23	29.29 ± 0.09	29.34 ± 0.14	29.37 ± 0.16	29.44 ± 0.05	29.19 ± 0.22	29.11 ± 0.07
C16:1	1.53 [a,b] ± 0.01	1.58 [a] ± 0.02	1.54 [a,b] ± 0.01	1.55 [a,b] ± 0.00	1.53 [a,b] ± 0.03	1.59 [a] ± 0.01	1.54 [a,b] ± 0.01	1.52 [b] ± 0.01
C17:0	0.97 ± 0.01	0.99 ± 0.01	0.98 ± 0.00	0.99 ± 0.00	1.00 ± 0.01	1.01 ± 0.02	0.99 ± 0.04	0.94 ± 0.05
C17:1	0.25 ± 0.00	0.26 ± 0.00	0.25 ± 0.00	0.25 ± 0.00	0.27 ± 0.03	0.26 ± 0.00	0.25 ± 0.00	0.25 ± 0.01
C18:0	15.31 [a,b] ± 0.09	15.32 [a,b] ± 0.01	15.16 [a,b] ± 0.02	15.28 [a,b] ± 0.06	15.32 [a,b] ± 0.08	15.46 [a] ± 0.03	15.16 [a,b] ± 0.25	14.96 [b] ± 0.15
C18:1n9t	0.49 ± 0.00	0.52 ± 0.00	0.46 ± 0.00	0.51 ± 0.02	0.51 ± 0.00	0.50 ± 0.00	0.49 ± 0.05	0.45 ± 0.05
C18:1n9c	21.30 [a,b] ± 0.12	21.29 [a,b] ± 0.13	21.39 [a] ± 0.13	21.19 [a,b] ± 0.09	21.19 [a,b] ± 0.01	21.00 [a,b,c] ± 0.03	20.96 [b,c] ± 0.16	20.77 [c] ± 0.00
C18:2n6t	0.52 ± 0.00	0.55 ± 0.01	0.52 ± 0.01	0.53 ± 0.01	0.54 ± 0.01	0.54 ± 0.00	0.55 ± 0.03	0.52 ± 0.02
C18:2n6c	3.63 [a,b] ± 0.02	3.67 [a] ± 0.05	3.64 [a,b] ± 0.01	3.69 [a] ± 0.02	3.61 [a,b] ± 0.00	3.65 [a,b] ± 0.01	3.58 [b] ± 0.02	3.61 [a,b] ± 0.01
C20:0	0.22 ± 0.00	0.24 ± 0.00	0.22 ± 0.00	0.23 ± 0.00	0.23 ± 0.00	0.24 ± 0.01	0.23 ± 0.01	0.22 ± 0.02
C18:3n3	0.66 ± 0.00	0.68 ± 0.01	0.66 ± 0.00	0.68 ± 0.01	0.66 ± 0.00	0.68 ± 0.01	0.66 ± 0.01	0.65 ± 0.02
C9t11	0.66 ± 0.00	0.67 ± 0.01	0.66 ± 0.01	0.66 ± 0.00	0.66 ± 0.01	0.67 ± 0.00	0.66 ± 0.01	0.64 ± 0.01
C20:3n6	0.18 [b] ± 0.00	0.19 [b] ± 0.00	0.18 [b] ± 0.00	0.19 [b] ± 0.00	0.18 [b] ± 0.00	0.27 [a] ± 0.00	0.18 [b] ± 0.00	0.18 [b] ± 0.00
C20:3n3	0.11 ± 0.00	0.12 ± 0.00	0.11 ± 0.00	0.12 ± 0.00	0.11 ± 0.00	0.12 ± 0.00	0.11 ± 0.00	0.11 ± 0.00
SFAs	69.48 [a,b] ± 0.17	69.29 [b] ± 0.22	69.39 [b] ± 0.12	69.45 [a,b] ± 0.13	69.56 [a,b] ± 0.08	69.52 [a,b] ± 0.05	69.77 [a,b] ± 0.27	70.08 [a] ± 0.12
MUFAs	24.62 [a] ± 0.13	24.70 [a] ± 0.14	24.70 [a] ± 0.11	24.56 [a] ± 0.09	24.55 [a] ± 0.06	24.41 [a,b] ± 0.04	24.35 [a,b] ± 0.18	24.07 [b] ± 0.05
PUFAs	5.86 [a,b] ± 0.03	5.97 [a,b] ± 0.08	5.87 [a,b] ± 0.01	5.95 [a,b] ± 0.03	5.85 [a,b] ± 0.01	6.03 [a] ± 0.01	5.84 [a,b] ± 0.09	5.81 [b] ± 0.07
UFAs	30.48 [a,b] ± 0.17	30.67 [a] ± 0.22	30.57 [a] ± 0.12	30.51 [a] ± 0.13	30.40 [a,b] ± 0.07	30.44 [a,b] ± 0.05	30.19 [a,b] ± 0.27	29.88 [b] ± 0.12
n3	0.79 ± 0.00	0.82 ± 0.01	0.79 ± 0.00	0.81 ± 0.01	0.79 ± 0.01	0.82 ± 0.01	0.79 ± 0.02	0.78 ± 0.02
n6	4.41 [a,b] ± 0.03	4.49 [a,b] ± 0.06	4.42 [a,b] ± 0.01	4.48 [a,b] ± 0.03	4.40 [b] ± 0.00	4.54 [a] ± 0.02	4.39 [b] ± 0.06	4.39 [b] ± 0.03
n3/n6	0.18 ± 0.00	0.18 ± 0.00	0.18 ± 0.00	0.18 ± 0.00	0.18 ± 0.00	0.18 ± 0.00	0.18 ± 0.00	0.18 ± 0.00
SCFAs(4–8)	5.78 [a,b] ± 0.01	5.78 [a,b] ± 0.38	5.80 [a,b] ± 0.35	5.77 [a,b] ± 0.15	5.89 [a,b] ± 0.16	5.50 [b] ± 0.21	6.16 [a,b] ± 0.41	6.83 [a] ± 0.32
MCFAs (10–16)	49.43 ± 0.04	49.28 ± 0.16	49.45 ± 0.18	49.42 ± 0.22	49.33 ± 0.16	49.60 ± 0.11	49.42 ± 0.17	49.39 ± 0.08
LCFAs (17–24)	44.78 [a,b] ± 0.03	44.94 [a] ± 0.23	44.75 [a,b] ± 0.17	44.81 [a,b] ± 0.07	44.78 [a,b] ± 0.00	44.89 [a] ± 0.10	44.42 [a,b] ± 0.58	43.78 [b] ± 0.40

a–c: Different letters in the same row correspond to significant differences ($p < 0.05$); C core portion, S under rind portion; SFAs: saturated fatty acids; MUFAs: monounsaturated fatty acids; PUFAs: polyunsaturated fatty acids; UFAs: unsaturated fatty acids; SCFAs: short-chain fatty acids; MCFAs: medium-chain fatty acids; LCFAs: long-chain fatty acids.

3.5. Volatile Organic Compounds

Volatile organic compounds (VOCs) identified in the Provolone del Monaco cheese samples were divided into 6 families, including organic acids, alcohols, ketones, aldehydes, esters, and other compounds (Table 3).

Organic acids were the most abundant VOCs detected. In particular, acetic acid (ranging from 18.25 to 2503.02 µg/kg), butanoic acid (ranging from 186.02 to 11,469.36 µg/kg), and hexanoic acid (ranging from 203.99 to 5022.33 µg/kg) were found. These molecules are typical of dairy products and come from the oxidative decarboxylation of α-keto acids, which originate from branched amino acids released during proteolysis [34]. Organic acids (and relative quantities) are the main responsibles for the Provolone del Monaco PDO cheese aroma, giving it its pungent, sour, cheesy, and buttery odors [35–38].

Alcohols and ketones are the second and third most abundant families detected, respectively, and arise from β-oxidation and decarboxylation of free fatty acids released during lipolysis [34]. Most likely, for this reason, no alcohols were revealed at the beginning of ripening. Among the alcohols, ethanol (ranging from 9.49 to 56.73 µg/kg), 1-butanol (ranging from 6.38 to 79.81 µg/kg), and 2-ethyl-1-hexanol (ranging from 8.20 to 80.20 µg/kg), mostly contribute to the Provolone del Monaco aroma, with characteristic alcoholic, fruity, and spicy aromas, respectively. Among ketones, acetoin (3-hydroxy-2-butanone, ranging from 3.77 to 61.91 µg/kg), which is a characteristic dairy product volatile compound, can be found only in the first days of ripening. Aldehydes (hexanal and nonanal) were found in small amounts during ripening. These compounds are produced from α-keto acids by decarboxylation [34] and can be quickly converted into alcohols or their respective acids [39] during ripening. This could explain why hexanal (ranging from 26.60 to 63.66 µg/kg) was found only at time 0 in the cheese samples. Esters were the less abundant family detected. Two molecules were identified, ethyl hexanoate (ranging from 35.24 to 123.33 µg/kg) and butyl butanoate (ranging from 33.92 to 47.24 µg/kg). Their origin could be attributed to the reaction between free fatty acids and alcohols [34].

An integrated and multidisciplinary system of analysis is becoming a valuable tool to analyze and study food, taking into account the various aspects of food quality and composition by employing innovative and emerging technologies with statistical analysis. Such a multi-technique approach to food research enables utilizing all the resources of quality, safety, and traceability in food systems. The joined approach of multidisciplinary analysis techniques of food matrices with the multivariate statistical approach enlarges the possibilities to exploit a wide range of food traits and aspects. The "Integrated Approach" is the key to modern food research and the innovative challenge for analyzing and modelling agro-food systems in their totality [40–43]. To better understand the VOC distribution during ripening, a Principal Components Analysis (PCA) was carried out using the recorded dataset, which is shown in Figure 2.

The PCA illustrates which families are characteristic at different ripening times. The F1 and F2 factors explain 97.8% of the total variability. In particular, the factor F1 explains 74.7%. The main contribution to F1 comes from alcohols (21.3%). The factor F2 explains 23.1% of the variability, with the main contribution given from esters (39.0%), followed by aldehydes (38.7%). As expected, the majority of molecules were found in the last months of ripening. In particular, after 90 and 180 days, cheese samples showed a large amount of organic acids and alcohols. Furthermore, cheese at the 270-day timepoint was characterized mainly by esters, ketones, and other compounds.

Table 3. Volatile organic compounds families (µg/kg ± SD) in cheese samples during 270 days of ripening.

Samples	Ripening Days	Acids	Alcohols	Ketones	Aldehydes	Esters	Others
C3	0	642.86 [c] ± 163.29	28.37 [c] ± 11.47	21.28 [a] ± 0.19	34.08 [a] ± 7.34	n.d.	12.88 [b,c] ± 0.97
	90	2658.35 [b,c] ± 502.10	49.53 [c] ± 3.88	3.77 [c] ± 0.06	n.d.	n.d.	7.44 [c] ± 1.29
	180	7105.93 [a] ± 2043.31	195.40 [a] ± 0.60	14.25 [a,b] ± 3.54	n.d.	n.d.	25.77 [a,b] ± 5.50
	270	5029.01 [a,b] ± 256.45	110.03 [b] ± 18.95	6.50 [b,c] ± 2.55	21.26 [a] ± 0.88	68.87 [a] ± 4.01	35.99 [a] ± 2.99
C5	0	529.72 [c] ± 54.28	42.47 [b] ± 5.77	26.00 [a] ± 1.17	39.84 [a] ± 3.17	n.d.	9.38 [b] ± 1.79
	90	9822.30 [a] ± 327.61	384.37 [a] ± 56.89	6.81 [b] ± 0.52	n.d.	46.78 [a,b] ± 12.61	26.48 [a] ± 7.10
	180	6459.75 [b] ± 290.65	279.10 [a] ± 43.78	10.03 [b] ± 1.58	n.d.	47.24 [a,b] ± 7.57	14.41 [a,b] ± 0.34
	270	5172.55 [b] ± 915.88	343.78 [a] ± 54.10	9.41 [b] ± 4.08	12.64 [b] ± 2.97	69.17 [a] ± 22.67	27.94 [a] ± 2.83
S3	0	1659.50 [b] ± 34.48	22.72 [c] ± 2.33	46.96 [b] ± 6.17	74.28 [a] ± 4.15	n.d.	22.30 ± 7.87
	90	9171.68 [a,b] ± 41.58	287.42 [a,b] ± 18.37	251.86 [b] ± 5.08	n.d.	n.d.	39.45 ± 15.89
	180	4147.72 [a,b] ± 526.24	154.72 [b,c] ± 10.36	76.98 [b] ± 12.68	n.d.	n.d.	14.45 ± 1.95
	270	16,051.11 [a] ± 6335.85	360.26 [a] ± 74.44	604.81 [a] ± 110.20	50.30 [b] ± 0.26	123.33 [a] ± 0.41	68.57 ± 20.44
S5	0	441.49 [b] ± 141.81	21.38 [b] ± 4.69	26.47 ± 1.97	34.74 [a] ± 2.80	n.d.	12.67 [b] ± 0.08
	90	17,539.56 [a] ± 2130.94	360.20 [a] ± 88.95	511.24 ± 253.21	n.d.	n.d.	49.49 [a] ± 6.08
	180	15,607.76 [a] ± 2303.45	271.80 [a] ± 28.88	219.30 ± 13.33	17.50 [a,b] ± 2.15	n.d.	44.82 [a,b] ± 6.04
	270	7128.88 [b] ± 2002.45	209.47 [a,b] ± 45.13	255.68 ± 67.22	22.09 [a] ± 9.31	62.71 [a] ± 5.76	26.57 [a,b] ± 13.52

a–c: Different letters in the same column correspond to significant differences ($p < 0.05$); C3 core portion of 3 kg size; C5 core portion of 5 kg size; S3 under rind portion of 3 kg size; S5 under rind portion of 5 kg size; n.d. = not detected.

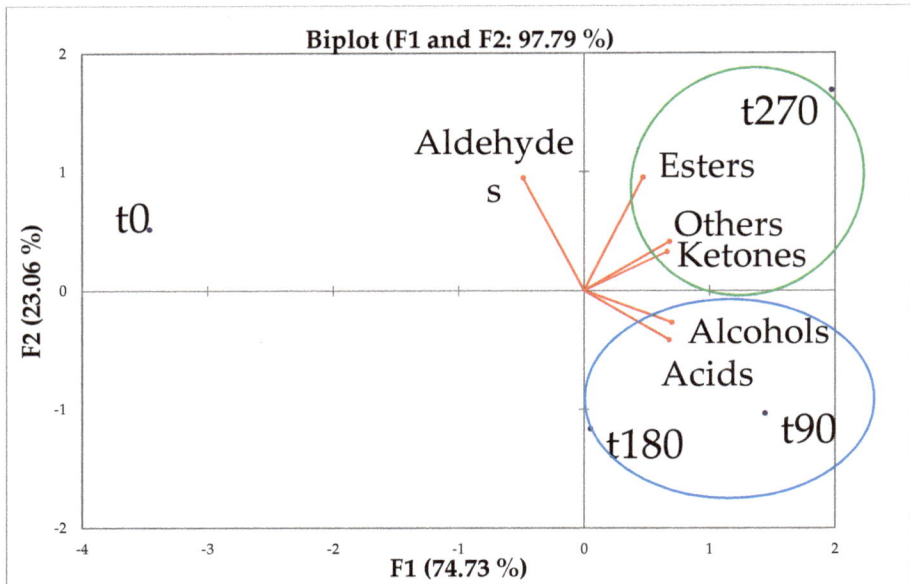

Figure 2. Principal components analysis (PCA) of volatile organic compounds (VOC) evaluated in Provolone del Monaco samples after 0 (t0), 90 (t90), 180 (t180), and 270 (t270) days of ripening.

4. Conclusions

Investigation and valorization of traditional products is essential for the optimization of their potential beneficial and healthy properties, for the preservation of agro-biodiversity, and sustainability promotion. According to this perspective view, our research supports quality in food and in the meantime helps to promote sustainable resource management through environmental sound farming systems linked to territorial characterization and to local cultural heritage. Provolone del Monaco PDO, an example of traditional cheese, showed important chemical changes during 270 days of ripening. As expected, the main significant differences involved the loss of water (ranging from 47% to 29%), the increase of acidity (from 0.16% to 0.22%), and the large production of nitrogen (from 3.8% to 5%). Cheese samples showed significant differences in maturation index (SN/TN) values (from 6% to 28.5%), indicating that a strong proteolytic activity occurred during ripening. No significant differences were found in the fatty acid profile. Provolone del Monaco cheese is mainly characterized by medium chain saturated fatty acids, with palmitic acid (C16:0) as the most abundant one (29%). Only a few significant differences in the fatty acid profile were found during the 270 days of ripening. With reference to the aromatic profile, the analysis of the data showed that acids are the principal volatile compounds contributing to the cheese aroma with pungent and buttery odors. PCA also showed that at 180 days of ripening, acids, alcohols and ketones represent the most characteristic volatile compound families, and at 270 days, mainly esters can be found. The comparison between the core samples and the portions below the rind samples showed some differences, primarily due to the low water content of the latter portion. Mainly, the S portions showed lower water content and a higher nitrogen percentage than the C portions. Concerning the two different sizes examined, namely, 3 and 5 kg, only small differences were observed, indicating that the size (weight) does not affect the quality of Provolone del Monaco cheese. These results, characterizing the nutritional profile of Provolone del Monaco PDO should be promoted throughout atlases, leaflets, brochures, handbooks at events at the national and local level, such as local restaurants, typical local products markets, street markets, and during gastronomic and local festivals with cultural and artistic events to promote the territory. Promotion and dissemination of knowledge of Provolone del Monaco PDO, should be encouraged from nutritional, cultural, and touristic point of views, in order to reinforce gastronomic heritage and promote food tourism in rural areas for rural and regional economic development, as well as to support short food chain at zero kilometers. Indeed, the results should represent a valid tool for promotion of socioeconomic development, enhancement of territories, biodiversity preservation, and sustainability.

Author Contributions: R.R., A.S., N.M. conceived the work. N.M., R.R., A.S., A.D., M.L., A.A. wrote the manuscript. N.M., F.P., A.M., G.M., A.A. carried out the experimental study and analyzed the data. All authors made a substantial contribution to revise the work and approved it for publication.

Funding: The research received no external funding.

Conflicts of Interest: The authors declare no conflict of interest.

References

1. DOP Italian Regulation 10A02770, 2010, n.56 of 09.03.2010. Available online: https://www.gazzettaufficiale. it/do/atto/serie_generale/caricaPdf?cdimg=10A0277000100010110012&dgu=2010-03-09&art. dataPubblicazioneGazzetta=2010-03-09&art.codiceRedazionale=10A02770&art.num=1&art.tiposerie=SG (accessed on 25 March 2019).
2. Aponte, M.; Fusco, V.; Andolfi, R.; Coppola, S. Lactic acid bacteria occurring during manufacture and ripening of Provolone del Monaco cheese: Detection by different analytical approaches. *Int. Dairy J.* **2008**, *18*, 403–413. [CrossRef]
3. Romano, R.; Boriello, I.; Macaldi, C.; Giordano, A.; Spagna Musso, S. Influenza del grado di stagionatura sulla composizione in acidi grassi, ω-3 e CLA del Provolone del Monaco. *Prog. Nutr.* **2008**, *10*, 165–173.
4. Romano, R.; Masucci, F.; Giordano, A.; Spagna Musso, S.; Naviglio, D.; Santini, A. Effect of tomato by-products in the diet of Comisana sheep on composition and conjugated linoleic acid content of milk fat. *Int. Dairy J.* **2010**, *20*, 858–862. [CrossRef]
5. Di Monaco, R.; Di Marzo, S.; Cavella, S.; Masi, P. Valorization of traditional foods: The case of Provolone del Monaco cheese. *Br. Food J.* **2005**, *107*, 98–110. [CrossRef]
6. Delgado, F.J.; González-Crespo, J.; Ladero, L.; Cava, R.; Ramírez, R. Free fatty acids and oxidative changes of a Spanish soft cheese (PDO 'Torta del Casaro') during ripening. *Int. J. Food Sci. Technol.* **2009**, *44*, 1721–1728. [CrossRef]
7. Romano, R.; Giordano, A.; Le Grottaglie, L.; Manzo, N.; Paduano, A.; Sacchi, R.; Santini, A. Volatile compounds in intermittent frying by gas chromatography and nuclear magnetic resonance. *Eur. J. Lipid Sci. Technol.* **2013**, *115*, 764–773. [CrossRef]
8. Thierry, A.; Collins, Y.F.; Mukdsi, M.C.A.; Mcsweeney, P.L.H.; Wilkinson, M.G.; Spinnler, H.E. Lipolysis and Metabolism of Fatty Acids in Cheese. In *Cheese, Chemistry, Physics and Microbiology*, 4th ed.; Academic Press—Elsevier: San Diego, CA, USA, 2017; ISBN 9780124170124.
9. Hassan, F.A.M.; Abd El-Gawad, M.A.M.; Enab, A.K. Flavour Compounds in Cheese (Review). *Res. Prec. Instr. Mach.* **2013**, *2*, 15–29. [CrossRef]
10. Aydin, C.M. Formation of volatile and non-volatile compounds in cheese. *Int. J. Sci. Technol. Res.* **2017**, *6*, 252–263.
11. McSweeney, P.L.; Sousa, M.J. Biochemical pathways for the production of flavour compounds in cheeses during ripening: A review. *Lait* **2000**, *80*, 293–324. [CrossRef]
12. Bradley, R.L.; Vanderwarn, A.A. Determination of moisture in cheese and cheese products. *AOAC Int.* **2001**, *84*, 570–592.
13. D.M. Official Method of Cheese Analysis. Ministry Regulation April 21 1986. Available online: https://www.gazzettaufficiale.it/do/gazzetta/serie_generale/3/pdfPaginato?dataPubblicazioneGazzetta= 19861002&numeroGazzetta=229&tipoSerie=SG&tipoSupplemento=SO&numeroSupplemento=88& progressivo=0&numPagina=4&edizione=0&elenco30giorni= (accessed on 27 March 2019).
14. Nota, G.; Spagna Musso, S.; Romano, R.; Naviglio, D.; Improta, C. Idrolisi rapida degli esteri degli steroli nei grassi. *Riv. Italiana Sostanze Grasse* **1995**, *72*, 315–316.
15. Lee, J.H.; Diono, R.; Kim, G.Y.; Min, D.B. Optimization of solid phase microextraction analysis for the headspace volatile compounds of Parmesan cheese. *J. Agric. Food Chem.* **2003**, *51*, 1136–1140. [CrossRef]

16. Fox, P.F.; Singh, T.K.; McSweeney, P.L.H. Proteolysis in cheese during ripening. *Food Rev. Int.* **2005**, *4*, 457–509.

17. McSweeney, P.L.H. Biochemistry of cheese ripening. *Int. J. Food Microbiol.* **2004**, *57*, 127–144. [CrossRef]

18. Akın, N.; Aydemir, S.; Koçak, C.; Yıldız, M.A. Changes of free fatty acid contents and sensory properties of white pickled cheese during ripening. *Food Chem.* **2003**, *80*, 77–83. [CrossRef]

19. Bintsis, T.; Papademas, P. An overview of the cheesemaking process. In *Global Cheesemaking Technology: Cheese Quality and Characteristics*, 1st ed.; Papademas, P., Bintsis, T., Eds.; John Wiley & Sons, Ltd.: New York, NY, USA, 2017; pp. 120–156.

20. Gobbetti, M.; Morea, M.; Baruzzi, F.; Corbo, M.R.; Matarante, A.; Considine, T.; Fox, P.F. Microbiological, compositional, biochemical and textural characterisation of Caciocavallo Pugliese cheese during ripening. *Int. Dairy J.* **2002**, *12*, 511–523. [CrossRef]

21. Almena-Aliste, M.; Mietton, B. Cheese classification, characterization and categorization: A global perspective. In *Cheese and Microbes*, 1st ed.; Donnelly, C.W., Ed.; ASM Press: Herndon, VA, USA, 2014; pp. 39–72.

22. Naviglio, D.; Romano, R.; Pizzolongo, F.; Santini, A.; De Vivo, A.; Schiavo, L.; Nota, G.; Spagna Musso, S. Rapid determination of esterified glycerol and glycerides in triglycerides fats and oils by means of periodate method after transesterficatior. *Food Chem.* **2007**, *102*, 399–405. [CrossRef]

23. Cornu, A.; Rabiau, N.; Kondjoyan, N.; Verdier-Metz, I.; Pradel, P.; Tournayre, P.; Berdagué, J.L.; Martin, B. Odour-active compound profiles in Cantal-type cheese: Effect of cow diet, milk pasteurization and cheese ripening. *Int. Dairy J.* **2009**, *19*, 588–594. [CrossRef]

24. Esposito, G.; Masucci, F.; Napolitano, F.; Braghieri, A.; Romano, R.; Manzo, N.; Di Francia, A. Fatty acid and sensory profiles of Caciocavallo cheese as affected by management system. *J. Dairy Sci.* **2014**, *97*, 1918–1928. [CrossRef]

25. Scerra, M.; Chies, L.; Caparra, P.; Cilione, C.; Foti, F. Effect of only pasture on fatty acid composition of cow milk and Ciminà caciocavallo cheese. *J. Food Res.* **2016**, *5*, 20–28. [CrossRef]

26. Rodríguez-Alcalá, L.M.; Castro-Gómez, M.P.; Pimentel, L.L.; Fontec, J. Milk fat components with potential anticancer activity—A review. *Biosci. Rep.* **2017**, *37*, BSR20170705. [CrossRef]

27. den Hartigh, L.J. Conjugated Linoleic Acid Effects on Cancer, Obesity, and Atherosclerosis: A Review of Pre-Clinical and Human Trials with Current Perspectives. *Nutrients* **2019**, *11*, 370. [CrossRef]

28. Zhang, J.; Yi, M.; Zha, L.; Chen, S.; Li, Z.; Li, C.; Gong, M.; Deng, H.; Chu, X.; Chen, J.; et al. Sodium butyrate induces endoplasmic reticulum stress and autophagy in colorectal cells: Implications for apoptosis. *PLoS ONE* **2016**, *11*, e0147218. [CrossRef]

29. Borycka-Kiciak, K.; Banasiewicz, T.; Rydzewska, G. Butyric acid—A well-known molecule revisited. *Gastroenterol. Rev.* **2017**, *12*, 83–89. [CrossRef]

30. Deeth, H.C. Lipoprotein lipase and lipolysis in milk. *Int. Dairy J.* **2006**, *16*, 555–562. [CrossRef]

31. Singh, T.K.; Drake, M.A.; Cadwallader, K.R. Flavor of Cheddar cheese: A chemical and sensory perspective. *Compr. Rev. Food Sci. Food Saf.* **2003**, *2*, 166–189. [CrossRef]

32. Karagul-Yuceer, Y.; Drake, M.A.; Cadwallader, K.R. Aroma-active components of liquid Cheddar whey. *J. Food Sci.* **2003**, *68*, 1215–1219. [CrossRef]

33. Karagul-Yuceer, Y.; Vlahovich, K.N.; Drake, M.A.; Cadwallader, K.R. Characteristic aroma compounds of rennet casein. *J. Agric. Food Chem.* **2003**, *51*, 6797–6801. [CrossRef]

34. Marilley, L.; Casey, M.G. Flavours of cheese products: Metabolic pathways, analytical tools and identification of producing strains. *Int. J. Food Microbiol.* **2004**, *90*, 139–159. [CrossRef]

35. Lecanu, L.; Ducruet, V.; Jouquand, C.; Gratadoux, J.J.; Feigenbaum, A. Optimization of headspace solid-phase microextraction (SPME) for the odour analysis of surface-ripened cheese. *J. Agric. Food Chem.* **2002**, *50*, 3810–3817. [CrossRef]

36. Kaminarides, S.; Stamou, P.; Massouras, T. Changes of organic acids, volatile aroma compounds and sensory characteristics of Halloumi cheese kept in brine. *Food Chem.* **2007**, *100*, 219–225. [CrossRef]

37. Varming, C.; Agerlin, M.; Thomas, P.; Ardö, S. Challenges in quantitative analysis of aroma compounds in cheeses with different fat content and maturity level. *Int. Dairy J.* **2013**, *29*, 15–20. [CrossRef]

38. Sádecká, J.; Kolek, E.; Pangallo, D.; Valík, L.; Kuchta, T. Principal volatile odorants and dynamics of their formation during the production of May Bryndza cheese. *Food Chem.* **2014**, *150*, 301–306. [CrossRef]

39. Di Marzo, S.; Di Monaco, R.; Cavella, S.; Romano, R.; Borriello, I.; Masi, P. Correlation between sensory and instrumental properties of Canestrato Pugliese slices packed in biodegradable films. *Trends Food Sci. Technol.* **2006**, *17*, 169–176. [CrossRef]

40. Romano, R.; Santini, A.; Le Grottaglie, L.; Manzo, N.; Visconti, A.; Ritieni, A. Identification markers based on fatty acid composition to differentiate between roasted Arabica and Canephora (Robusta) coffee varieties in mixtures. *J. Food Compos. Anal.* **2014**, *35*, 1–9. [CrossRef]

41. Pannico, A.; Schouten, R.E.; Basile, B.; Romano, R.; Woltering, E.J.; Cirillo, C. Non-destructive detection of flawed hazelnut kernels and lipid oxidation assessment using NIR spectroscopy. *J. Food Eng.* **2015**, *160*, 42–48. [CrossRef]

42. Durazzo, A.; Kiefer, J.; Lucarini, M.; Marconi, S.; Lisciani, S.; Camilli, E.; Gambelli, L.; Gabrielli, P.; Aguzzi, A.; Finotti, E.; et al. An innovative and integrated food research approach: Spectroscopy applications to milk and a case study of a milk-based dish. *Braz. J. Anal. Chem.* **2018**, *5*, 12–27. [CrossRef]

43. Durazzo, A.; Kiefer, J.; Lucarini, M.; Camilli, E.; Marconi, S.; Gabrielli, P.; Aguzzi, A.; Gambelli, L.; Lisciani, S.; Marletta, L. Qualitative Analysis of Traditional Italian Dishes: FTIR Approach. *Sustainability* **2018**, *10*, 4112. [CrossRef]

Communication

Qualitative Analysis of Traditional Italian Dishes: FTIR Approach

Alessandra Durazzo [1,*], Johannes Kiefer [2], Massimo Lucarini [1], Emanuela Camilli [1], Stefania Marconi [1], Paolo Gabrielli [1], Altero Aguzzi [1], Loretta Gambelli [1], Silvia Lisciani [1] and Luisa Marletta [1]

[1] CREA Research Centre for Food and Nutrition, Via Ardeatina 546, 00178 Rome, Italy; massimo.lucarini@crea.gov.it (M.L.); emanuela.camilli@crea.gov.it (E.C.); stefania.marconi@crea.gov.it (S.M.); paolo.gabrielli@crea.gov.it (P.G.); altero.aguzzi@crea.gov.it (A.A.); loretta.gambelli@crea.gov.it (L.G.); silvia.lisciani@crea.gov.it (S.L.); luisa.marletta@crea.gov.it (L.M.)

[2] Technische Thermodynamik, Universität Bremen, Badgasteiner Str. 1, 28359 Bremen, Germany; jkiefer@uni-bremen.de

* Correspondence: alessandra.durazzo@crea.gov.it; Tel.: +39-065-149-4430

Received: 29 September 2018; Accepted: 5 November 2018; Published: 9 November 2018

Abstract: Italian cuisine and its traditional recipes experience an ever-increasing popularity around the world. The "Integrated Approach" is the key to modern food research and the innovative challenge for analyzing and modeling agro-food systems in their totality. The present study aims at applying and evaluating Fourier Transformed Infrared (FTIR) spectroscopy for the analysis of complex food matrices and food preparations. Nine traditional Italian recipes, including *First courses*, *One-dish meals*, *Side courses*, and *Desserts*, were selected and experimentally prepared. Prior to their analysis via FTIR spectroscopy, the samples were homogenized and lyophilized. The IR spectroscopic characterization and the assignment of the main bands was carried out. Numerous peaks, which correspond to functional groups and modes of vibration of the individual components, were highlighted. The spectra are affected by both the preparation procedures, the cooking methods, and the cooking time. The qualitative analysis of the major functional groups can serve as a basis for a discrimination of the products and the investigation of fraud. For this purpose, the FTIR spectra were evaluated using Principal Component Analysis (PCA). Our results show how the utilization of vibrational spectroscopy combined with a well-established chemometric data analysis method represents a potentially powerful tool in research linked to the food sector and beyond. This study is a first step towards the development of new indicators of food quality.

Keywords: FTIR-ATR; traditional Italian recipes; chemometrics; PCA

1. Introduction

Currently, food research is focusing not only on the chemical characteristics and functions of individual foods and/or food components, but is also aimed at understanding their combination in dishes, meals, and diets for a more complete and real assessment of their nutritional impact. In fact, studies on the interactions between individual food components and/or between the different ingredients of cooked and composite dishes seem to play a greater role in the concept of "food synergy" and in reconsidering the total food matrix as a variable of health interest [1].

Italian cuisine and its traditional recipes represent a great legacy of tradition and culture for the country becoming more and more popular all over the world for its great quality, abundance, and variety of food preparations. The Italian culinary traditions are typically Mediterranean, based on cereals, vegetables, fruit, fish, and, in particular, olive oil. Strongly influenced by gastronomic traditions, history, local availability, and seasonality, the typical Italian foods and traditional recipes once were closely linked to the production area and were mostly consumed locally [2]. Today, they represent a real heritage and are part of the national food distribution network contributing to the economic and cultural improvement of the entire country. In addition, traditional knowledge passed on from generation to generation and held by local communities, including the ancient practices of the preparation and conservation of the foods and sustainable and natural resources, often played an integral role in achieving sustainable development goals [3].

Despite this, the traditional dishes remain an integral part of a nation's culture that is worth investigation. This needs to consider that the current Italian eating habits are part of the Mediterranean diet and its variants, although they are influenced by the globalization of the market and the changes in lifestyle. Nevertheless, modern Italian cuisine seems to fit well with the Italian culinary traditions [4].

A significant fraction of the dietary intake includes mixed dishes usually based on multiple ingredients/foods. They can be prepared at home or at a restaurant, or they can represent industrially processed foods. To preserve the cultural elements by providing the related dietary information, many researchers have paid special attention to the study of the nutritional characteristics of traditional foods and recipes over the last decade [5–10]; traditional foods are in fact key elements in the dietary patterns of a country. Ramdath et al. [7] have studied the nutritional composition of commonly consumed dishes in Trinidad to determine the associations between diet and the risk of chronic diseases. The aim was to improve the dietary quality among the population. Costa et al. [9] and Lombardi Boccia e Marletta [5] emphasized the importance of knowing the nutritional composition. Their aim was to contribute to the promotion of the local biodiversity and sustainable diets by maintaining healthy dietary patterns within local cultures. They used several traditional foods from the Black Sea area and from Italy as a case study. However, detailed characteristics of traditional full dishes are usually not included in the national food composition databases, which are commonly used to make dietary recommendations. Therefore, it is difficult to make a meaningful comparison between what we eat and the recommendations made. Costa et al. [8] and Durazzo et al. [10] made attempts to include such a holistic approach to food and food analysis.

There have been large research projects in this area, see, for example, the EuroFIR project (www.eurofir.net). For instance, Guerrero et al. [11] have identified the main different aspects that European consumers seem to perceive when thinking about traditional food products: habits and nature, origin and locality, processing and elaboration, and sensory properties. The authors concluded that when innovations are applied to traditional food products, their degree of acceptance is strongly dependent on the product and on the type of innovation [11].

The "Integrated Approach" is the key to modern food research and the innovative challenge for analyzing and modelling agro-food systems in their totality [12,13]. As is generally true in the food sector, the recent developments and advancements in instrumentation as well as chemometric pattern recognition techniques, including principal component analysis (PCA), hierarchical cluster analysis, and partial least squares regression (PLSR), have amplified the range of the IR spectroscopy applications to food products. Thus, the IR spectra can be used to identify or differentiate between samples [14]. The list of applications includes evaluating and determining components [15,16], monitoring contaminants and adulterants [17,18], classification [19,20], discrimination [21,22], authentication [23,24], etc. In this context, the review of Rodriguez-Saona and Allendorf [25] is worth mentioning. It gives an overview of the main applications of near infrared (NIR) and mid-infrared (MIR) spectroscopy for the rapid authentication and detection of adulteration of food.

FTIR spectroscopy coupled with multivariate statistical data analysis has an enormous potential to facilitate the measurement of quality attributes. Generally, the application of statistical methods in food science allows the highlighting of effective trends, to study and investigate relationships, and to draw conclusions from experimental data [26]. Chemometrics is often referred to as the discipline that uses mathematical and statistical methods to obtain relevant information from collected data [27]. From the chemometric analysis of FTIR spectra, a large amount of information can be derived [28].

FTIR spectroscopy represents a rapid, less destructive, and high-throughput method for the analysis of food products. It provides simplified handling and it enables the samples being examined directly in their original state. The FTIR method has successfully been used for food composition determination, food distinction/differentiation, and food authentication in several food matrices [29,30]. Karoui et al. [31] described the application of IR spectroscopy and chemometric data analysis to different food groups: dairy products, meat and meat products, fish, edible oils, cereals and cereal products, sugar and honey, fruit and vegetable, and coffee. Their focus was on the analysis of intact food systems and the exploration of their molecular structure-quality relationships. Franca and Olivera [32] gave a critical overview of the application of FTIR in food processing and engineering. However, only a few studies have used FTIR spectroscopy to analyze composite dishes [31,32]. To make an important step beyond the published work, the present study aims at applying and evaluating FTIR spectroscopy for the analysis of some Italian complex food matrices and food preparations. The objectives include (1) qualitative analysis of the major functional groups and (2) discrimination analysis by applying PCA to the FTIR data set in order to test whether or not a classification is possible. In previous studies [10,33], the focus was on traditional analytical methods to determine the proximate composition and the dietary intake evaluation. In contrast, the present work aims at applying an integrated analytical approach for classification of these dishes as reported for the case study of the Béchamel sauce [13].

2. Materials and Methods

2.1. Selection of Recipes and Experimental Preparations

Nine traditional Italian recipes, including *First courses*, *One-dish meals*, *Side courses*, and *Desserts*, as shown in Table 1, were selected and experimentally prepared [10,33] in a dedicated lab-kitchen following a validated and standardized protocol developed within the EuroFIR Network [34].

Table 1. Ingredients, preparation procedures, cooking methods, and cooking times of some Italian traditional dishes.

Original Name	Food Name	Ingredients (g/100 g)	Preparation Procedures and Cooking Methods	Timing (min.)
		First Courses		
Spaghetti alle vongole [10,33]	*Spaghetti with clams*	Spaghetti (48.8), clams without shell (44.8), extra virgin olive oil (3.3), garlic (1.4), salt (0.4), parsley chopped (1.2), chili pepper (0.1).	Boil the spaghetti in salted water. Fry the garlic and pepper in olive oil in a pan, then add the clams and leave on the heat until the valves open. Add cooked spaghetti and parsley.	16
Pasta alla amatriciana [10]	*Amatriciana pasta*	Pasta (tortiglioni) (37.5), diced tomatoes (37.5), Amatrice cheek lard diced (16), Roman Pecorino cheese PDO (hard cheese from sheep milk) (7.5), extra virgin olive oil (1.1), salt (0.3), chili pepper (0.1).	Cut the cheek lard into strips and brown it in a pan with olive oil and chili pepper. Add the tomato and salt, simmer for about 20 min. Meanwhile, boil the pasta in salted water, then drain it and toss with the sauce, and, lastly, add the Pecorino cheese.	25
Pasta alla carbonara [10]	*Carbonara Pasta*	Short pasta (47.3), bacon cubes (20.3), Roman Pecorino cheese PDO (hard cheese from sheep milk) (13.6), eggs (16.6), extra virgin olive oil (1.4), salt (0.4), black pepper (0.4).	Mix eggs, Pecorino cheese, salt, and pepper to become a creamy mixture. Meanwhile, brown the bacon in a pan. Boil the pasta in salted water, drain and add it to the bacon, and then remove the pan from the heat. Finally, pour the egg mixture onto the pasta and mix until the cream thickens slightly.	13

Table 1. *Cont.*

Original Name	Food Name	Ingredients (g/100 g)	Preparation Procedures and Cooking Methods	Timing (min.)
		One-dish meals		
Parmigiana di melanzane [33]	*Eggplant parmesan*	Eggplants (49.8), vegetable oil (for frying), extra virgin olive oil (0.5), diced tomatoes (35.7), two garlic cloves, mozzarella cheese (uncured cheese) (9.6), Parmigiano Reggiano cheese PDO (hard cheese) (4.2), leaves of basil (0.1), salt (0.1).	Cut the eggplants into 1 cm thick slices. Fry the slices for about 5 min in boiling oil, remove and drain on kitchen paper. Fry the garlic in pan with olive oil, then add the tomato. Simmer the sauce over medium heat. Cut the mozzarella into small cubes. Sprinkle the bottom of a baking pan with the tomato sauce. Make layers of fried eggplants, tomato sauce, Parmigiano Reggiano cheese, and mozzarella. Continue until the ingredients are exhausted, and finish with the sauce and Parmigiano Reggiano cheese. Bake in a preheated static oven at 180 °C.	25
Gâteau di patate [10,33]	*Potato flan*	Baked potatoes crushed (61.5), ham (7.9), mozzarella cheese (uncured cheese) (6.7), scamorza cheese (semisoft cheese) (6.6), Parmigiano Reggiano cheese PDO (hard cheese) (5.3), butter (1.3), milk (2.3), pepper (0.01), nutmeg (0.01), eggs (6.9), salt (0.1), bread crumbs (1.1).	Finely crush the boiled potatoes. Mix the potatoes in a large bowl with butter, milk, and Parmigiano Reggiano cheese, and cool the mixture. Add the eggs, the mozzarella, and the diced scamorza, the ham cooked in strips, the salt, the pepper, and the nutmeg. Pour the mixture in the baking dish and sprinkle breadcrumbs on the well-flat surface. Bake in a traditional oven at 160 °C. Lightly grate the surface.	20
		Side courses		
Carciofi alla romana [10,33]	*Roman artichokes*	"Roman" artichokes PGI (74.4), chopped mint (0.1), chopped parsley (2), finely chopped garlic (0.6), extra virgin olive oil (6.1), salt (0.4), water (42.5), pepper (0.05).	Stuff the artichokes with mint, parsley, garlic, pepper, and salt, and place them in a pan. Add water and olive oil, and stew over medium heat.	20
		Desserts		
Pan di Spagna [10,33]	*Sponge cake*	Eggs (49.6), sugar (27.1), flour (22.6), salt (0.04).	Beat the egg yolks with the sugar in a bowl until pale and fluffy. Whisk the egg whites to stiff peaks in another bowl then gently pour into the egg yolks mixture. Add a little at a time to the sifted flour. Pour the mixture into the mold and bake in a traditional oven at 180 °C.	20
Torta di mele [33]	*Apple pie*	Apples cultivar "Renetta" (37.7), sugar (15), wheat flour (18), butter (9.3), eggs (9.2), whole milk (7.5), baking powder (1), vanilla (0.03), grated lemon peel (0.2), lemon juice (2.4).	Whip the butter in a bowl with the granulated sugar until you get a frothy mixture, add the eggs, one at a time, the milk, and the grated lemon peel. Add the sifted flour and baking powder, stirring with a spatula, and add a pinch of salt. Peel the apples, cut them into slices that are not too large. Pour the mixture into a properly greased and floured mold, leveling well. Carefully place the apple slices on top of the dough. Bake in a preheated oven at 180 °C.	30
Pastiera [33]	*Pastiera cake*	Shortcrust pastry (31.5), ricotta cheese (soft cheese from sheep milk) (22.3), sugar (16.7), wheat cooked canned (13.4), whole milk (3.4), eggs (10.1), butter (1), citron candied (1.2), orange flower water (0.3) vanilla (0.02).	Melt the butter in a pan with the grated lemon and orange peels. Add the cooked wheat and milk, simmer until it becomes a cream. Allow to cool. In a bowl, mix the ricotta with the sugar and the egg yolks, add to the mixture the cream of cooked wheat, the candied citron, orange flower water, and vanillin, and finally the beaten egg whites. Roll out the pastry and place it in a buttered cake tin. Pour the mixture and decorate with crossed strips of shortcrust pastry. Bake in a preheated oven at 180 °C until it becomes a nice amber color.	40

In detail, for every selected recipe, a document collection was carried out from the most popular and traditional cookbooks in Italy (Il cucchiaio d'argento; La cucina italiana, etc). A "standard recipe" was identified and one "preparation protocol" was elaborated to just establish ingredients, amounts, preparation, and cooking techniques (time, temperature, utensils, etc.). The sampling plan considered the collection of single ingredients in different retail stores and supermarkets. Ingredients were purchased by collecting the main food brands and/or varieties of the same product. Each individual food brand (primary sample) for every ingredient (secondary sample) was properly weighed and then combined to make an aggregate sample (pool) before using it for the preparation of the final dish (laboratory sample). The dish was assembled and cooked by trained personnel according to the "preparation protocol" of the "standard recipe". For this purpose, common household methods and utensils were used. Two independent batches (about 2 kg for each composite dish) were produced in different periods. After cooking, the prepared dishes were weighed once more. Then, they were homogenized, frozen at −30 °C, and then lyophilized for the FTIR analysis. The lyophilization guarantees the homogenization and uniformity of the samples. In addition, this method allowed an optimal storage to protect the sample from oxidation, contamination, or other potential effects [34,35].

2.2. FTIR-ATR Analysis

The FTIR spectra were recorded on a Nicolet iS10 FT-IR spectrometer equipped with a diamond crystal cell for attenuated total reflection (ATR) operation. The spectra were acquired (32 scans per sample or background) in the range of 4000–650 cm^{-1} at a nominal resolution of 4 cm^{-1}. The spectra were corrected using the background spectrum of air. The analysis was carried out at room temperature. For a measurement, a lyophilized sample was placed onto the surface of the ATR crystal. Before acquiring a spectrum, the ATR crystal was carefully cleaned with wet cellulose tissue and dried using a flow of nitrogen gas. The cleaned crystal was checked spectrally to ensure that no residue was retained from the previous sample. The spectrum of every sample was collected 5 times to check the reproducibility and do a statistical analysis. It should be noted that the individual spectra of the same dish varied to some extent, up to variations in the absorbance values of 30%. However, the characteristic signatures remained very similar so that averaged spectra are shown in the following.

2.2.1. Qualitative Analysis of the Spectra

As a first step, the FTIR spectra were analyzed with respect to the spectral band positions to identify the signatures of the major functional groups. An assignment of the main bands was carried out by analyzing the acquired spectra and by comparing them with those in the literature.

2.2.2. Multivariate Analysis of the Spectra

The application of PCA to the FTIR spectra was carried out in the second step. The evaluation algorithms were implemented in Matlab using the *princomp* routine. PCA is a purely mathematical and unsupervised procedure, which is aimed at the most meaningful representation of a data set. For this purpose, the dimensionality of the data matrix is reduced and the relevant information is extracted. In a first step, the covariance matrix of the initial data matrix is calculated. Then, the eigenvalues and eigenvectors of this covariance matrix are determined. Thirdly, the eigenvectors are sorted with respect to their variance. The eigenvector with the largest eigenvalue shows the most variance in the data set and is thus called the first principal component (PC1), and so on. In this procedure, the signal variance is a measure for the relevance of the information stored in the data set. The implemented diagonalization of the covariance matrix results in a decomposition of the original data matrix into the loadings and the scores. The scores represent the initial data in the new dimensional space spanned by the principal components. The loadings are the eigenvectors of the matrix. In other words, they represent the contribution of the original data to creating the principal components. Samples displayed in the new space show closer positions when they have a strong relationship. The resulting scores plot can be used for a classification of the data set.

In practice, the purpose of unsupervised methods is to find groups of samples, which have related features, to allow their separation into different classes. PCA is the most widespread unsupervised tool in the field of food [36].

3. Results and Discussion

3.1. Qualitative Analysis

FTIR spectroscopy can be defined as a "fingerprint analytical technique" for the structural identification of compounds considering that no two chemical structures will have the same FTIR spectrum [37]. FTIR provides a characteristic signature of chemical or biochemical substances present in the sample by featuring their molecular vibrations (stretching, bending, and torsions of the chemical bonds) in specific infrared regions.

Figures 1–4 show the FTIR spectra of the experimental preparations, grouped as *First Course* (*Spaghetti alle vongole, Pasta alla amatriciana, Pasta alla carbonara*), *One-dish meals* (*Parmigiana di melenzane, Gâteau di patate*), *Side courses* (*Carciofi alla romana*), and *Desserts* (*Pan di Spagna, Torta di mele, Pastiera*).

The spectra contain a multitude of bands that are more or less characteristic of food samples. The spectra are affected by the preparation procedures, the cooking methods, and the cooking time. In the spectra displayed, the specific bands characteristic of the dishes are highlighted and assigned as follows.

The FTIR spectra show bands associated with the main classes of organic molecules. The band at 3289 cm^{-1} is characteristic of NH and OH stretching vibrations. The region of 2923–2853 cm^{-1} can be assigned to the symmetrical and asymmetric stretching modes of the CH$_3$ and CH$_2$ groups. The 1745 cm^{-1} band corresponds to the stretching of the C=O ester carbonyl or carboxylic acid groups, which are characteristic of fatty acids. The amide I band at 1645 cm^{-1} results from the C=O stretching in the amides I, II, and III, while the amidic band II at 1545 cm^{-1} originates from the bending vibrations of the N-H groups. The IR region from 1500 to 800 cm^{-1} is commonly referred to as the "fingerprint" region, which includes bands corresponding to the vibrations of the C-O, C-C, C-H, and C-N bonds [38]. This region is, on the one hand, very rich in information, but, on the other hand, difficult to analyze due to its complexity. This area provides important information about organic compounds, such as sugars, alcohols, and organic acids, present in the sample.

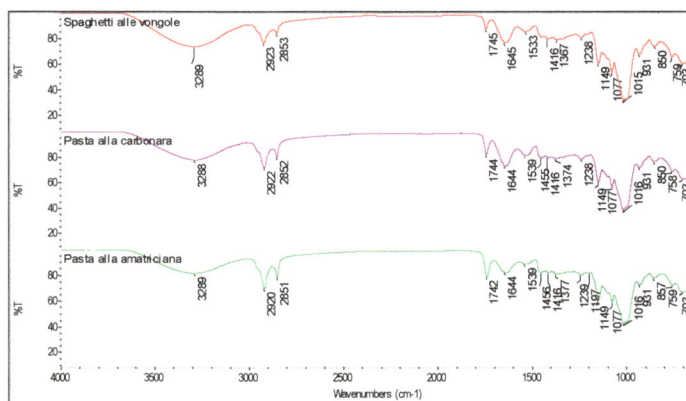

Figure 1. Averaged FTIR spectra of experimental *First courses* dishes (*Spaghetti alle vongole, Pasta alla amatriciana, Pasta alla carbonara*) in the mid-infrared region (4000–650 cm^{-1}).

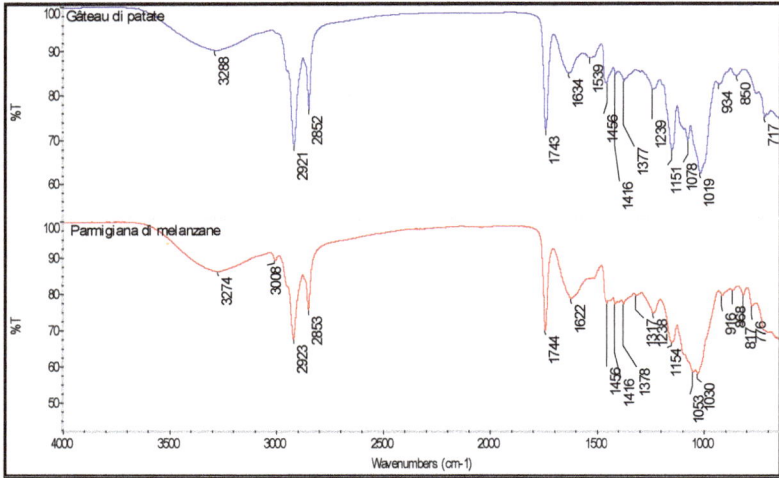

Figure 2. Averaged FTIR spectra of *One dish meals* (*Parmigiana di melanzane, Gâteau di patate*) in the mid-infrared region (4000–650 cm^{-1}).

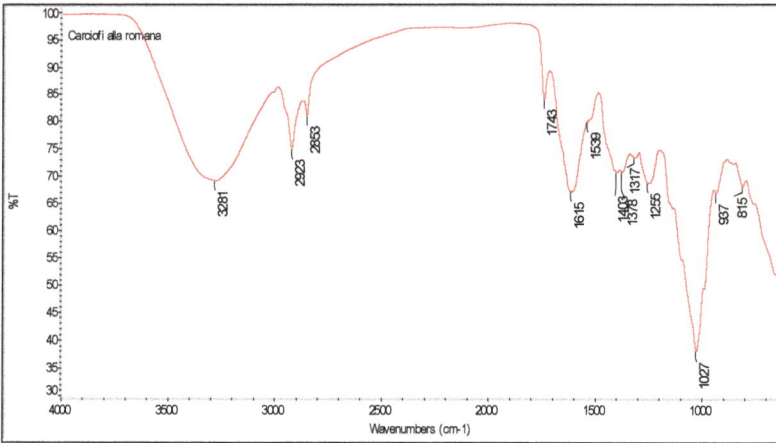

Figure 3. Averaged FTIR spectra of *Side courses* (*Carciofi alla romana*) in the mid-infrared region (4000–650 cm^{-1}).

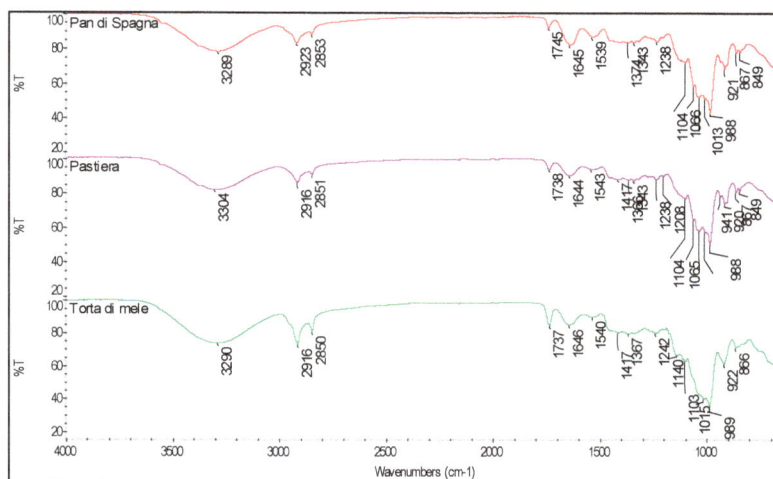

Figure 4. Averaged FTIR spectra of *Desserts* (*Pan di Spagna*, *Torta di mele*, *Pastiera*) in the mid-infrared region (4000–650 cm^{-1}).

Concerning the *First courses*, the peaks observed (Figure 1) at 1015, 1077, and 1149 cm^{-1} in all three dishes are related to the C-O-C stretching and CO(-COH) stretching of starch [39]. Li et al. [39] describes how these three peaks were observed to have increased significantly in gluten proteins of cooked noodles. This reflects the presence of residual starch in the samples, and it is potentially induced by the increased connectivity between proteins and starch during cooking. The peak found at 850 cm^{-1} is assigned to the aromatic C-H out-of-plane deformation, and the peak at 759 cm^{-1} probably corresponds to the S-N stretching mode [40]. Basically, FTIR spectroscopy can also be used to observe changes in the secondary structure of gluten proteins [41,42]. For all *First Courses* samples, there are peaks at 1645 and 1539 cm^{-1}, which are likely assigned as the amide bands I and II, respectively. Interestingly, Li et al. [42] described how the absorption intensity of the peak at 1655 cm^{-1} increased during the mixing, resting, and sheeting processes of noodles, whereas it decreased again during cooking. Another typology of FTIR application on pasta, and widely described for cereal-based products [43], is given by Kamil et al. [44], who used FTIR spectroscopy as a tool for detecting the adulteration of pasta.

As reported in Figure 2, *one-dish meals* were represented by *Gâteau di patate* and *Parmigiana di melenzane*. Concerning potato-based food products, other authors have applied FTIR spectroscopy to potato chips [45–47] and mashed potatoes [48]. In addition, studies of the surface of potato tubers [49] and a potato peel extract [50] or its starch-component [51,52] have been reported. Sivakesava and Irudayaraj, [45] applied Fourier transform infrared photoacoustic spectroscopy (FTIR-PAS) for analyzing potato chips. They monitored the changes in chemical groups related to the fat and oil treatment of home-made chips during heating at 80 °C. The authors showed that when oxidation proceeded, the band near 3005 cm^{-1} decreased at a rate that was dependent on the nature of the oil used to fry the potato chips. This band was assigned CH groups at a cis double bond. It disappeared completely with the progressing oxidation in the samples fried for 30 s, suggesting isomerisation of the oils in the potato chips. Under oxidative conditions, a peculiar behavior was observed. There was a decrease of the band of the ester carbonyl functional group of the triglycerides (1749 cm^{-1}). In addition, a decrease of the bands from the bending vibrations of the CH$_2$ and CH$_3$ aliphatic groups (1464 and 1375 cm^{-1}) was observed. Against the intuition, there was no change of the bands at 1239 and 1160 cm^{-1} during the oxidation process. The band at 1099 cm^{-1}, on the other hand, decreased in the presence of oleic acyl groups, while the overlapping of the methylene rocking vibration and the out-of plane bending vibration of the cis-disubstituted olefins observed at 721 cm^{-1} did not change after heating. Concerning the main ingredient of *Parmigiana di melanzane*, some authors have previously investigated the FTIR application on eggplant peel extract [53].

The only representative of *Side dishes* is the *Carciofi alla romana* dish. Concerning artichoke, the main ingredient of *Carciofi alla romana*, previous studies were focused on artichoke fiber [54], inulin [55], and pectin [56]. Fiore et al. [54] have described the characteristic bands of artichoke fiber, extracted from the stem of a plant. Recently, Ceylan et al. [56] have examined structural features of isolated pectin from different sections of globe artichoke and the industrial waste using FTIR spectroscopy, monitoring the process-dependent structural changes in the molecular structure of pectin molecules. In a recent work of Wang et al. [57], the authors monitored the effects of the heating rate on the fast pyrolysis behavior and on the product distribution of Jerusalem artichoke stalk by using TG-FTIR and Py-GC/MS.

As shown in Figure 3, the peaks observed at 1615, 1539, 1403, 1027, and 815 cm^{-1} are attributed to the presence of chlorogenic acid and caffeoylquinic acid. Other research groups [58–60] underlined how spectral features in the wavenumber region of 1700−1600 cm^{-1} and 1300−800 cm^{-1} have an important role in further characterizing chlorogenic acid isomer composition. In particular, Liang et al. [60] assigned the spectral features of pure chlorogenic acid isomer standards as follows: The peak at 809 cm^{-1} was attributed to cyclohexane C-O twisting; the peaks at 1120 and 1165 cm^{-1} are related to cyclohexane CH, C-OH bending, and the phenyl ring bending vibration, respectively; the peak at 1276 cm^{-1} is due to phenyl CH rocking vibrations; the peak at 1605 cm^{-1} was assigned to the phenyl ring stretching; and the peak at 1627 cm^{-1} was attributed to C=C ethylenic stretching.

In the spectra of the desserts (Figure 4), the region of 1500–900 cm^{-1} contains the signals of the main monosaccharides, such as fructose and glucose, and disaccharides, such as sucrose. These peaks exhibit high intensities as can be expected. In particular, the region of 900–750 cm^{-1} corresponds to an abnormal saccharide configuration [61]. Several authors studied the peaks characterizing the sugar fraction [62–64]. In detail, the peak at 921 cm^{-1} corresponds to the typical bending of C-H of carbohydrates, and the peaks at 1038 cm^{-1} and 1238 cm^{-1} correspond to the stretching of C-O in the C-OH group and to the C-C stretching of the carbohydrate structure, respectively. Moreover, the peak at 1104 cm^{-1} corresponds to the C-O stretching of the C-O-C bond and the peak at 1343 cm^{-1} is due to the O-H bending of the C-OH group. The peak at 1417 cm^{-1} is a combination of the O-H bending of the C-OH group and the C-H bending of the alkenes.

An example of a previous application of FTIR to desserts is given by Syahariza et al. [31], who detected lard adulteration in cake formulations.

3.2. Multivariate Data Analysis

In the second part of this study, the FTIR spectra were analyzed using PCA. This approach is commonly utilized for a classification of the data set. It must be admitted that the different dishes investigated in this study can also be discriminated by the naked eye before they are homogenized and lyophilized. However, as an important step towards using FTIR and PCA for the detection of potential fraud, e.g., the use of low quality ingredients in expensive products, we tested if the method is capable of distinguishing between the different dishes. In future work, the discrimination of samples from one type of dish prepared in different ways will be the aim.

In the first step, the full spectra were analyzed. The resulting score plot (not shown) revealed that most of the dish types grouped together. However, the spectra of the *Spaghetti alle vongole* appeared among the data point cloud of the desserts, and hence an unambiguous classification was not possible. Therefore, we applied the PCA to selected wavenumber ranges in the second step. Considering the CH/OH stretching range (4000–2700 cm^{-1}), the resulting score plot revealed a similar behavior as the full spectra. In contrast, feeding the fingerprint region (2000–650 cm^{-1}) into the PCA algorithm leads to a different situation. Figure 5 displays the resulting score plot, in which a clear classification can be observed. Models developed from FTIR spectra using the fingerprint region resulted in models with superior discriminative performance. In our analysis, only the first three Principal Components (PCs) were considered as they cover most of the variance of the data set and their loading plots exhibit a reasonable signal-to-noise ratio, which indicated that they are meaningful. It should be noted that the wavenumber ranges were not optimized in any way to avoid the need for a priori knowledge in the data evaluation.

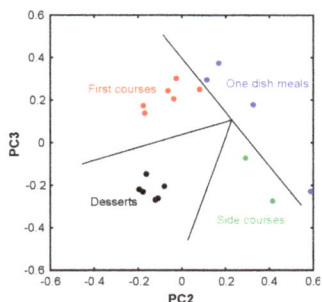

Figure 5. PCA applied to the data set of FTIR spectra feeding the fingerprint region (2000–650 cm^{-1}).

4. Conclusions

This study provided spectroscopic fingerprint signatures of the most representative functional groups of some traditional Italian recipes. Numerous peaks, which correspond to functional groups and vibrational modes of the individual ingredients, were identified. Moreover, effects of the preparation procedures, cooking methods, and cooking time were observed. In detail, this study reported an FTIR approach for the discrimination of nine different traditional Italian dishes, which can be clearly grouped into four subtypes using principal component analysis (PCA). According to their characteristic FTIR signatures, an unambiguous classification of the food groups was possible, enabled by their distinctive ingredients. This underlines that FTIR spectroscopy in the fingerprint region combined with chemometric data analysis in terms of PCA represents a powerful tool for the reliable discrimination between composite dishes.

This is a very promising outcome as it provides further support for our hypothesis that the combination of FTIR and multivariate data analysis is a suitable means of product discrimination, a tool for continuous product quality monitoring in the food processing industry, and a method for the detection and investigation of fraud. Since the rather simple PCA algorithm already shows a satisfactory performance, other, more complex approaches, such as artificial neural networks (ANN), will likely perform even better.

The mid-term aim of our research is to utilize spectroscopic data sets and the information extracted from them by multivariate methods as new indicators of food quality. This approach will certainly not be limited to food as it is easily transferable to other sectors.

Author Contributions: A.D., J.K., M.L. and L.M. have conceived the work and wrote the manuscript. All authors have carried out the experimental study and analyzed the data. All authors have made a substantial contribution to revise the work, and approved it for publication.

Funding: This work was supported by the project QUALIFU-SIAGRO (MiPAAF D.M. 2087/7303/09, 28/01/2009).

Conflicts of Interest: The authors declare no conflict of interest.

References

1. Jacobs, D.R., Jr.; Tapsell, L.C. Food, not nutrients, is the fundamental unit in nutrition. *Nutr. Rev.* **2007**, *65*, 439–450. [CrossRef] [PubMed]
2. Capatti, A.; Montanari, M. *La Cucina Italiana. Storia di Una Cultura*, 7th ed.; Editori Laterza: Bari, Italy, 2006.

3. Cordonier Segger, M.C.; Phillips, F. Indigenous Traditional Knowledge for Sustainable Development: The Biodiversity Convention and Plant Treaty Regimes. *J. For. Res.* **2015**, *20*, 430–437. [CrossRef]
4. Camilli, E.; Lisciani, S.; Marconi, S.; Gabrielli, P.; Durazzo, A.; Gambelli, L.; Aguzzi, A.; Lucarini, M.; Turrini, A.; Zanoni, B.; et al. Some traditional Mediterranean dishes in the Italian diet. In Proceedings of the 1st World Conference on the Mediterranean diet: Revitalizing the Mediterranean Diet, Milan, Italy, 6–8 July 2016.
5. Lombardi-Boccia, G.; Marletta, L. *La Carne Bovina nella Gastronomia Italiana. Valore Nutritivo di Alcune Ricette Tipiche*; Istituto Nazionale di Ricerca per gli Alimenti e la Nutrizione: Rome, Italy, 2008.
6. Vasilopoulou, E.; Trichopoulou, A. The micronutrient content of traditional Greek foods. *Med. J. Nutr. Metab.* **2009**, *2*, 97–102. [CrossRef]
7. Ramdath, D.D.; Hilaire, D.G.; Brambilla, A.; Sharma, S. Nutritional composition of commonly consumed composite dishes in Trinidad. *Int. J. Food Sci. Nutr.* **2011**, *62*, 34–46. [CrossRef] [PubMed]
8. Costa, H.S.; Vasilopoulou, E.; Trichopoulou, A.; Finglas, P. New nutritional data on traditional foods for European food composition databases. *Eur. J. Clin. Nutr.* **2010**, *64*, 73–81. [CrossRef] [PubMed]
9. Costa, H.S.; Albuquerque, T.G.; Sanches-Silva, A.; Vasilopoulou, E.; Trichopoulou, A.; D'Antuono, L.F.; Alexieva, I.; Boyko, N.; Costea, C.; Fedosova, K.; et al. New nutritional composition data on selected traditional foods consumed in Black Sea Area countries. *J. Sci. Food Agric.* **2013**, *93*, 3524–3534. [CrossRef] [PubMed]
10. Durazzo, A.; Lisciani, S.; Camilli, E.; Gabrielli, P.; Marconi, S.; Gambelli, L.; Aguzzi, A.; Lucarini, M.; Maiani, G.; Casale, G.; et al. Nutritional composition and antioxidant properties of traditional Italian dishes. *Food Chem.* **2017**, *218*, 70–77. [CrossRef] [PubMed]
11. Guerrero, L.; Guàrdia, M.D.; Xicola, J.; Verbeke, W.; Vanhonacker, F.; Zakowska-Biemans, S.; Sajdakowska, M.; Sulmont-Rossé, C.; Issanchou, S.; Contel, M.; et al. Consumer-driven definition of traditional food products and innovation in traditional foods. A qualitative cross-cultural study. *Appetite* **2009**, *52*, 345–354. [CrossRef] [PubMed]
12. Romano, R.; Santini, A.; Le Grottaglie, L.; Manzo, N.; Visconti, A.; Ritieni, A. Identification markers based on fatty acid composition to differenziate between roasted *Arabica* and *Canephora* (*Robusta*) coffee varieties in mixtures. *J. Food Compos Anal.* **2014**, *35*, 1–9. [CrossRef]
13. Durazzo, A.; Kiefer, J.; Lucarini, M.; Marconi, S.; Lisciani, S.; Camilli, E.; Gambelli, L.; Gabrielli, P.; Aguzzi, A.; Finotti, E.; et al. An innovative and integrated food research approach: Spectroscopy applications to milk and a case study of a milk-based dish. *Braz. J. Anal. Chem.* **2018**, *5*, 12–27. [CrossRef]
14. Diem, M. *Introduction to Modern Vibrational Spectroscopy*; Wiley: New York, NY, USA, 1993.
15. Lucarini, M.; Durazzo, A.; Sánchez Del Pulgar, J.; Gabrielli, P.; Lombardi-Boccia, G. Determination of fatty acid content in meat and meat products: The FTIR-ATR approach. *Food Chem.* **2018**, *267*, 223–230. [CrossRef] [PubMed]
16. Nesakumar, N.; Baskar, C.; Kesavan, S.; Rayappan, J.B.B.; Alwarappan, S. Analysis of Moisture Content in Beetroot using Fourier Transform Infrared Spectroscopy and by Principal Component Analysis. *Sci. Rep.* **2018**, *8*, 7996. [CrossRef] [PubMed]
17. De Girolamo, A.; Cervellieri, S.; Cortese, M.; Porricelli, A.C.R.; Pascale, M.; Longobardi, F.; von Holst, C.; Ciaccheri, L.; Lippolis, V. Fourier transform near-infrared and mid-infrared spectroscopy as efficient tools for rapid screening of deoxynivalenol contamination in wheat bran. *J. Sci. Food Agric.* **2018**. [CrossRef] [PubMed]
18. Li, Z.; Suslick, K.S. Portable optoelectronic nose for monitoring meat freshness. *ACS Sens.* **2016**, *1*, 1330–1335. [CrossRef]
19. Christou, C.; Agapiou, A.; Kokkinofta, R. Use of FTIR spectroscopy and chemometrics for the classification of carobs origin. *J. Adv. Res.* **2017**, *10*, 1–8. [CrossRef] [PubMed]
20. Manfredi, M.; Robotti, E.; Quasso, F.; Mazzucco, E.; Calabrese, G.; Marengo, E. Fast classification of hazelnut cultivars through portable infrared spectroscopy and chemometrics. *Spectrochim. Acta A Mol. Biomol. Spectrosc.* **2018**, *189*, 427–435. [CrossRef] [PubMed]
21. Cai, J.-X.; Wang, Y.F.; Xi, X.-G.; Li, H.; Wei, X.L. Using FTIR spectra and pattern recognition for discrimination of tea varieties. *Int. J. Biol. Macromol.* **2015**, *78*, 439–446. [CrossRef] [PubMed]

22. Lee, B.-J.; Zhou, Y.; Lee, J.S.; Shin, B.K.; Seo, J.-A.; Lee, D.; Kim, Y.S.; Choi, H.K. Discrimination and prediction of the origin of Chinese and Korean soybeans using Fourier transform infrared spectrometry (FT-IR) with multivariate statistical analysis. *PLoS ONE* **2018**, *13*, e0196315. [CrossRef] [PubMed]

23. Alamprese, C.; Casiraghi, E. Application of FT-NIR and FT-IR spectroscopy to fish fillet authentication. *LWT Food Sci. Technol.* **2015**, *63*, 720–725. [CrossRef]

24. Jiménez-Sotelo, P.; Hernández-Martínez, M.; Osorio-Revilla, G.; Meza-Márquez, O.G.; García-Ochoa, F.; Gallardo-Velázquez, T. Use of ATR-FTIR spectroscopy coupled with chemometrics for the authentication of avocado oil in ternary mixtures with sunflower and soybean oils. *Food Addit. Contam. Part A Chem. Anal. Control Expos. Risk Assess.* **2016**, *33*, 1105–1115. [CrossRef] [PubMed]

25. Rodriguez-Saona, L.E.; Allendorf, M.E. Use of FTIR for Rapid Authentication and Detection of Adulteration of Food. *Annu. Rev. Food Sci. Technol.* **2011**, *2*, 467–483. [CrossRef] [PubMed]

26. Granato, D.; de Araújo, V.M.; Jarvis, B. Observations on the use of statistical methods in Food Science and Technology. *Food Res. Int.* **2014**, *55*, 137–149. [CrossRef]

27. Otto, M. *Chemometrics: Statistics and Computer Application in Analytical Chemistry*, 3rd ed.; Wiley: New York, NY, USA, 2016; ISBN 978-3-527-34097-2.

28. Kumar, N.; Bansal, A.; Sarma, G.S.; Rawal, R.K. Chemometrics tools used in analytical chemistry: An overview. *Talanta* **2014**, *123*, 186–199. [CrossRef] [PubMed]

29. Karoui, R.; Downey, G.; Blecker, C. Midinfrared spectroscopy coupled with chemometrics: A tool for the analysis of intact food systems and the exploration of their molecular structure-quality relationships—A review. *Chem. Rev.* **2010**, *110*, 6144–6168. [CrossRef] [PubMed]

30. Franca, A.S.; Oliveira, L.S. Potential uses of fourier transform infrared spectroscopy (FTIR) in food processing and engineering. In *Food Engineering*; Siegler, B.C., Ed.; Nova Science Publishers Inc.: Hauppauge, NY, USA, 2011; ISBN 978-1-61728-913-2.

31. Syahariza, Z.A.; Che Man, Y.B.; Selamat, J.; Bakar, J. Detection of lard adulteration in cake formulation by Fourier transform infrared (FTIR) spectroscopy. *Food Chem.* **2005**, *92*, 365–371. [CrossRef]

32. Shaikh, T.H.; Mahesar, S.A.; Shah, S.N.; Kori, A.H.; Sherazi, S.T.H.; Lakho, S.A. FTIR spectroscopy combined with chemometric: A versatile tool for quality evaluation of fried vermicelli. *Ukrainian Food J.* **2017**, *6*, 61–76. [CrossRef]

33. Durazzo, A.; Camilli, E.; Marconi, S.; Lisciani, S.; Gabrielli, P.; Gambelli, L.; Aguzzi, A.; Lucarini, M.; Kiefer, J.; Marletta, L. Nutritional composition and dietary intake of composite dishes traditionally consumed in Italy. *J. Food Compos. Anal.* **2018**. under review.

34. Finglas, P.M.; Berry, R.; Astley, S. Assessing and improving the quality of food composition databases for nutrition and health applications in Europe: The contribution of EuroFIR. *Adv. Nutr.* **2014**, *5*, 608–614. [CrossRef]

35. Greenfield, H.; Southgate, D.A.T. Food composition data. In *Production, Management, and Use*, 2nd ed.; Food and Agriculture Organization of the United Nations: Rome, Italy, 2003.

36. Abdi, H.; Williams, L.J. Principal component analysis, Wiley Interdiscip. *Rev. Comput. Stat.* **2010**, *2*, 433–459.

37. Yap, K.Y.L.; Chan, S.Y.; Lim, C.S. Infrared-based protocol for the identification and categorisation of ginseng and its products. *Food Res. Int.* **2007**, *40*, 643–652. [CrossRef]

38. Smith, B.C. *Infrared Spectra Interpretation. A Systematic Approach*, 1st ed.; CRC Press LLC: Boca Raton, FL, USA, 1999.

39. Li, W.; Dobraszczyk, B.J.; Dias, A.; Gil, A.M. Polymer conformation structure of wheat proteins and gluten subfractions revealed by ATR-FTIR. *Cereal Chem.* **2006**, *83*, 407–410. [CrossRef]

40. Deepa, M.; Agnihotry, S.A.; Gupta, D.; Chandra, R. Ion-pairing Effects and Ion-solvent-polymer Interactions in Lin(CF3SO2)2-PC-PMMA Electrolytes: A FTIR Study. *Electrochim. Acta* **2004**, *49*, 373–383. [CrossRef]

41. Bock, J.E.; West, R.; Iametti, S.; Bonomi, F.; Marengo, M.; Seetharaman, K. Gluten structural evolution during pasta processing of refined and whole wheat pasta from hard white winter wheat: The influence of mixing, drying, and cooking. *Cereal Chem.* **2015**, *92*, 460–465. [CrossRef]

42. Li, Y.; Chen, Y.; Li, S.; Gao, A.; Dong, S. Structural changes of proteins in fresh noodles during their processing. *Int. J. Food Prop.* **2017**, *20*, S202–S213. [CrossRef]

43. Ambrose, A.; Cho, B.K. A Review of Technologies for Detection and Measurement of Adulterants in Cereals and Cereal Products. *J. Biosyst. Eng.* **2014**, *39*, 357–365. [CrossRef]

44. Kamil, M.M.; Hussien, A.M.S.; Ragab, G.H.; Khalil, S.K.H. Detecting Adulteration of Durum Wheat Pasta by FT-IR Spectroscopy. *J. Am. Sci.* **2011**, *7*, 573–578.

45. Sivakesava, S.; Irudayaraj, J. Analysis of potato chips using FTIR photoacoustic spectroscopy. *J. Sci. Food Agric.* **2000**, *80*, 1805–1810. [CrossRef]

46. Rein, A.; Rodriguez-Saona, L.E. *Measurement of Acrylamide in Potato Chips by Portable FTIR Analyzers*; Application Note; Agilent Technologies, Inc.: Santa Clara, CA, USA, 2013.

47. Ayvaz, H.; Rodriguez-Saona, L.E. Application of Handheld and Portable Spectrometers for Screening Acrylamide Content in Commercial Potato Chips. *Food Chem.* **2015**, *174*, 154–162. [CrossRef] [PubMed]

48. Dankar, I.; Haddarah, A.; Omar, F.E.L.; Pujolà, M.; Sepulcre, F. Characterization of food additive-potato starch complexes by FTIR and X-ray diffraction. *Food Chem.* **2018**, *260*, 7–12. [CrossRef] [PubMed]

49. Erukhimovitch, V.; Tsror (Lahkim), L.; Hazanovsky, M.; Huleihel, M. Direct identification of potato's fungal phyto-pathogens by Fourier-transform infrared (FTIR) microscopy. *J. Spectrosc.* **2010**, *24*, 609–619. [CrossRef]

50. Ibrahim, T.H.; Chehade, Y.; Abouzour, M. Corrosion Inhibition of Mild Steel using Potato Peel Extract in 2M HCl Solution. *Int. J. Electrochem. Sci.* **2011**, *6*, 6542–6556.

51. Bartosova, A.; Soldán, M.; Sirotiak, M.; Blinová, L.; Michaliková, A. Application of FTIR-ATR spectroscopy for determination of glucose in hydrolysates of selected starches. *Res. Pap. Fac. Mater. Sci. Technol. Slovak Univ. Technol.* **2013**, *21*, 116–121.

52. Abdullah, A.H.D.; Chalimah, S.; Primadona, I.; Hanantyo, M.H.G. Physical and chemical properties of corn, cassava, and potato starchs. *IOP Conf. Ser. Earth Environ. Sci.* **2018**, *160*, 012003. [CrossRef]

53. Darvanjooghi, M.H.K.; Davoodi, S.M.; Dursun, A.I.; Ehsani, M.R.; Karimpour, I.; Ameri, E. Application of treated eggplant peel as a low-cost adsorbent for water treatment toward elimination of Pb^{2+}: Kinetic modeling and isotherm study. *Adsorpt. Sci. Technol.* **2018**, *36*, 1112–1143. [CrossRef]

54. Fiore, V.; Valenza, A.; Di Bella, G. Artichoke (*Cynara scolymus* L.) fibres as potential reinforcement of composite structures. *Compos. Sci. Technol.* **2011**, *71*, 1138–1144. [CrossRef]

55. López-Molina, D.; Navarro-Martínez, M.D.; Rojas Melgarejo, F.; Hiner, A.N.; Chazarra, S.; Rodríguez-López, J.N. Molecular properties and prebiotic effect of inulin obtained from artichoke (*Cynara scolymus* L.). *Phytochemistry* **2005**, *66*, 1476–1484. [CrossRef] [PubMed]

56. Ceylan, C.; Bayraktar, O.; Atci, E.; Sarrafi, S. Extraction and characterization of pectin from fresh globe artichoke and canned artichoke waste. *GIDA J. Food* **2017**, *42*, 568–576. [CrossRef]

57. Wang, B.; Xu, F.; Zong, P.; Zhang, J.; Tian, Y.; Qiao, Y. Effects of heating rate on fast pyrolysis behavior and product distribution of Jerusalem artichoke stalk by using TG-FTIR and Py-GC/MS. *Renew. Energy* **2019**, *132*, 486–496. [CrossRef]

58. Ribeiro, J.S.; Salva, T.J.; Ferreira, M.M.C. Chemometric studies for quality control of processed Brazilian coffees using drift. *J. Food Qual.* **2010**, *33*, 212–227. [CrossRef]

59. Nallamuthu, I.; Devi, A.; Khanum, F. Chlorogenic acid loaded chitosan nanoparticles with sustained release property, retained antioxidant activity and enhanced bioavailability. *Asian J. Pharm. Sci.* **2015**, *10*, 203–211. [CrossRef]

60. Liang, N.; Lu, X.; Hu, Y.; Kitts, D.D. Application of Attenuated Total Reflectance–Fourier Transformed Infrared (ATR-FTIR) Spectroscopy to determine the Chlorogenic acid isomer profile and antioxidant capacity of coffee Beans. *J. Agric. Food Chem.* **2016**, *64*, 681–689. [CrossRef] [PubMed]

61. Tulchinsky, V.M.; Zurabiab, S.F.; Asankozhoev, K.A.; Kogan, G.A.; Khorlin, A.V. Study of the infrared spectra of oligosaccharides in the region 1000–400 cm^{-1}. *Carbohydr. Res.* **1976**, *51*, 1–8.

62. Bureau, S.; Ruizb, D.; Reich, M.; Gouble, B.; Bertrand, D.; Audergon, J.M.; Renard, C.M.G.C. Application of ATR-FTIR for a rapid and simultaneous determination of sugars and organic acids in apricot fruit. *Food Chem.* **2009**, *115*, 1133–1140. [CrossRef]

63. Velázquez, T.G.; Revilla, G.O.; Loa, M.Z.; Espinoza, Y.R. Application of FTIR-HATR spectroscopy and multivariate analysis to the quantification of adulterants in Mexican honeys. *Food Res. Int.* **2009**, *42*, 313–318. [CrossRef]

64. Anjos, O.; Campos, M.G.; Ruiz, P.C.; Antunes, P. Application of FTIR-ATR spectroscopy to the quantification of sugar in honey. *Food Chem.* **2015**, *169*, 218–223. [CrossRef] [PubMed]

MDPI

Article

Wild Plants Potentially Used in Human Food in the Protected Area "Sierra Grande de Hornachos" of Extremadura (Spain)

José Blanco-Salas *, Lorena Gutiérrez-García, Juana Labrador-Moreno and Trinidad Ruiz-Téllez

Department of Vegetal Biology, Ecology and Earth Science, University of Extremadura, 06071 Badajoz, Spain; zoolorena5@hotmail.com (L.G.-G.); labrador@unex.es (J.L.-M.); truiz@unex.es (T.R.-T.)
* Correspondence: blanco_salas@unex.es; Tel.: +34-924-289-300 (ext. 89052)

Received: 29 November 2018; Accepted: 11 January 2019; Published: 16 January 2019

Abstract: Natura 2000 is a network of protected spaces where the use of natural resources is regulated through the Habitat Directive of the European Union. It is essential for the conservation of biodiversity in Europe, but its social perception must be improved. We present this work as a demonstration case of the potentialities of one of these protected areas in the southwest (SW) Iberian Peninsula. We show an overview of the catalog of native wild plants of the place, which have nutritional and edible properties, having been used in human food by the peasant local population over the last century, and whose consumption trend is being implemented in Europe mainly through the haute cuisine and ecotourism sectors. What is offered here is a study of the case of what kind of positive contribution systematized botanical or ethnobotanical scientific knowledge can make toward encouraging innovative and sustainable rural development initiatives. A total of 145 wild plants that are potentially useful for leading tourism and consumers toward haute cuisine, new gastronomy, enviromentally-friendly recipes, and Natura 2000 Conservation are retrieved. The methodology used for our proposal is based on recent proposals of food product development and Basque Culinary Center initiatives.

Keywords: local foods; bioactive components; traditional recipes; traditional dietary patterns; edible plants; Mediterranean; innovative gastronomy

---

1. Introduction

1.1. Plant Resources Global Availability and Potential Use

From approximately 250,000 plant species, it has been estimated that up to 75,000 could be edible [1], and some 7000 are regularly eaten worldwide [2]. Even though the average Western citizen has now access to more species of edible plants than ever, a common household shopping list will not include more than 45 plant species as food supplies. In fact, Western societies tend to use much fewer species than indigenous communities. Regarding wild resources (plant species that grow spontaneously in populations in natural or managed habitats, thriving independently of direct human action, following the definition of [3], 73 plant species are reported to be used in Álava (Spain, [4]), while 66 plant species are reported to be used in Emilia (Italy, [5]), the inhabitants of the Quinling Mountains of Saanxi (China) use up to 185 species [6], and the indigenous peoples of southern Ecuador use up to 354 [7]. The main actor responsible for this steady loss of culinary traditions is globalization [8]; wild edible plants are also often stigmatized due to associations with food shortage and poverty, climatic extreme events, or politically conflictive times ([9] and references therein), leading to the abandonment of wild edible resources and the consequent decline of knowledge in plant uses. What has become increasingly clear is that any approach to biodiversity and landscape conservation must incorporate traditional knowledge in order to achieve success.

1.2. Combined Conservation Trends Focusing on the Natura 2000 Network

In rural communities of southern Spain, a close relationship between biodiversity, local culture, and dialectal diversity has evolved over centuries of history. There is a genuine biocultural heritage [10] that has been specifically developed on a local spatial scale and has been until recent times transmitted through generations. There, significant areas belong to the Natura 2000 Network. This is the organized area for the conservation of biodiversity in Europe. It is regulated through the Habitats Directive of the European Union, which enables compatible development, knowledge, bioeconomy, and innovation [11]. However, sometimes, the measures proposed by the environmental organisms are seen by the local population with suspicion or rejection, and they are challenged [12]. The problem has worried the European authorities to the point of having implemented specific action programs aimed at improving its social perception [13]. Most are oriented to new trends and development opportunities to favor biodiversity and Natura 2000 Network acceptance.

1.3. The Particular Case of the Sierra Grande de Hornachos: Ecological and Cultural Relevance

Hornachos is a small town (3777 inhabitants) [14] of the central Badajoz province (Extremadura) in the southwest of the Iberian Peninsula that has this problematic. It provides an optimal study case for biocultural heritage whose record and divulgation is timely. In addition to this, there is a sociological reality in the territory of unpopularity related to the Natura 2000 Network. In this framework, we are actually carrying out a project supported by the European Union and the regional government. It is intended in the medium term to solve real problems of rejection and blockage that have occurred in the past. An SAC and SPA (Special Area for Conservation and Special Protection Area for Birds) have been developed in the "Sierra Grande de Hornachos" within the Natura 2000 Network [15]. Its forests and thickets represent an ideal environment for the life of abundant Mediterranean wild fauna of high value with elements such as the griffon vulture, the golden eagle, and more than 220 species of vertebrates. There is the cartography of the predominance of holm oaks in flat areas, hillsides, and foothills; of cork oaks in shady and watercourses; of mixed forest; of wild olive trees; of summit junipers; and of *Nerium oleander (adelfares)* and *Flueggea tinctoria (tamujares)* in the riparian vegetation. Surprisingly, the flora has been much less inventaried, and no specific floristic catalog of the area has been published to date. For ethnobotanical studies, there are also no specific systematized works, despite the potential of the place. It has a rich historical past that includes a mixture of cultures, such as Christian, Jewish, and Muslim. This circumstance has produced a strong imprint in the agriculture of the area. It influenced the maintenance of agroecosystems, including many activities related to nature, hunting, fishing, medicine, gastronomy, and even traditional crafts, which are as interesting as those from Andalucía [16].

1.4. Local Traditional Knowledge on Food Plants and Resource Management at Sierra Grande de Hornachos

In this region, as well as in others of the Mediterrenean [17], economic marginality, historical factors, and recurrent political crises in old times forced its inhabitants to make the most out of the wild resources found in the vicinity of their settlement. However, over the last decades, the development of agriculture and the globalization phenomena have led to the progressive abandonment of wild edible resources [9,17], and nowadays, it is still a rich bio-heritage that holds value.

Taking into consideration that the regional flora of Extremadura has 1938 species [18] and that the "Sierra Grande de Hornachos" comprises 12,469 ha of well-conserved Mediterranean vegetation units, its floristic resources were expected to be interesting. In a first aproximation, a list that did not reach 50 species was the only published result [19], featuring threatened plants such as *Lavatera triloba*, *Erodium mouretii*, *Orchis italica* and *Ophrys fusca* subsp. *dyris* as standouts. The complete floristic inventory of the area was approached by our team in 2016/17 Project IB16003 with the expectation of finding at least 400 species. We made field prospections, and to date, 1301 taxa make up the checklist of the wild vascular plants of Sierra Grande de Hornachos, according to our campaigns and investigations. This systematical biodiversity information (of 1301 taxa) can be crossed with another source (database) that is not open access, but is available in the Ministry of the Environment of Spain: The Spanish Inventory of Traditional Knowledge Relating to Biodiversity. Doing this, all of the vegetal resources can be studied and presented to the local inhabitants of the region as possibilities for the development and progress toward a sustainable society of 21st century. The real challenge relating to the conservation of rural areas and phytodiversity in Europe is to find new trends and opportunities for plants and people. That seems to be the most sucessful way to provide social support for the Natura 2000 Network. This is the main focus of our work. Constructing the catalog of the wild plants of the area, we can look for the uses of these species and explore their possibilities. We can study how they were employed in other places in the country. This means that they can effectively be also utilized in the same way. Thus, the traditional used and conserved knowledge will drive innovative applications in the area. In summary, a positive view of the sustainable exploitation of natural resources in the Natura 2000 Network can promote a friendly face of conservation measures. For this reason, it is important to explore this method of management.

1.5. Innovation for Conservation: Wild Gathering, Agro-Tourism, Gourmet Foods, and Local Memories

New initiatives are starting in Europe. Agro-tourism is rapidly increasing its popularity through urban Western societies, ultimately enabling the survival of traditional knowledge and the gathering of wild plants itself [20]. Nevertheless, a renaissance of interest in traditional plant knowledge is beginning. Gathering wild edible plants has become popular as a means of enjoying outdoor contact with nature and granting alternative, high-end culinary experiences [9]. Wild plant consumption is not unfainly seen as "an occupation for the poor, which increased during times of bad crops and famine", but also a fashionable part of 21st century haute cuisine [17]. It is even increasingly seen as a form of reaffirming cultural identity, sophistication, and commitment to the community. In brief, innovative strategies suggested for natural resources are excellent tools for promoting the culture of biological conservation among social sectors, and hopefully serve to mitigate local rejects to Natura 2000 regulations.

In that framework and with the underlying characteristics of the case of Hornachos, we state the specific objectives for this work.

1.6. Objectives

The objective of this work is to record the threatened biocultural heritage of a Natura 2000 area (Hornachos) on wild edible plants with the double purpose of preserving it and setting the basis for a valorization and divulgation program focused on the local and visiting population.

2. Materials and Methods

1. The database of the Spanish Inventory of Traditional Knowledge Relating to Biodiversity (IECTB) [21–24] comprising of 3156 record (taxa) was taken as a basis. To date, it is not yet open access. So, permission to consult it was formally required by the authors (J. Blanco-Salas) and allowed by the IECTB national coordinator (M. Pardo de Santayana).

   - IECTB is the Inventory of Traditional Knowledge that has been published in Spain but dispersed in a difficult literature. The realization of this inventory arose from a legislative requirement (Article 70 of Law 42/2007, of December 13, the Natural Heritage and Biodiversity of Spain). In this inventory, the preparation of plants for food uses is written in the same way as it has been reported from local informants.

2. In parallel, our research group created a database of Hornachos plants. Basic information was taken from the Global Biodiversity Information Facility (GBIF) [25], the open-access program ANTHOS [26], and the Herbarium UNEX (University of Extremadura) collections.

- GBIF [25] stands for the Global Biodiversity Information Facility. It is an international network and research infrastructure funded by the world's governments that is aimed at providing anyone, anywhere, open access to data about all types of life on Earth.
- ANTHOS [26] is an open-access program that was developed to divulge and show society information about the biodiversity of Spanish plants on the Internet. The initiative was born under the umbrella of the Flora Iberica research project (Ministry of Agriculture, Food, and the Environment) and the Royal Botanical Garden of Madrid, Spain.
- UNEX Herbarium is the herbarium of the University of Extremadura, which was formed by 36,451 specimens of vascular plants whose main origin is the autonomous region of Extremadura (Spain) and Portugal. Here is the place where we included the voucher speciments of our field collections.

3. The Hornachos list finally resulted in 1301 plants. We looked for 1301 names in the IECTB Inventory official database, and 834 matched (Figure 1). This means that we firstly had 834 useful taxa and $1301 - 834 = 467$ taxa without use.

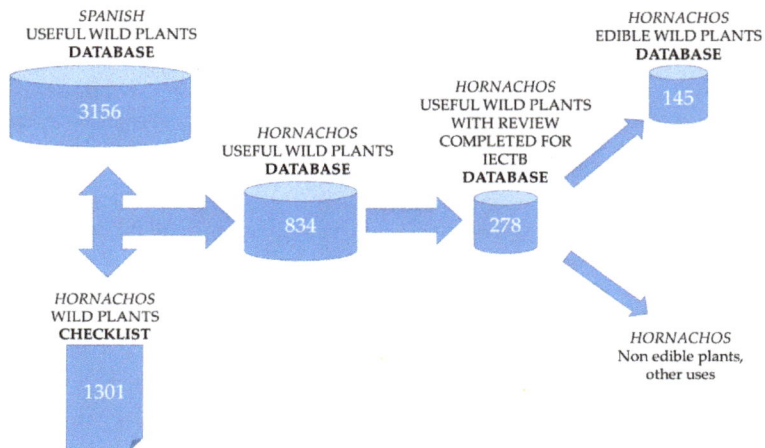

Figure 1. Methodology scheme.

4. The IECTB inventory database information compiled in "one file for plant" format has not yet been completely published. Of the 834 useful Hornachos plants included in the IECTB, to date, the IECTB has made information on 278 species available in published books [21–24] (the remainder of the 834 species will be published in the future, when the IECTB project is finished). Of these 278 useful Hornachos plants, 145 species were edible plants. These names of edible plants were listed.

5. A provisional database of the edible plants of Hornachos was built, with 145 records (corresponding to the provisional 145 edible plants in Hornachos).

    - Botanical nomenclature was adapted to the Angiosperm Phylogeny Group (APG) IV System [27] and included vernacular names.
    - Part of the plant used (upon the information taken from the IECTB).
    - Subcategory of food use. The ones considered were: A = Vegetables; B = Roots, bulbs, tubers, and rhizomes; C = Sweet fruits; D = Dry and oleaginous fruits; E = Cereals and pseudocereals; F = Fat; G = Alcoholic drinks; H = Non-alcoholic drinks; I = Condiments; J = Sugars and sweeteners; K = Candies and chewing; L = Other food uses. They are the same as considered by the IECTB. The information for filling the corresponding fields was literally transferred from the IECTB.
    - Preparation. It was filled upon the following classification: 1 = Soups; 2 = "gofio" (= roasted cereal stirred into liquid); 3 = Broths and purees; 4 = Omelettes; 5 = Cakes; 6 = Rices; 7 = Salads; 8 = Cold vegetable soups; 9 = Pies/Patties; 10 =Potages; 11 = Stews; 12 = Scrambled; 13 = Sautéed/boiled/toasted; 14 = Milky; 15 = Fried/breaded; 16 = Sauces; 17 = Oil; 18 = Vinegar; 19 = Renner; 20 = Preservative; 21 = Dressings/condiments; 22 = Flour; 23 = Sugar; 24 = Desserts; 25 = Ice creams; 26 = Sorbets; 27 = Cakes/Cakes; 28 = Sweet; 29 = Jams; 30 = Marmalades; 31 = Jellies; 32 = To the natural; 33 = Brine; 34 = Pickles; 35 = In syrup; 36 = In alcohol; 37 = Wines; 38 = Liquors; 39 = Syrups for cocktails; 40 = Juices; 41 = Brandies; 42 = Infusions; 43 = Coffee/tea; 44 = Diluted syrups. This classification is based on that of Bertrand [28]. We assigned the information about preparation, which was recorded in IECTB format, to the corresponding culinary categories of Bertrand.

6. Finally, tables and graphs were made in order to facilitate discussing the results.

3. Results

The constructed database had a total of 145 records. For each record, we considered the nomenclatural field (family, scientific, and vernacular), 15 sub-categories of food uses according to the IECTB, 42 Bertrand food classifications, and 22 possible parts of the the plant used. The total matrix of managed results included $145 \times 3 \times 15 \times 42 \times 22 = 6,029,100$ possible raw data. The raw data results are summarized in Table A1 of Appendix A, and the corresponding boxes of Appendix B (Boxes A1–A8). The resulting 145 species belong to 49 families. Asteraceae is the best represented, with 25 different vegetables, and Lamiaceae stands out for the preparation of alcoholic and non-alcoholic drinks, as well as condiments. Figure 2 summarizes the diversity of uses among the taxonomic groups' plant uses, which covered up to 12 subcategories of the IECTB, mostly involving brewing and the aromatization of beverages (alcoholic, 51 species, and non-alcoholic, 28 species). Up to 39 species were used as a form of "wild candy", being picked and chewed as mere entertainment. In the following paragraphs, we discuss the scope of the applications found.

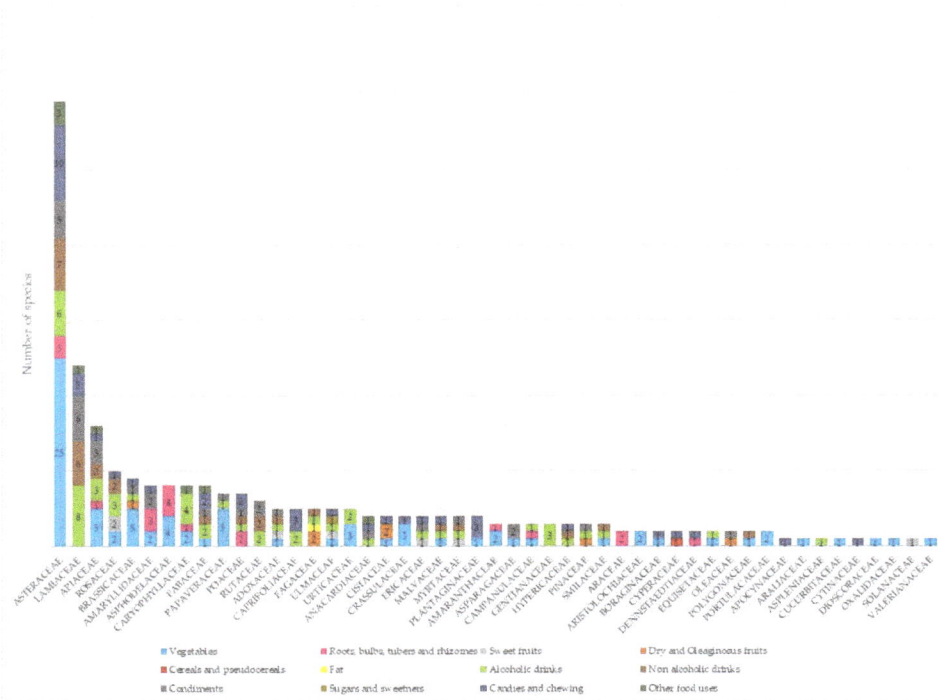

Figure 2. In colors, the diversity of uses among the families. Each bar shows its corresponding number of species.

4. Discussion

4.1. Hornachos Natura 2000 Area and Wild Edible Plants: Diversity and General Possibilities

The diversity of edible plants in the Hornachos community catalogued to date is high in comparison with other territories of Europe. It is more than double the number of edible plants used in similar but more urbane environments of the Basque Country (northern Spain [4]) or Emilia (central Italy [5]) and is approaching figures that have been published for human settlements in rural Asia [6]. However, it is still far from those that have been published for the indigenous communities of South America [7]. Apparently, the rich biocultural heritage of Hornachos is still well preserved. This can be explained by the overall lower level of industrialization of Extremadura in comparison with the more developed areas of the country, where contact with nature is much lower, causing the progressive demise of traditional knowledge.

Asteraceae is the most useful family of our catalogue, but this is also one of the three floristically richest families of the regional flora [18]. As in other Mediterranean countries (see a compilation in [9]), the diversity of wild greens correlates with the presence of bitter, pungent, or acrid secondary metabolites that allow for a variety of nuances in the perception of taste. Another nutritional aspect that has been studied recently is the influence of micronutrients in the correct performance of many metabolic processes, even at the prenatal stage [29]. Experimental studies on women have demonstrated the influence and significance of a wild vegetable diet in folate levels, and similar nutritional parameters [30] and specific reviews have focused on the adolescents of developed and underdeveloped countries. [31] Traditional indigenous foods have been studied to evaluate their levels of micronutrients [32], and the suitability of wild vegetables for alleviating human dietary deficiencies has been addressed [33]. Specific studies of plants that have been traditionally consumed in Spain [34] have also been made.

In addition to this, previous studies in the field of haute cuisine have been published on the possibilities of innovation in this field [35–38]. In Northern Europe [39,40], interesting new design initiatives have been developed related to high-level gastronomy [41–43]. Important professionals of the sector with recognized international prestige have proposed cooking with wild plants [28]. The avant-garde cuisine has been linked to the world of art through the creative universe of great chefs from southern Europe such as Ferrán Adriá, El Bulli, and Mugaritz. Specialized official universities have also become subjects, such as the Basque Culinary Centre in Mondragón (Spain) [44]. Following the artistic concept of Chinese cuisine [45], the model has become directed toward the elaboration of dishes [46,47] that are a mixture of art and science. Wild biodiversity has started to be considered as an interesting resource for the development of new gastronomic and edible products [48,49]. Ethnobotany and the study and cataloguing of traditional knowledge by the local population constitute an important resource for implementing new food product development strategies [50].

Figure 3 shows the number of species of our study, which can be organized according to the different culinary uses of Bertrand classification [28], based on the raw data compiled in Table A1. Apart from the individual potential of each species, we consider that the examples that we expose next have an added value.

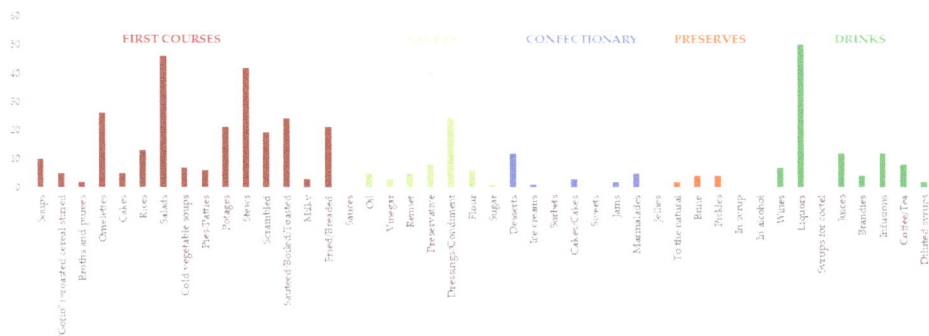

Figure 3. Distribution of the 145 species of the study according to the culinary classification proposed by Bertrand [28].

4.2. Wild Edible Plants: Innovative Gastronomy and Healthy Promotion

First courses and dishes can be creatively improved by the incorporation of many of the above-mentioned examples. Table A1 summarizes which vegetables have been experimented by traditional cooks, to prepare all kind of soups, broths, purees, omelettes, pies, patties, and salt cakes. The most common way of using the plants of our catalogue has been in stewing rices or potages, or adding them to salads. There is an immense arsenal of creativity at the disposal of the talent of the future professionals of haute cuisine. In Table A1, the species assigned to the Bertrand classification numbers 7, 10, and 11 are direct examples (last column). Another interesting new way of exploring haute cuisine is the new formulation of fried, breaded, sautéed, boiled, or toasted culinary preparations made out of the plants that we have assigned in Table A1 as belonging to the Bertrand classification numbers 13 and 15.

Soups can be made with *Chenopodium* or *Urtica*, which also carry *Allium* or *Foeniculum*, for example. The first one provides minerals, fiber, vitamins, and essential fatty acids that enhance the sensory and functional value of the food [51]. The latter is especially significant in the last two cases. Omelettes recipes can be based on *Crepis vesicaria* subsp. *haenseleri*, *Scorzonera laciniata*, *Sonchus oleraceus*, *Borago officinalis*, or *Eruca vesicaria*. There is even a tradition of using the young shoots of *Tamus communis* and *Brionia dioica* as a substitute for *Asparagus* for this same purpouse. Their particularly bitter taste makes them peculiar, and precaution about the preparation must be taken in account in order to avoid the unwanted effects that the bibliography describes [17]. Scrambled *Capsella bursa-pastoris* and cold vegetable soups (gazpachos) where *Sisymbrium orientale* or *Allium ampeloprasum* are added have a bromatological explanation. These plants are rich in sulfur components belonging to the glucosinolate group (for the *Brassicaceae*) and the non-proteic amino acids (for the species of *Allium*).

In any case, they are very odorous molecules with a high biological activity [52]. In a similar direction, the use of other species of *Allium* or elements of the mentioned familiy (e.g., *Rapistrum rugosum*, *Sisymbrium irio*), as proposed in Table A1, for the recipes for stew presentations, is worthwhile because it enables the incorporation of the phyochemical variability of the systematics to the cuisine. For the preparation of rice, *Sisymbrium orientale* is recommended. The same can be said about *Sonchus asper*, which has an excelent nutritional profile, being very rich in fiber and ω-3 acids [53]. We emphasize as well the singularity of the proposal of using previously cooked *Cytinus hypocistis* because it is a healthy and antioxidant food, but it is also very astringent [54]. Salad is a very fashionable group for gourmets in agro-tourism. New developments can be made by taking any of the plants that are codified with the number seven from from Table A1. Some of them (e.g., *Apium, Chenopodium, Chondrilla juncea, Bellis perennis, Montia fontana, Silene vulgaris, Stellaria media, Sonchus tenerrimus, Tragopogon porrifolius, Umbilicus rupestris, Valerianella microcarpa*, and *Veronica anagallis-aquatica*) are already entering the high Spanish hotel industry and occupying small selectede market niches. They have organoleptic and nutritional values [54] mainly as antioxidants [55], which is very interesting.

Another group that is worthy of mention for its variability is the drinks category. We have found a large list of plants that remain at the moment underexploited in the Hornachos local area, having a great potential in the industry of liquors, spirits, and soft drinks. Table A1 summarizes the potentialities of the flora of this Natura 2000 area in this sector (Bertrand classification numbers 36 to 44). Our proposal can really be transferred to land, because Hornachos is situated next to Tierra de Barros (Badajoz), which is an active vinegrowing district, with a developed liqueur industry. There is a possibility of making vinegar with rosemary oil and brandy with a few plants that could be explored. They are *Sambucus nigra, Foeniculum vulgare,* and *Arbutus unedo*, respectively, which are very rich [56–58] in phenolics derivatives. Diluted syrups with *Achillea ageratum* and *Smilax aspera* are surely based on the chemical composition of these two plants, for they have esteroidic and saponine components [59] that favor those sorts of viscous formulations. Table A1 also shows the infusions that could be prepared with the autochthonous elements of the territory. In the case of chamomiles, which are recommended for their stomachic and digestive properties, the different species (*Chamaemelum nobile, Chamaemelum fuscatum, Matricaria chamomilla*, and *Heychrysum stoechas*) provide different qualities and let the consumer choose according to their individual preferences. This is related to the chemical profile of the essential oil, which has been characterized by chromatography [60,61]. Other digestive genera include *Mentha, Origanum, Sideritis*, and *Thymus* [62]. The antiinflamatory profile of *Phlomis* [63] and *Malva* [64] or the sedative of other cases such as *Melissa, Hypericum*, or *Crataegus* [64] are very interesting from the nutraceutic point of view.

The Confectionery and Preserves groups have discrete representations (see Figure 3), which do not detract from the importance of particular species. The sweetening power of *Celtis australis* would be worth investigating for its improvement and exploitation of low-calorie programs, which are in the interest of the overweight public [65]. Others can be objects of creativity, such as ice creams, sorbets, or marmalades made out of the mature fruits of *Myrtus communis*, which have excellent antioxidant properties [66]. The brines of *Sedum album* (which is very rich in flavonol glycosides [67]) or the pickles of *Portulaca oleracea* (which have large amounts of ω-3 and ω-6 fatty acids [68] all represent really high gourmet challenges.

We must finally emphasize the importance of the Sauces group, since the most abundant list of species corresponds to those used as condiments. They are plants with essential oils, which are well-known in popular gastronomy and mostly belong to the *Lamiaceae* and *Apiaceae* families. In Table A1, these are assigned to number 21. However, we must emphasize them over the relevance of natural rennets such as *Scolymus hispanicus*, *Cynara cardunculus*, and *Cynara humilis*, because they open many possibilities of introducing new taste sensations and presentations. Something similar can be said about a potential flour use of *Elymus repens* or *Scirpoides holoschoenus*, [69] based in traditional knowledge about vegetal biodiversity. The latter is another food technology and scientific challenge, because the starchy rhizome is very rich in resveratrol and other antioxidant molecules that should be better studied for its best utilization.

5. Conclusions

The envisaged promotion of local products throughout environmentally sustainable techniques further contributes to environmental protection. The valorization of traditional foods and recipes is necessary in order to simultaneously preserve the local agro-biodiversity, sustainability, and productive systems. In this context, it is becoming important to address the consumer folk knowledge toward innovative applications. This strategy should represent a valid tool for the promotion of socioeconomic development, the enhancement of territories, and biodiversity preservation.

Author Contributions: Conceptualization, T.R.-T.; Methodology, J.B.-S.; Investigation, J.B.-S., L.G.-G.; Resources, J.L.-M.; Data Curation, L.G.-G.; Writing-Original Draft Preparation, T.R.-T.; Writing-Review and Editing, L.G.-G.; Validation, J.L.-M.; Visualization, L.G.-G., J.L.-M.; Supervision, T.R.-T.; Project Administration, J.B.-S.; Funding Acquisition, T.R.-T., J.L.-M.

Funding: This research was by Junta de Extremadura (Spain) and European Regional Development Fund, through grant number IB16003.

Acknowledgments: To Manuel Pardo de Santayana (Universidad Autónoma de Madrid) and the Research Group "Proyecto Inventario Español de Conocimientos Tradicionales Relativos a la Biodiversidad". To Pedro Escobar García (Natural History Museum, Vienna, Austria) and two anonymous referees for their helpful corrections in the draft versions of the manuscript.

Conflicts of Interest: The authors declare no conflict of interest.

Appendix A.

Table A1. Catalog of edible plants present in the protected area " Sierra Grande de Hornachos".

Family	Scientific Name	Vernacular Name	PP	Subc	Pr	CL F
Adoxaceae	*Sambucus nigra* L.	Saúco	I, Fr	A, C, G, H	a, b	4, 24, 29, 30, 33, 37, 38, 40
	Viburnum tinus L.	Durillo	L	I	b	21
Amaranthaceae	*Chenopodium album* L.	Cenizo	Wp, R	A, B	a, b	1, 4, 6, 7, 8, 10, 11, 12, 22
	Chenopodium murale L.	Cenizo	Wp	A	c	7
Amaryllidaceae	*Narcissus bulbocodium* L. *	Campanitas	B, F	B, K	b	n.d.
	Allium ampeloprasum L.	Ajo porro	Wp, B, Bp	B, I	a, b	1, 4, 6, 7, 8, 11, 12, 13, 14, 15, 21
Anacardiaceae	*Allium roseum* L.	Ajo porro	L, B	A, B, I	a, b	2, 7, 10, 11, 21, 24
	Pistacia lentiscus L.	Lentisco	Ap, Wp, L, Sv	G, I, K, L	b	20, 21, 38
	Apium graveolens L.	Apio	Wp	A	a, b	n.d.
	Apium nodiflorum (L.) Lag.	Berra	St, Bs	A	a, b	2, 7, 8, 11
Apiaceae	*Foeniculum vulgare* Mill.	Hinojo	L, Bs, I, St, Wp, S, Fr	A, G, H, I, K, L	a, b	1, 4, 6, 7, 9, 10, 11, 13, 21, 37, 38, 40, 41, 42
	Scandix australis L.	Quijones	Wp, Ap	A, G, H, I	a, b	11, 21, 38, 40, 41
	Scandix pecten-veneris L.	Alfileres	L, Fr	A	a, b	11
	Thapsia villosa L. *	Cañaheja	R, Wp	B, G, I	a, b	21, 38

Table A1. *Cont.*

Family	Scientific Name	Vernacular Name	PP	Subc	Pr	CL F
Apocynaceae	*Vinca difformis* Pourr. *	Alcandorea	F	K	b	n.d.
Araceae	*Arisarum simorrhinum* Dur. *	Candil	Tu	B	c	2
	Arum italicum Mill. *	Aro	Tu	B	a	5, 10, 14
Araliaceae	*Hedera helix* L. *	Hiedra	St	A	c	n.d.
Aristolochiaceae	*Aristolochia paucinervis* Pomel *	Candil	St	A	a	11
	Aristolochia pistolochia L. *	Candil	St	A	a	11, 12
Asparagaceae	*Ruscus aculeatus* L.	Rusco	Bs, St	A, I	a, b	4, 21, 24
	Urginea maritima (L.) Baker *	Cebolla	L	I	b	20
Asphodelaceae	*Asphodelus aestivus* Brot.	Cebollino	L, Bs, Tu	A, B	a	11
	Asphodelus albus Mill.	Gamón	L, Bs, Tu	A, B	a	11
	Asphodelus fistulosus L.	Cebollino	L, Bs, Tu	A, B	a	11
	Asphodelus ramosus L.	Gamón	L, Bs, Tu	A, B	a	11
Aspleniaceae	*Asplenium ceterach* L.	Doradilla	Wp	G	b	38
Asteraceae	*Achillea ageratum* L.	Árnica	Wp	A, H	c	44
	Anacyclus clavatus (Desf.) Pers.	Magarza	L, Bs	A	c	n.d.
	Andryala integrifolia L.	Árnica	L, St	A	a, b	7
	Andryala laxiflora DC.	Árnica	L, St	A	b	7
	Andryala ragusina L.	Ajonje	St, Sv	K	b	n.d.
	Anthemis cotula L.	Magarza	Wp	G	c	38
	Bellis perennis L.	Margarita	L, F	A, K	b	7
	Calendula arvensis L.	Caléndula	Bs, F	A	a, b	11
	Carthamus lanatus L.	Cardo	L, F	A, I	a, b	7, 10, 11, 19
	Chamaemelum fuscatum (Brot.) Vasc.	Magarza	F	H	a	42
	Chamaemelum nobile (L.) All.	Manzanilla	I, F	G, H	a, b	38, 42
	Chondrilla juncea L.	Ajonjera	L, St, Wp, Sv	A, K	a, b	1, 7, 11
	Cichorium intybus L.	Achicoria	L, Bs, R, F	A, H, K	a, b	1, 7, 9, 11, 12, 13, 15, 43
	Crepis capillaris (L.) Wallr.	Almirón	Wp	A	c	n.d.
	Crepis foetida L.	Almirón		A	c	n.d.
	Crepis vesicaria subsp *haenseleri* (Boiss. ex DC.) P.D.Sell	Achicoria	L	A	a,b	4, 7, 10, 11
	Cynara cardunculus L.	Cardo	L, St, F, Wp, Pn	A, I, K	a, b	6, 7, 10, 11, 15, 19, 24, 33
	Cynara humilis L.	Alcachofa	I, F	A, L	a, b	19
	Dittrichia graveolens (L.) Greuter	Hierba matapulgas	Ap	I	a	21
	Dittrichia viscosa (L.) Greuter	Olivarda fina	I	G	b	38
	Galactites tomentosa Moench	Cardo	St	A	c	n.d.
	Helichrysum stoechas (L.) Moech	Manzanilla	Pe, F, Ap	G, H, L	a	38, 43
	Mantisalca salmantica (L.) Briq. & Cavillier	Escoba	L, St	A	a, b	7, 10, 11, 12
	Matricaria chamomilla L.	Manzanilla	I, F	G, H	a, b	40, 42
	Reichardia intermedia (Schultz Bip) Samp.	Lechuguilla		A	a, b	7, 13, 15, 24
	Scolymus hispanicus L.	Cardillo	L, St, R, F	A, I	a	4, 11, 12, 13, 15, 19, 20, 21, 32
	Scorzonera angustifolia L.	Teta de vaca	L, Pf, F	A, K	a, b	7, 11
	Scorzonera hispanica L.	Alcarcionera	Bl, R	K	b	
	Scorzonera laciniata L.	Berbaja	L, F, R, St, F	A, B, K	a, b	4, 7, 11, 13
	Silybum marianum (L.) Gaertner	Cardo mariano	Wp, S, Fr	A, I, K	a, b	7, 10, 11, 13, 15, 19
	Sonchus asper (L.) Hill	Cerraja	L	A	a, b	6, 7, 9, 13, 15
	Sonchus oleraceus L.	Cerraja	L, Bs, R	A, B	a, b	2, 4, 5, 6, 7, 9, 10, 11, 12, 13, 14, 15, 24
	Sonchus tenerrimus L.	Cerraja	L	A	a, b	5, 7
	Tragopogon porrifolius L.	Teta de vaca	L, I, R	A, K	a, b	7, 11
	Urospermum picroides (L.) Scop. ex F.W. Schmidt	Cerrajón	L, Bs, R	A, B, H	a, b	7, 9, 13 ,15
	Xanthium spinosum L.	Cadillo	Wp	G	c	38
Boraginaceae	*Borago officinalis* L.	Borraja	L, St, F	A, K	a, b	4, 7, 11, 12, 15, 24
Brassicaceae	*Capsella bursa-pastoris* (L.) Medik.	Bolsa de pastor	L, Ap, Fr	A, G, K	a, b	1, 12, 38
	Eruca vesicaria (L.) Cav.	Oruga	L, Bs, S	A, D, I	a, b	4, 7, 8, 10, 11, 15, 21
	Rapistrum rugosum (L.) All.	Ravaniza	L	A	a	6, 10, 11
	Sisymbrium irio L.	Rabaniza	L	A	a	11
	Sisymbrium orientale L.	Tamarilla		A	a, b	6, 8

Table A1. *Cont.*

Family	Scientific Name	Vernacular Name	PP	Subc	Pr	CL F
Campanulaceae	*Campanula rapunculus* L.	Vara de San José	Bs, R, Fr	A,B,G	a,b	7,13,38
	Lonicera etrusca G. Santi	Madreselva	F	K	b	n.d.
Caprifoliaceae	*Lonicera implexa* Aiton	Madreselva	F, Ap	G, K	b	38
	Lonicera periclymenum subsp *hispanica* (Boiss & Reuter) Nyman	Madreselva	I, F	G, K	c	38
	Herniaria cinerea DC.	Arenaria	Wp	G	c	38
	Herniaria glabra L.	Arenaria	Wp	G	c	38
Caryophyllaceae	*Herniaria lusitanica* Chaudhri	Arenaria	Wp	G	c	38
	Herniaria scabrida Boiss	Arenaria	Wp	G	c	38
	Silene vulgaris (Moench) Garcke	Colleja	L, St, R, Wp	A, B, L	a, b	4, 6, 7, 9, 10, 11, 12, 15, 34
	Stellaria media (L.) Vill.	Pamplina	L	A	a, b	7, 11
Cistaceae	*Cistus albidus* L.	Estepa	S	D	c	n.d.
	Cistus ladanifer L.	Jara	F, S, Sv	A, D, K	b	n.d.
Crassulaceae	*Sedum album* L.	Arroz	L	A, K	b	33, 34
	Umbilicus gaditanus Boiss.	Campanitas		A	c	n.d.
	Umbilicus rupestris (Salisb.) Dandy	Ombligo de Venus	L	A	a, b	7, 11
Cucurbitaceae	*Bryonia dioica* Jacq.	Nueza	Bs, Z	A	a, b	1, 4, 7, 8, 11, 12
Cyperaceae	*Scirpoides holoschoenus* (L.) Soják	Junco	S, Bl	E, K	a, b	2, 22
Cytinaceae	*Cytinus hypocistis* (L.) L.	Colmenita	F	K	b	13, 15
Dennstaedtiaceae	*Pteridium aquilinum* (L.) Kuhn	Helecho	Rz, L	B, K	a, b	5, 22
Dioscoreaceae	*Tamus communis* L.	Espárrago	Bs	A	a, b	4, 11, 13, 15
Equisetaceae	*Equisetum ramosissimum* Desf.	Cola de caballo	Bs, St	A, G	a, b	15, 38
Ericaceae	*Arbutus unedo* L.	Madroño	Fr, L, St, Ba	C, G, I, K	a, b	21, 24, 29, 30, 37, 38, 41
	Bituminaria bituminosa (L) Stirton	Angelota	F	K	b	n.d.
Fabaceae	*Lygos sphaerocarpa* (L.) Boiss	Retama	St, Bs, L	I, L	b	20
	Medicago sativa L.	Alfalfa	Bs, Wp	A, G, H	a	1, 4, 6, 7, 12, 13, 15, 38, 43
	Onobrychis humilis (L.) G. López	Carretilla	F	K	b	n.d.
	Spartium junceum L.	Retama	F	G	c	38
Fagaceae	*Quercus rotundifolia* Lam.	Encina	Fr	D, F, G, H	a, b	5, 13, 17, 22, 24, 27, 38, 40, 43
	Quercus suber L.	Alcornoque	Fr	D	a	13
Gentianaceae	*Centaurium erythraea* subsp *grandiflorum* (Biv) Melderis	Centaura	Wp	G	c	38
	Centaurium maritimum (L.) Fritsch	Centaura	Wp	G	c	38
	Centaurium pulchellum (Swartz) Druce	Centaura	Wp	G	c	38
Hypericaceae	*Hypericum perforatum* L.	Hipérico	F, Ap	G, H, K	a, b	38, 42
	Calamintha nepeta (L.) Savi	Hierba nieta	Wp, Ap, F	G, H, I, L	a	21, 38, 43
	Marrubium vulgare L.	Marrubio	F. Ap	G	c	38
	Melissa officinalis subsp *altissima* (Sibth & Sm) Arcangeli	Toronjil	Fr, Ap, L, F	G, H, I	a, b	21, 37, 38, 42
Lamiaceae	*Mentha pulegium* L.	Poleo	F, Ap, L	G, H, I	a	21, 38, 42
	Origanum vulgare subsp. *vires* (Hoffmanns. & Link) Bonnier & Layens	Orégano	Wp, Ap, F	G, H, I	a, b	21, 37, 38, 42
	Phlomis lychnitis L.	Candilera	L, F	H, K	a, b	42, 43
	Phlomis purpurea L.	Matagallos	L	K	b	
	Rosmarinus officinalis L.	Romero	Bs, L, St, F	G, I, K	b	17, 18, 21, 37, 38
	Sideritis hirsuta L.	Rabo de gato	F, Ap	G, H	a	38, 42
	Thymus mastichina (L.) L.	Mejorana	F, Ap	G, I	b	21, 38
Malvaceae	*Malva sylvestris* L.	Malva	L, P, F, Fr	A, G, H, K	a, b	7, 11, 15, 38, 40, 42
Myrtaceae	*Myrtus communis* L.	Arrayán	Fr, F, Ap, L	C, G, H, I	a, b	20, 21, 25, 30, 37, 38
Oleaceae	*Olea europaea* L.	Acebuche	Fr	D	b	17, 33
	Phillyrea angustifolia L.	Labiérnago	Wp	I	a	20

Table A1. *Cont.*

Family	Scientific Name	Vernacular Name	PP	Subc	Pr	CL F
Oxalidaceae	*Oxalis corniculata* L.	Trébol	L	A	c	n.d.
	Papaver dubium L.	Amapola	L, P	A	a, b	6, 7, 27
	Papaver hybridum L.	Amapola	L, St	A	a, b	7, 10, 11
Papaveraceae	*Papaver pinnatifidum* Moris	Amapola	L	A	a, b	6, 7, 27
	Papaver rhoeas var. *rhoeas* L.	Amapola	L, St, P, F, S	A, G, I	a, b	4, 7, 8, 10, 11, 12, 13, 15, 21, 38
	Roemeria hybrida (L.) DC.	Amapola morada	Bs	A	a, b	4, 7, 10, 12, 13
Pinaceae	*Pinus pinea* L.	Pino piñonero	S, Fr	D, G, I	a, b	11, 20, 21, 24, 34, 38
	Digitalis purpurea subsp *toletana* (Font Quer) Hinz *	Digital	F	K	b	n.d.
Plantaginaceae	*Digitalis thapsi* L. *	Biloria	F	K	b	n.d.
	Veronica anagallis-aquatica L.	Berro	L, St	A, K	a, b	7
	Arundo donax L.	Caña	St, Bl	I, K	b	20, 21
Poaceae	*Cynodon dactylon* (L.) Pers.	Grama	Rz, Wp	B, G, H, K	a, b	13, 22, 38, 40, 42
	Elymus repens (L.) Gould	Grama	Rz	B	a	4, 22
Polygonaceae	*Rumex pulcher* L.	Romaza	L, Ap	A, H	a	4, 6, 7, 10, 13, 15, 40, 43
Portulacaceae	*Montia fontana* subsp. *amporitana* Sennen	Pamplina	L, St	A	b	7
	Portulaca oleracea L.	Verdolaga	L, St, Bs	A	a, b	4, 7, 10, 11, 15, 34
	Agrimonia eupatoria L.	Agrimonia	Wp	G	c	38
Rosaceae	*Crataegus monogyna* Jacq.	Espino albar	L, St, Bs, Fr, S	A, C, G, H	a, b	30, 32, 41, 43
	Rubus ulmifolius Schott	Zarzamora	L, Bs, F, Fr, A	A, C, G, H, K	a, b	4, 13, 24, 30, 38, 40
Rutaceae	*Ruta angustifolia* Pers.	Ruda	Wp, L, F, S	G, H, I	b	4, 12, 17, 18, 21, 38, 40
	Ruta montana L.	Ruda	Wp, L, F, S	G, H, I	b	4, 12, 17, 18, 21, 38, 40
Smilaceae	*Smilax aspera* var. *altissima* Moris & De Not	Zarzaparrilla	Bs, R	A, G, H	a, b	12, 38, 40, 44
Solanaceae	*Solanum nigrum* L.	Hierba mora		C	b	n.d.
Ulmaceae	*Celtis australis* L.	Almez	Fr	C, G, J	b	23, 38
	Ulmus minor Miller	Olmo	L, Bs, F	A, K	b	n.d.
Urticaceae	*Urtica dioica* L.	Ortiga	Bs, Wp	A, G	a, b	1, 3, 4, 7, 11, 12, 13, 38
	Urtica membranacea Poiret	Ortiga	L, St	A	a	3, 4, 7, 13
	Urtica urens L.	Ortiga	L, St, Wp	A, G	a	1, 4, 7, 10, 11, 12, 13, 15, 24, 38
Valerianaceae	*Valerianella microcarpa* Desv.	Canónigo	L	A	c	7

PP = Parts of the used plant (BL = Base of leaf/stem; Bp = Base of the pseudostem; Bs = Tender shoots; B = Bulb; Ba = Bark; F = Flowers; Fr = Fruits; L = Leaves; I = Inflorescence; Ap = Aerial part; Fp = Floral peduncle; Pn = "Penca"; P = Petals; Wp = Whole plant; R = Root; Rz = Rhizome; Sv = Sap/resin/exudate/latex; S = Seeds; St = Stems; Tu = Tubers; Z = tendrils). Subc = Subcategories of food uses proposed by the Spanish Inventory of Traditional Knowledge Relating to Biodiversity (A = Vegetables; B = Roots, bulbs, tubers, and rhizomes; C = Sweet fruits; D = Dry and oleaginosus fruits; E = Cereals and pseudocereals; F = Fat; G = Alcoholic drinks; H = Non-alcoholic drinks; I = Condiments; J = Sugars and sweeteners; K = Candies and chewing; L = Other food uses); Pr = Preparation (a = cooked, b = raw, c = unknown); Cl F = Classification of foods according to Bertrand (2015) (1 = Soups; 2 = "Gofio" (= roasted cereal stirred into liquid); 3 = Broths and purees; 4 = Omelettes; 5 = Cakes; 6 = Rices; 7 = Salads; 8 = Cold vegetable soups; 9 = Pies/Patties; 10 = Potages; 11 = Stews; 12 = Scrambled; 13 = Sautéed/boiled/toasted; 14 = Milky; 15 = Fried/breaded; 16 = Sauces; 17 = Oil; 18 = Vinegar; 19 = Renner; 20 = Preservative; 21 = Dressings/condiments; 22 = Flour; 23 = Sugar; 24 = Desserts; 25 = Ice creams; 26 = Sorbets; 27 = Cakes/Cakes; 28 = Sweet; 29 = Jams; 30 = Marmalades; 31 = Jellies; 32 = To the natural; 33 = Brine; 34 = Pickles; 35 = In syrup; 36 = In alcohol; 37 = Wines; 38 = Liquors; 39 = Syrups for cocktails; 40 = Juices; 41 = Brandies; 42 = Infusions; 43 = Coffee/tea; 44 = Diluted syrups). * Toxic plants. n.d. = no data.

Appendix B.

Box A1. Vegetables (85 taxa).

The most common:

Asteraceae (30.59%), Liliaceae (9.41%), and Apiaceae (8.24%).

Part consumed:

Leaves (65.88%), followed by the tender shoots (35.29%) or the stems (24.71%). Other parts as the flowers or fruits are less used (10.59%, 5.88%). More seldom are tendrils, (2.35%), roots and bulbs, petals, or specifically the floral receptacle or peduncule (1.18%).

Consumption mode:

Five peculiar cases—*Andryala laxiflora, Bellis perennis, Crataegus monogyna, Montia fontana* subsp. *Amporitana,* and *Sedum album*—are exclusively consumed in crude for brine, entertainment, or salads.

The largest group of useful vegetable plants (50 species) is composed by those than can be consumed either raw or cooked in omelettes, stews, purées, soups, rice, and pastries; regarding their preparation, they can be fried, scrambled, or boiled. These are: *Allium roseum, Andryala integrifolia, Apium graveolens, Apium nodiflorum, Bryonia dioica, Calendula arvensis* subsp. *arvensis, Calendula arvensis* subsp. *macroptera, Campanula rapunculus, Carthamus lanatus, Chenopodium album, Chondrilla juncea, Cichorium intybus, Crepis vesicaria* subsp. *haenseleri, Cynara cardunculus* subsp. *cardunculus, Equisetum ramosissimum, Eruca vesicaria, Foeniculum vulgare, Foeniculum vulgare* subsp. *piperitum, Malva sylvestris, Mantisalca salmantica, Papaver dubium, Papaver hybridum, Papaver pinnatifidum, Papaver rhoeas* var. *rhoeas, Portulaca oleracea, Portulaca oleracea* subsp. *granulato stellulata, Reichardia intermedia, Roemeria hybrida, Rubus ulmifolius, Sambucus nigra, Scandix australis* subsp. *australis, Scandix australis* subsp. *microcarpa, Scandix pecten-veneris, Scorzonera angustifolia, Scorzonera laciniata, Silene vulgaris, Silybum marianum, Sisymbrium orientale, Smilax aspera* var. *altissima, Smilax aspera* var. *aspera, Sonchus asper, Sonchus oleraceus, Sonchus tenerrimus, Stellaria media, Tamus communis, Tragopogon porrifolius, Umbilicus rupestris, Urospermus picroides, Urtica dioica,* and *Veronica anagallis-aquatica.*

Eighteen species have been popularly consumed stewed: *Aristolochia paucinervis* and *Aristolochia pistolochia* (whose fresh stems are used to make a stew in which they are fried with oil and garlic, salt, bread crumbs, paprika, and eggs), *Asphodelus aestivus, Asphodelus albus* subsp. *albus, Asphodelus albus* subsp. *villarsii, Asphodelus fistulosus, Asphodelus ramosus, Borago officinalis* (although the flowers may be eaten raw), *Capsella bursa-pastoris, Cynara humilis* (from which the inflorescences are consumed), *Medicago sativa, Rapistrum rugosum, Rumex pulcher, Ruscus aculeatus* (in omelettes), *Scolymus hispanicus* subsp. *hispanicus, Sisymbrium irio, Urtica membranacea,* and *Urtica urens* (in salad after being scalded or cooked).

Box A2. Roots, bulbs, tubers, and rhizomes (21 taxa).

Sweet fruits (seven taxa)

Families:

Belonging to Rosaceae and five more families: Caprifoliaceae, Ericaceae, Myrtaceae, Solanaceae, and Ulmaceae.

Part consumed:

Fruits

Consumption mode:

Arbutus unedo, Crataegus monogyna, Myrtus communis, and *Rubus ulmifolius* have been used both for direct consumption or as raisins, as for elaborations such as liquors, jams, ice cream, and various desserts.

In the case of *Sambucus nigra,* although in some areas it was said that its fruits were poisonous, once ripe, they have been used to make jams.

Only in two cases—*Celtis australis* and Solanum *nigrum*—is the raw consumption of fruits mentioned as a treat.

Dry and oleaginous fruits (eight taxa)

Families:

Fagaceae, Oleaceae, Pinaceae, Brassicaceae, and Cistaceae belong to this group.

Part consumed:

Seeds and fruits

Consumption mode:

For the species *Pinus pinea* and *Quercus rotundifolia,* seeds can be eaten raw, but they are toasted, roasted, or cooked to make them more palatable. Regarding Quercus suber, its oak is eaten cooked, and it has stringent properties.

For the species *Cistus ladanifer, Eruca vesicaria,* and *Olea europaea,* raw seeds or fruits are consumed directly from the plant or after processing, such as in brines or oils.

Finally, we have reported the mention of elaborated *Cistus albidus* seeds as being very appetizing.

Cereals and pseudocereals; legumes; fat (two taxa)

Families:

Cyperacerae and Fagaceae.

Part consumed:

Seeds and fruits

Consumption mode:

Scirpoides holoschoenus, whose seeds can form a possible base for the preparation of flours.

Quercus rotundifolia, which takes advantage of the food fats in their fruits, extracting them by cooking.

Box A3. Alcoholic drinks (58 taxa).

The most common:

Lamiaceae (13.79%), Asteraceae (10.34%), Caryophyllaceae (10.34%), Apiaceae (8.62%), Gentianaceae (6.90%), and 20 more families.

Part consumed:

Flowers (25.86%) or the fruit/seeds (15.52%) of the whole of the aerial part (51.79%).

Consumption mode:

All are used as ingredients in the elaboration of liquors.

These species are: *Agrimonia eupatoria* subsp. *eupatoria, Anthemis cotula, Arbutus unedo, Asplenium ceterach, Calamintha nepeta* subsp. *nepeta = Satureja calamintha, Campanula rapunculus, Capsella bursa-pastoris, Celtis australis, Centaurium erythraea* subsp. *grandiflorum, Centaurium erythraea* subsp. *erythraea, Centaurium maritimum, Centaurium pulchellum, Chamaemelum nobile, Crataegus monogyna, Cynodon dactylon* var. *dactylon, Cynodon dactylon* var. *villosum, Dittrichia viscosa, Equisetum ramosissimum, Foeniculum vulgare, Foeniculum vulgare* subsp. *piperitum, Helichrysum stoechas, Herniaria cinerea, Herniaria glabra, Herniaria lusitanica, Herniaria lusitanica* var. *gaditana, Herniaria scabrida, Herniaria scabrida* subsp. *guadarramica, Hypericum perforatum, Lonicera implexa, Lonicera peryclimenum* subsp. *hispanica, Malva sylvestris, Marrubium vulgare, Matricaria chamomilla, Medicago sativa, Melissa officinalis* subsp. *altissima, Mentha pulegium, Myrtus communis, Origanum vulgare* subsp. *virens = Origanum virens, Papaver rhoeas* var. *rhoeas, Pinus pinea, Pistacia lentiscus, Quercus rotundifolia, Rosmarinus officinalis, Rubus ulmifolius, Ruta angustifolia, Ruta montana, Sambucus nigra, Scandix australis* subsp. *australis, Scandix australis* subsp. *microcarpa, Sideritis hirsuta, Smilax aspera* var. *altissima, Smilax aspera* var. *aspera, Spartium junceum, Thapsia villosa, Thymus mastichina, Urtica dioica, Urtica urens,* and *Xanthium spinosum.*

Box A4. Non-alcoholic drinks (32 taxa).

The most common:
Mostly Asteraceae (21.88%), Lamiaceae (18.75%) and Apiaceae (12.50%).
Part consumed:
It is prepared with the plant or its flowers (90.63%), and less frequently (9.38%) with the fruits or seeds.
Consumption mode:
Elaboration uses include by decoction, maceration, or infusions, and to a lesser extent for syrups or juices (as in *Sambucus nigra*).

Seven of these plants are considered medicinal: *Achillea ageratum, Foeniculum vulgare, Foeniculum vulgare* subsp. *piperitum, Matricaria chamomilla, Scandix australis* subsp. *australis, Scandix australis* subsp. *Microcarpa,* and *Sideritis hirsuta.* The rest (*Chamaemelum fuscatum, Chamaemelum nobile, Cynodon dactylon* var. *dactylon, Cynodon dactylon* var. *villosum, Hypericum perforatum, Malva sylvestris, Melissa officinalis* subsp. *altissima, Mentha pulegium, Myrtus communis, Origanum vulgare* subsp. *virens, Phlomis lychnitis, Rubus ulmifolius, Rumex pulcher, Ruta angustifolia, Ruta montana, Sambucus nigra, Smilax aspera* var. *Altissima,* and *Smilax aspera* var. *aspera*) are associated with refreshing properties, and whose drinks were taken for the mere pleasure of tasting them.

Finally, seven species have been cited as older everyday drinks: *Calamintha nepeta, Cichorium intybus, Crataegus monogyna, Helichrysum stoechas, Medicago sativa, Quercus rotundifolia,* and *Urospermus picroides.*

Box A5. Condiments (32 taxa).

The most common:
Lamiaceae (18.75%), Apiaceae (15.63%), and Asteraceae (15.63%).
Part consumed:
The plant (31.25%), the leaf (28.13%), the flower (21.88%), the fruit/seed (19.01%), the stem (15.63%), and less frequently, the bark or the bulb (3.13%).
Consumption mode:
Twenty-two of these plants are specially significant for the dressing of olives, stews, and meats: *Allium ampeloprasum, Allium roseum, Calamintha nepeta* subsp. *nepeta = Satureja calamintha, Dittrichia graveolens, Eruca vesicaria, Foeniculum vulgare, Foeniculum vulgare* subsp. *piperitum, Melissa officinalis* subsp. *altissima, Mentha pulegium, Myrtus communis, Origanum vulgare* subsp. *virens = Origanum virens, Papaver rhoeas* var. *rhoeas, Rosmarinus officinalis, Ruscus aculeatus, Ruta angustifolia, Ruta montana, Scandix australis* subsp. *australis, Scandix australis* subsp. *microcarpa, Scolymus hispanicus* subsp *hispanicus, Thapsia villosa, Thymus mastichina,* and *Viburnum tinus.*

Six species (*Arbutus unedo, Arundo donax, Lygos sphaerocarpa = Retama sphaerocarpa, Phillyrea angustifolia, Pinus pinea,* and *Urginea maritima*) are appreciated for their preservative properties, and are used to maintain the quality of game meats, healing hams, olives, and cheeses.

Only in the case of *Pistacia lentiscus* do known uses include it both as a seasoning and as a preservative for olives and game meat. Finally, the flowers and fruits of *Carthamus lanatus, Cynara cardunculus* subsp. *Cardunculus,* and *Silybum marianum,* can be used for curdling milk and making cheeses.

Box A6. Sugars and sweeteners (one taxon).

Family:
Ulmaceae.
Part consumed:
Fruit.
Consumption mode:
Celtis australis, Ulmaceae, whose ripe fruits were boiled to be used as a sugar substitute in times of scarcity.

Box A7. Candies and chewing (43 taxa).

The most common:
Asteraceae (23.26%) and 26 more families.
Part consumed:
The parts of the plants of this group that have been consumed as sweets and masticatories are very diverse, from the flowers (48.84%), leaves (9.30%), stems (6.98%), fruits (6.98%), and roots/rhizome (11.63%), to the sap, latex (2.33%), or the gills (2.33%).
Consumption mode:
Except in specific cases in which the original product underwent a slight transformation (for example in the production of gum with the sap of Pistacia lentiscus), consumption was directly from the plant, and without subjecting them to any alteration of their properties and natural state.
The most appreciated part is the flowers, since in the case of 21 plants, this is the part that is most consumed (*Bellis perennis, Bituminaria bituminosa, Borago officinalis, Cichorium intybus, Cytinus hypocistis, Cytinus hypocistis* subsp *macranthus, Digitalis purpurea* subsp *toletana, Digitalis thapsi, Hypericum perforatum, Lonicera etrusca, Lonicera implexa, Lonicera peryclimenum* subsp. *hispanica, Narcissus bulbocodium, Onobrychis humilis, Onobrychis humilis* var. *glabrescens, Phlomis lychnitis, Rosmarinus officinalis, Scorzonera angustifolia, Scorzonera laciniata, Ulmus minor,* and *Vinca difformis*). The resting 22 species of this group are: *Andryala ragusina, Arbutus unedo, Arundo donax, Capsella bursa-pastoris, Chondrilla juncea, Cistus ladanifer, Cynara cardunculus* subsp. *cardunculus, Cynodon dactylon* var. *dactylon, Cynodon dactylon* var. *villosum, Foeniculum vulgare, Foeniculum vulgare* subsp. *piperitum, Malva sylvestris, Phlomis purpurea, Pistacia lentiscus, Pteridium aquilinum, Rubus ulmifolius, Scirpoides holoschoenus = Scirpus holochoenus, Scorzonera hispanica, Sedum album, Silybum marianum, Tragopogon porrifolius,* and *Veronica anagallis-aquatica.*

Box A8. Other food uses (nine taxa).

Families:
Diverse and singular (Anacardiaceae, Apiaceae, Asteraceae, Caryophyllaceae, Fabaceae, and Lamiaceae).
Part consumed:
In most cases (66.67%), the whole plant is applied.
Consumption mode:
Calamintha nepeta and *Helichrysum stoechas* improve the quality of wine; the florets of *Cynara humilis* can be used as curd milk for obtaining cheese; *Silene vulgaris* can be used to clean the cheeses; *Foeniculum vulgare* and/or *Foeniculum vulgare* subsp *piperitum* can be used as cooking water to wash the pig guts (in the slaughters) to disinfect them and neutralize the smell; the tender shoots and buds of *Lygos sphaerocarpa* and *Pistacia lentiscus* can be used to purify water tanks. Finally, *Matricaria chamomilla* removes the bad taste after drinking a bitter almond.

References

1. Şerban, P.; Wilson, J.R.U.; Vamosi, J.C.; Richardson, D.M. Plant Diversity in the Human Diet: Weak Phylogenetic Signal Indicates Breadth. *Bioscience* **2008**, *58*, 151–159. [CrossRef]
2. Myers, N. *A Wealth of Wild Species: Storehouse for Human Welfare*; Westview press: Boulder, CO, USA, 1983.
3. Maurer, M.; Schueckler, A. *Use and Potential of Wild Plants in Farm Households*; Food and Agriculture Organization of the United Nations: Rome, Italy, 1999.
4. Alarcón, R.; Pardo-De-Santayana, M.; Priestley, C.; Morales, R.; Heinrich, M. Medicinal and local food plants in the south of Alava (Basque Country, Spain). *J. Ethnopharmacol.* **2015**, *176*, 207–224. [CrossRef] [PubMed]
5. Sansanelli, S.; Tassoni, A. Wild food plants traditionally consumed in the area of Bologna (Emilia Romagna region, Italy). *J. Ethnobiol. Ethnomed.* **2014**, *10*, 69–80. [CrossRef] [PubMed]
6. Kang, Y.; Łuczaj, Ł.; Kang, J.; Wang, F.; Hou, J.; Guo, Q. Wild food plants used by the Tibetans of Gongba Valley (Zhouqu county, Gansu, China). *J. Ethnobiol. Ethnomed.* **2014**, *10*, 20–33. [CrossRef] [PubMed]
7. Van den Eynden, V.; Cueva, E.; Cabrera, O.; Eynden, V. Van den Wild Foods from Southern Ecuador. *Econ. Bot.* **2003**, *57*, 576–603. [CrossRef]
8. Walsh, B. How the world eats. Available online: www.time.com/time/specials/2007/article/0,28804, 1628191_1626317_1626671,00.html (accessed on 27 September 2018).
9. Carvalho, A.M.; Barata, A.M. The Consumption of Wild Edible Plants. In *Wild Plants, Mushrooms and Nuts*; Ferreira, I.C.F.R., Morales, P., Barros, L., Eds.; John Wiley & Sons: Hoboken, NJ, USA, 2016; pp. 169–198.

10. Davidson-Hunt, I.J.; Turner, K.L.; Te Pareake Mead, A.; Cabrera-Lopez, J.; Bolton, R.; Idrobo, C.J.; Miretski, I.; Morrison, A.; Robson, J.P. Biocultural design: A new conceptual framework for sustainable development in rural indigenous and local communities. *Sapiens* **2012**, *5*, 33–45.

11. Evans, D. Building the European Union's Natura 2000 network. *Nat. Conserv.* **2012**, *1*, 11–26. [CrossRef]

12. Jackson, A.L.R. Renewable energy vs. biodiversity: Policy conflicts and the future of nature conservation. *Glob. Environ. Chang.* **2011**, *21*, 1195–1208. [CrossRef]

13. Cancino, S.J.; Torres, B.M.J.; Jiménez, E.O.; Garay, A.H.; Pérez, J.P.; Quiroz, J.F.E.; Carrillo, A.R.Q. Evaluación de la distancia entre plantas sobre el rendimiento y calidad de semilla de Brachiaria brizantha. *Tec. Pecu. Mex.* **2010**, *48*, 297–310.

14. INE. Instituto Nacional de Estadística. Available online: https://www.ine.es/ (accessed on 19 December 2018).

15. Natura 2000 Standard Dataform. Sierra Grande de Hornachos. Available online: http://natura2000.eea.europa.eu/Natura2000/SDF.aspx?site=ES0000072 (accessed on 19 December 2018).

16. García-Sánchez, E.; Hernández Bermejo, J.E.H.-B. Ornamental Plants in Agricultural and Botanical Treatises from Al-Andalus. In *Middle East Garden Traditions: Unity and Diversity*; Conan, M., Ed.; Dumbarton Oaks Research Library and Collection: Washington D.C., Washington, USA, 2008; pp. 75–94.

17. Jug-Dujaković, M.; Łukasz, Ł. The contribution of Josip Bakić' s research to the study of wild edible plants of the adriatic coast: a military project with ethnobiological and anthropological implications. *Slovak Ethnol.* **2016**, *2*, 158–168.

18. Devesa Alcaraz, J.A. *Vegetación y Flora de Extremadura*; Universitas Editorial: Badajoz, España, 1995.

19. Blanco-Salas, J.; Ruiz-Téllez, T. Propuesta innovadora de valorización de la biodiversidad vegetal del espacio protegido "ZIR Sierra Grande de Hornachos". In Proceedings of the VI Congreso Forestal Español, Extremadura, Spain, 26–30 Junio 2017; pp. 399–404.

20. Łuczaj, Ł.; Pieroni, A.; Tardío, J.; Pardo-De-Santayana, M.; Sõukand, R.; Svanberg, I.; Kalle, R. Wild food plant use in 21st century Europe: The disappearance of old traditions and the search for new cuisines involving wild edibles. *Acta Soc. Bot. Pol.* **2012**, *81*, 359–370. [CrossRef]

21. Pardo-De-Santayana, M.; Aceituno-Mata, L.; Molina, M. *Inventario Español de Conocimientos Tradicionales Relativos a la Biodiversidad. Primera Fase: Introducción, metodología y fichas*; Ministerio de Agricultura, Alimentación y Medio Ambiente: Madrid, Spain, 2014.

22. Pardo de Santayana, M.; Morales, R.; Tardío, J.; Molina, M. *Inventario Español de Conocimientos Tradicionales Relativos a la Biodiversidad. Fase II (1)*; Ministerio de Agricultura y Pesca, Alimentación y Medio Ambiente: Madrid, Spain, 2018.

23. Pardo de Santayana, M.; Morales, R.; Tardío, J.; Aceituno-Mata, L.; Molina, M. *Inventario Español de Conocimientos Tradicionales Relativos a la Biodiversidad. Fase II (2)*; Ministerio de Agricultura y Pesca, Alimentación y Medio Ambiente: Madrid, Spain, 2018.

24. Pardo de Santayana, M.R.; Aceituno-Mata, L. Metodología para el Inventario Español de los Conocimientos Tradicionales relativos a la Biodiversidad. In *Inventario Español de los Conocimientos Tradicionales Relativos a la Biodiversidad*; Ministerio de Agricultura, Alimentación y Medio Ambiente: Madrid, Spain, 2014; pp. 31–49.

25. GBIF. Global Biodiversity Information Facility. Available online: https://www.gbif.org (accessed on 1 July 2018).

26. Anthos. Sistema de Información sobre las Plantas en España. Available online: http://www.anthos.es/ (accessed on 1 July 2018).

27. Chase, M.W.; Christenhusz, M.J.M.; Fay, M.F.; Byng, J.W.; Judd, W.S.; Soltis, D.E.; Mabberley, D.J.; Sennikov, A.N.; Soltis, P.S.; Stevens, P.F.; et al. An update of the Angiosperm Phylogeny Group classification for the orders and families of flowering plants: APG IV. *Bot. J. Linn. Soc.* **2016**, *181*, 1–20. [CrossRef]

28. Bertrand, B. *Cocinar con Plantas Silvestres*; Editorial la Fertilidad de la Tierra: Madrid, Spain, 2015.

29. Persson, L.Å. Prenatal nutrition, socioenvironmental conditions, and child development. *Lancet Glob. Health* **2017**, *5*, 127–128. [CrossRef]

30. Ogle, B.M. Wild vegetables and micronutrient nutrition. Studies on the significance of wild vegetables in women's diets in Vietnam. *Acta Univ. Ups. Diss. Fac. Med.* **2001**, *1056*, 1–59.

31. Das, J.K.; Lassi, Z.S.; Hoodbhoy, Z.; Salam, R.A. Nutrition for the Next Generation: Older Children and Adolescents. *Ann. Nutr. Metab.* **2018**, *72*, 56–64. [CrossRef] [PubMed]

32. Ghosh-Jerath, S.; Singh, A.; Magsumbol, M.S.; Kamboj, P.; Goldberg, G. Exploring the Potential of Indigenous Foods to Address Hidden Hunger: Nutritive Value of Indigenous Foods of Santhal Tribal Community of Jharkhand, India. *J. Hunger Environ. Nutr.* **2016**, *11*, 548–568. [CrossRef] [PubMed]

33. Flyman, M.V.; Afolayan, A.J. The suitability of wild vegetables for alleviating human dietary deficiencies. *South African J. Bot.* **2006**, *72*, 492–497. [CrossRef]

34. Tardío, J.; Molina, M.; Aceituno-Mata, L.; Pardo-de-Santayana, M.; Morales, R.; Fernández-Ruiz, V.; Morales, P.; García, P.; Cámara, M.; Sánchez-Mata, M.C. Montia fontana L. (Portulacaceae), an interesting wild vegetable traditionally consumed in the Iberian Peninsula. *Genet. Resour. Crop Evol.* **2011**, *58*, 1105–1118. [CrossRef]

35. Larrea, J.L. *Innovación Abierta y Alta Cocina: Aprender a Innovar con Mugaritz*; Piramide: Madrid, Spain, 2011.

36. Corcuera, M. *25 años de la Nueva Cocina Vasca*; Askorri: Bilbao, Spain, 2003.

37. Harrintong, R.J. The Culinary Innovation Process: A barrier to Imitation. *J. Foodserv. Bus. Res.* **2004**, *7*, 35–37. [CrossRef]

38. Balaguer, O. *La Nouvelle Cuisine Des Desserts*; Montagud: Zaragoza, Spain, 2013.

39. Kiialainen, S. *Kuura New Nordic Cuisine*; Sima Productions: Zaragoza, Spain, 2014.

40. Gardoni, A.G. *Food by Design*; Booth-Clibborn: London, UK, 2002.

41. Myhrvold, N.; Young, C.; Bilet, M.; Smith, R.M.; Cooking Lab. *Modernist Cuisine: The Art and Science of Cooking*; Cooking Lab: Köln, Germany, 2011.

42. Stummerer, S.; Hablesreiter, M. *Food Design XL*; Springer: Vienna, Austria, 2009.

43. Weir, C.; Weir, R. *Ice Creams, Sorbets and Gelati: The Definitive Guide*; Grub Street: London, UK, 2010.

44. Hamilton, R.; Todoli, V. *Comida Para Pensar, Pensar Sobre El Comer: Una Reflexión Sobre el Universo Creativo de Ferrán Adriá, la Cocina de Vanguardia y su Relación con el Mundo del Arte*; Actar: Barcelona, Spain, 2009.

45. Dong, D. *Artistic Conception of Chinese Cuisine*; Chemical Industry Press: Pekin, China, 2013.

46. Puglisi, C.F. *Relæ: A Book of Ideas*; Ten Speed Press: New York, NY, USA, 2014.

47. Fernandez Guadaño, M. *Reinventores: Descubra Creatividad Y Nuevos Modelos De Negocio Con Los Mejores Cocineros*; Conecta: Barcelona, Spain, 2012.

48. Aduriz, L. *Diccionario botánico para cocineros*; Gourmandia: Donostia, Spain, 2006.

49. Brown, M. *Cereales, Semillas y Legumbres*; Lunwerg: Madrid, Spain, 2015.

50. Earle, M.; Earle, R.; Anderson, A. *Food Product Development: Maximising Success*; Wodohead Publishing: Cambridge, UK, 2011.

51. Poonia, A.; Upadhayay, A. Chenopodium album Linn: Review of nutritive value and biological properties. *J. Food sci. Technol.* **2015**, *52*, 3977–3985. [CrossRef] [PubMed]

52. Yarnell, E. Plant Chemistry in Veterinary Medicine: Medicinal Constituents and Their Mechanisms of Action. In *Veterinary Herbal Medicine*; Elsevier: Amsterdam, The Netherlands, 2007; pp. 159–182.

53. Guil-Guerrero, J.L.; Giménez-Giménez, A.; Rodríguez-García, I.; Torija-Isasa, M.E. Nutritional composition of Sonchus species (S. asper L., S. oleraceus L. and S. tenerrimus L.). *J. Sci. Food Agric.* **1998**, *76*, 628–632. [CrossRef]

54. Zucca, P.; Pintus, M.; Manzo, G.; Nieddu, M.; Steri, D.; Rinaldi, A.C. Antimicrobial, antioxidant and anti-tyrosinase properties of extracts of the Mediterranean parasitic plant Cytinus hypocistis. *BMC Res. Notes* **2015**, *8*, 562–580. [CrossRef] [PubMed]

55. Morales, P.; Fernández-Ruiz, V.; Sánchez-Mata, M.; Cámara, M.; Carvalho, A.M.; Pardo de Santayana, M.; Tardío, J.; Ferreira, I.C.F.R. Valoración de la actividad antioxidante de verduras silvestres. In Proceedings of the VII Congr. Ibérico AgroIngeniería y Ciencias Hortícolas, Madrid, Spain, 26–29 Agosto 2013.

56. Veberic, R.; Jakopic, J.; Stampar, F.; Schmitzer, V. European elderberry (Sambucus nigra L.) rich in sugars, organic acids, anthocyanins and selected polyphenols. *Food Chem.* **2009**, *114*, 511–515. [CrossRef]

57. Diao, W.R.; Hu, Q.P.; Zhang, H.; Xu, J.G. Chemical composition, antibacterial activity and mechanism of action of essential oil from seeds of fennel (Foeniculum vulgare Mill.). *Food Control* **2014**, *35*, 109–116. [CrossRef]

58. Morgado, S.; Morgado, M.; Plácido, A.I.; Roque, F.; Duarte, A.P. Arbutus unedo L.: From traditional medicine to potential uses in modern pharmacotherapy. *J. Ethnopharmacol.* **2018**, *225*, 90–102. [CrossRef]

59. De Cortes Sánchez-Mata, M.; Tardío, J. *Mediterranean Wild Edible Plants: Ethnobotany and Food Composition Tables*; Springer: Berlin, Germany, 2016.

60. De Santayana, M.P.; Morales, R. Manzanillas ibéricas: Historia Y Usos Tradicionales. *Rev. Fitoter.* **2006**, *6*, 143–153.

61. Pardo-de-Santayana, M.; Morales, R. Chamomiles in Spain. The Dynamics of Plant Nomenclature. In *Ethnobotany in the New Europe: People, Health and Wild Plant Resources*; Pardo-de-Santayana, M., Pieroni, A., Puri, R.K., Eds.; Berghahn Books: New York, NY, USA, 2010; pp. 282–306.
62. Stanković, M.S.; Radić, Z.S.; Blanco-Salas, J.; Vázquez-Pardo, F.M.; Ruiz-Téllez, T. Screening of selected species from Spanish flora as a source of bioactive substances. *Ind. Crops Prod.* **2017**, *95*, 493–501. [CrossRef]
63. Amor, I.L.B.; Boubaker, J.; Sgaier, M.B.; Skandrani, I.; Bhouri, W.; Neffati, A.; Kilani, S.; Bouhlel, I.; Ghedira, K.; Chekir-Ghedira, L. Phytochemistry and biological activities of Phlomis species. *J. Ethnopharmacol.* **2009**, *125*, 183–202. [CrossRef]
64. Barros, L.; Dueñas, M.; Carvalho, A.M.; Ferreira, I.C.F.R.; Santos-Buelga, C. Characterization of phenolic compounds in flowers of wild medicinal plants from Northeastern Portugal. *Food Chem. Toxicol.* **2012**, *50*, 1576–1582. [CrossRef]
65. World Health Organization Obesity and overweight. Available online: https://www.who.int/news-room/fact-sheets/detail/obesity-and-overweight (accessed on 19 December 2018).
66. Tuberoso, C.I.G.; Rosa, A.; Bifulco, E.; Melis, M.P.; Atzeri, A.; Pirisi, F.M.; Dessì, M.A. Chemical composition and antioxidant activities of Myrtus communis L. berries extracts. *Food Chem.* **2010**, *123*, 1242–1251. [CrossRef]
67. Wolbiś, M. Flavonol glycosides from Sedum album. *Phytochemistry* **1989**, *28*, 2187–2189. [CrossRef]
68. Petropoulos, S.; Karkanis, A.; Fernandes, Â.; Barros, L.; Ferreira, I.C.F.R.; Ntatsi, G.; Petrotos, K.; Lykas, C.; Khah, E. Chemical Composition and Yield of Six Genotypes of Common Purslane (Portulaca oleracea L.): An Alternative Source of Omega-3 Fatty Acids. *Plant Foods Hum. Nutr.* **2015**, *70*, 420–426. [CrossRef] [PubMed]
69. Popescu, A.; Pavalache, G.; Pirjol, T.N.; Istudor, V. Antioxidant Comparative Activity and Total Phenolic Content of Scirpus holoschoenus L. (Holoschoenus vulgaris Link) Depending on Extraction Condition and the Solvent Used. *Rev.Chim* **2016**, *67*, 255–259.

sustainability

MDPI

Article

Cultural Perspective of Traditional Cheese Consumption Practices and Its Sustainability among Post-Millennial Consumers

Zanete Garanti * and Aysen Berberoglu

Department of Business Administration, Faculty of Economics and Administrative Sciences,
Cyprus International University, Lefkosa 99258, North Cyprus; aysenb@ciu.edu.tr
* Correspondence: zgaranti@ciu.edu.tr; Tel.: +90-533-850-5809

Received: 17 July 2018; Accepted: 3 September 2018; Published: 6 September 2018

Abstract: (1) Background: The consumption of traditional foods has been linked to economic, social, and environmental sustainability; therefore, the main challenge of a changing marketplace is to ensure that young generations continue consuming traditional products. The current study uses a consumer culture theory (CCT) perspective to examine the following: (1) the way individuals use their traditional products to identify themselves with the culture and to feel that they are a part of the community, (2) the underlying values that turn young consumers into loyal customers of hellim/halloumi cheese, and (3) its implications to hellim/halloumi producers. (2) Methods: A qualitative research method is applied to study the perceptions of post-millennials towards traditional cheese products from a cultural perspective. (3) Results: The results of the study reveal that loyalty towards traditional food products amongst post-millennials is build based on (1) the memories that surround the food, (2) the rituals that preparing and eating a food involve, and (3) the identity that it builds, allowing people feel sense of belonging to their ethnic group. (4) Conclusions: Loyalty amongst post-millennials towards traditional food products tends to be emotional, rather than rational or behavioral. It allows us to present both theoretical and managerial implications. It also calls for more empirical research to understand the changing marketplace and post-millennials' consumption habits.

Keywords: consumer culture theory; post millennials; cheese; loyalty; Cyprus; traditional food

1. Introduction

Traditional food products are often recognized by means of association with their regional identity [1] and have always played an important role in the European, especially Mediterranean, culture, diet, and economy. As the new consumer generation of post-millennials is emerging, the biggest challenge for the marketplace [2] is to ensure progress that is proficient, so as to ensure that the current needs for an intact environment, social justice, and economic prosperity are met, without restraining the capacity of future generations to satisfy their desires [3]. The social sustainability [4] of traditional food is addressed in the current article, which focuses on traditional cheese product consumption amongst young consumers.

"How do you like your hellim/halloumi?" might sound like a silly question to some, but for Cypriots it is not a joke. With so many options available, everyone has their own favorite. Known as hellim in the Turkish Republic of Northern Cyprus, and halloumi in the Republic of Cyprus, this hard, rubbery, salty cheese represents the history and culture of the Mediterranean island. Recently, a battle to register hellim/halloumi cheese as protected designation of origin (PDO) in the European Union has led to many discussions to clarify who does the hellim/halloumi belong to. Being associated either with Greek or Turkish name, etymologically, the term halloumi points to an Arabic root, and cultural historians insist that the Venetian sources that had encountered halloumi in the pre-Ottoman period [5], originated from Egyptian and Roman civilizations [6].

Traditional food products can be defined based on the following four dimensions: (1) the key production process must be performed in a certain geographical area (national, regional, or local); (2) the recipe of the traditional food product must be authentic in terms of ingredients, raw materials, and the production process; (3) the traditional food product must have been traded for at least 50 years; and (4) it must be part of the gastronomic heritage [7]. Interestingly, hellim/halloumi cheese has many varieties available on the market, which is a rare phenomenon accruing in a case of a traditional food. Innovations in traditional food first might seem like a controversial phenomenon [8], but both the internal and external market demands, as well as industrial developments, impose innovations in traditional food products [9–11]. The innovations in food production in general and hellim/halloumi production in particular, did not interfere with the authenticity of hellim/halloumi, still allowing us to consider hellim/halloumi a Cypriot traditional food product. Firstly, it can only be produced in Cyprus. Secondly, while reaching an agreement on PDO, both sides of the island agreed that, historically, there are two major types of halloumi, fresh and mature. To secure the originality of hellim/halloumi cheese, it is defined that materials used in the production are fresh milk (sheep, goat, and cow), rennet, and fresh or dried Cypriot mint leaves and salt. It is also defined that the proportion of sheep or goat's milk, or the mixture thereof, must always be greater than the proportion of cow's milk [12]. The regulation allows for many varieties of cheese to be produced, for example, hellim/halloumi from 100% goat milk or 100% sheep milk, or mixed milk, or hellim/halloumi with or without mint, as well as differentiating salt levels, to name a few, without losing the authenticity of the traditional hellim/halloumi cheese. Thirdly, it has it has been produced and sold in island for centuries. Traditionally produced in rural areas by groups of women, hellim/halloumi production was seen as a social communal collaboration up until the 1970s [13], but as a result of industrial developments, hellim/halloumi producing acts in villages are rare. There are 53 industrial halloumi producers in southern Cyprus and 34 producers in the north [14], with the first factory producing hellim/halloumi in north being opened in 1958 [15], meaning that hellim/halloumi was commercialized around 60 years ago. Finally, it is also a gastronomic heritage with high consumption rates.

Traditional food products, especially the ones applying qualification and indication schemes, are known for the following: to promote the culture and traditions of the place of origin by using traditional practices of production [16–18]; to secure the originality of local foods [19,20] that are assumed to have greater health benefits [21]; are anticipated to act as a guarantee of the quality and safety by linking the quality attributes to foods 'locality' [16,22–25]; and can promote local communities, and support rural development [22,26–28], protecting rural areas from depopulation [1]. Hellim/halloumi cheese production is the major industry in Northern Cyprus, and is the main export product, according to the Cyprus Turkish Chamber of Industry [29]. We argue than hellim/halloumi cheese is one of the heathiest cheese options, firstly, because its recipe [12] does not allow for the use of any preservatives and additives, and secondly, because it uses fresh, local milk [30] that has traces of the wild plants that the animals have eaten [31]. More importantly, hellim/halloumi production uses an estimated 89% of the sheep and goat milk that is produced in the island [32], which gives hellim/halloumi its specific taste and aroma. As the price of the milk in Cyprus is the highest in the region [33], producing any other types of dairy products would not be as sustainable, because its price would not be competitive.

Traditional foods are typically consumed by middle-aged, ethnocentric, health-conscious select locals [34], which is not a case in Cyprus, where hellim/halloumi cheese is characterized by very high local consumption rates [14], even higher than the modern imported cheeses. In this paper, the authors aim to understand how loyalty towards traditional foods are built amongst young consumers, as previous research shows that younger customers are more concerned about emotional attachment and experiences [35,36] rather than the attributes of products and services. Therefore, the authors use a consumer culture theory (CCT) perspective to examine the following: (1) the way individuals use their traditional products to identify themselves with the culture and feel that they are a part of the community, (2) underlying values that turns young consumers into loyal customers of hellim/halloumi cheese, and (3) its implications to hellim/halloumi producers.

This research contributes to the existing knowledge in multiple ways. Firstly, it uses the perspective of CCT to understand more the complex aspect of loyalty that builds as a result of social and cultural experiences. Previously, Collin-Lachaud and Kjeldgaard [37] underwent one of the first studies that looked into a loyalty concept from a culturally informed perspective, using the attendees of music festivals as their study sample, but to the best of the authors' knowledge, there was no empirical evidence using a CCT perspective to explain loyalty in the traditional food industry. Moreover, the current study specifically explores the cultural aspect of loyalty building amongst young consumers, as previous research has identified the need for additional research [38] in order to understand transcend boundaries of consumption amongst different generations. In this study, we focused on post-millennials, also called Gen Z, who are classified as individuals are born after 1997 [39], mainly because they are the emerging market share, recently entering adulthood, and not much is known about their consumption habits. Finally, it brings out the managerial implications for the hellim/halloumi producers in Cyprus who are trying to establish long term loyalty relationships with their customers, and focusing on Gen Z as a large, emerging market segment.

2. Theoretical Background

In the field of consumer research, increasing interest is devoted to studying the marketplace phenomena from individuals' cultural perspectives [40], rather than focusing solely on the economic and psychological motivators and drivers of consumption. CCT explains how consumption experiences are shaped by "dynamic relationships between consumer actions, the marketplace, and cultural meanings" [41], and emphasizes the importance of consumption in the "construction of individual identities" [42]. When it comes to consuming traditional foods, consumers are driven by multiple factors of influence, far beyond the economic and psychological motivators known in the traditional literature of consumer behavior. Consumer culture theory summarizes the sociocultural, experiental, symbolic, and ideological aspects of consumption.

When the literature from the last 20 years is reviewed, it is possible to conclude that this concept has been sought over the years, but its conceptualization was made by Arnould and Thompson [41]. Contemporary consumer culture theory is a concept unifying the consumer actions, the marketplace, and cultural meanings together, while analyzing consumer behavior; in other words, taking into consideration all of the variables like the sociocultural, experiential, symbolic, and ideological aspects of consumption [41].

Throughout the years, consumption has retained its importance in terms of sociology, economics, and business literature, however, now the term is evaluated on the basis of the relations between the society and the individual, as a form of social classification, and communications systems, as a way to identity formation, ritualistic and community building processes, the relationship between the individual and the state, or the search for existentially fulfilling experiences [43]. However, over the years, marketing and consumer research literature had been limited to macro level perspectives, and marketing research did not focus on broader social, cultural, and macro-level considerations [44].

Early CCT suggestions objected the limitation of consumption to demographic or psychographic properties of consumers, like their gender, age, personality, or lifestyle traits, however, after the 1980s the CCT definition began to involve humanistic and experientialist paradigms, and CCT became attached to the "consumption experiences and the personal meanings" that consumers develop to products and experiences [45].

According to Schouten and McAlexander [38], consumer culture theory plays a very important role in the marketing discipline, because the interpersonal relationships people undertake to give their lives meaning is very crucial, for instance, by choosing how to spend their money and their time—while doing this they create their own categories, which is called the subcultures of consumption.

Consumer culture theory is significant in terms of consumer behavior, because its focal point is the meaning and identity creation in terms of products, brands, and experiences; it is related to the mentality created by the consumers in the marketplace, and the underlying reason for this mentality can be used to engage the attention of buyers and sellers.

In general business studies, culture is defined as a homogenous structure, including the meanings, ways of life, and shared values that affect the behavior of its members [45]. However, on the other hand, CCT does not acknowledge culture as its common definition, which is the "homogenous system of collectively shared meanings, ways of life, and unifying values shared by a member of society", but CCT is interested in investigating the heterogeneous allocation of meanings and the reasons for overlapping cultural groupings. Thus, consumer culture theory indicates a social arrangement as a result of the bonds between lived culture and social resources, meaningful ways of life, and the symbolic and material resources on which they depend [41]. Continuously, CCT does not detect and structure the behavior, but rather interprets the consumers' reactions or feelings based on patterns of behavior [45].

An important aspect within CCT is identity building by consumers. According to Patterson et al. [46], CCT claims that the marketplace involves ample amounts of commodities and symbolic resources to be used by consumers in terms of identity construction. Moreover, the study of culture in the marketplace indicates that the collective identity of consumers is created through an engagement in common consumption interests [46].

Consecutively, another important dimension of consumer culture theory is the 'experience', which refers to the phenomenon of the story behind the consumers relations with the products or brands. According to Ahola [47], it is possible to conceptualize the 'experience' in three dimensions within CCT. The first dimension is named the 'humanistic frame', in which the experience of the consumer is a personally unique event that is associated with significant emotional experiences, a kind of more inner phenomenon. The second one is called 'extraordinariness', where an "extraordinary experience is activated by unusual events and is characterized by high levels of emotional intensity and experience". The last framed is named the 'marketplace frame', where the main question is "how are the meanings of experience formed in the marketplace?" [47].

When the extant literature is reviewed, it is possible to come across many studies regarding consumer behavior, however, not many of the studies are grounded on consumer culture theory. Usually, the studies are looking at the individual consumer experiences or meanings rather than trying to reveal the culture created by a distinctive group of consumers.

One of the earliest studies in the literature was carried out by Schouten and McAlexander [38]. In their study, the authors considered the subcultures of consumption "as a distinctive subgroup of society that self-selects on the basis of a shared commitment to a particular product class, brand, or consumption activity, including an identifiable, hierarchical social structure; a unique ethos, or set of shared beliefs and values; and unique jargons, rituals, and modes of symbolic expression".

A later study by Arnould and Thompson [41], was the first to clearly define the consumer culture theory, which had a significant contribution to the literature.

Lastly, another recent study that was carried out by Collin-Lachaud and Kjeldgaard [37], grounded their research on the theory of consumer culture. The study considered the concept of loyalty from a consumer culture theory perspective. While evaluating the annual (French) music festivals and their ritualized meanings for consumers, they addressed the loyalty from perspectives of (1) social rather than individual, (2) an outcome of a social evaluation of emotional experiences rather than individual satisfaction, (3) temporally, and (4) spatially structured and structuring.

3. Research Context

The world milk production has increased by 64% from 1970 to 2012. A large part of this expansion has been used to increase the cheese production in all regions of the world. Innovations, new product development, and increased demand from emerging economies [48] have led to a higher than ever cheese consumption. Germany, the Netherlands, and France are the main cheese exporters in the world, and each exported more than $3 billion worth of cheese in 2015 [49]. The top cheese consumer is Denmark, where the people consume 28.1 kg of cheese per capita in a year. The second highest consumer is Iceland, followed by Finland, with 27.7 kg and 27.3 kg cheese consumption per capita, respectively [50].

As a product that holds geographical indication, hellim/halloumi cheese can be only produced in Cyprus (both northern and southern Cyprus). As the focus of this paper is the consumers in northern Cyprus, the growth of hellim/halloumi cheese production has been analyzed. The total milk product produced in northern Cyprus has significantly increased since 1995 (Figure 1), mostly due to the demand from domestic and external markets, as extensive marketing campaigns towards hellim/halloumi cheese were undertaken by the Cyprus Turkish Chamber of Industry, producers, and other stakeholders.

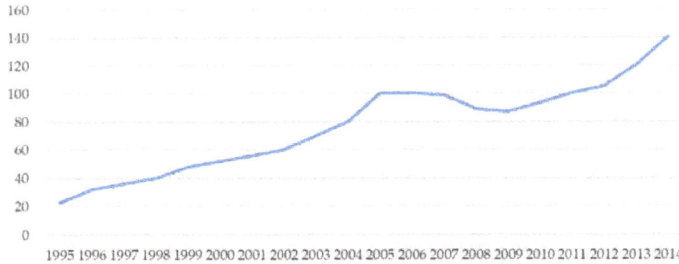

Figure 1. Total milk product production in Northern Cyprus (million tons). Source: the Cyprus Turkish Chamber of Industry.

Hellim/halloumi exports made up 22% of the total exports from Northern Cyprus in 2014. The export values have increased by 33% in the analyzed time period (Figure 2), mainly due to price increases and an increased demand from the market.

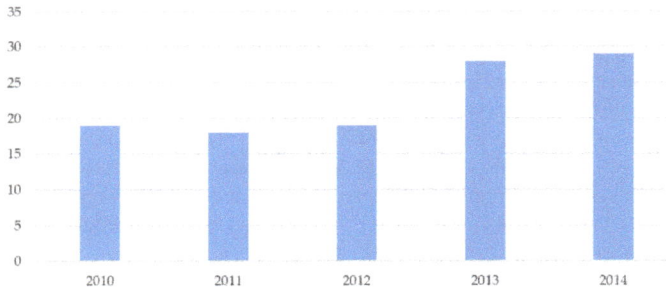

Figure 2. Hellim/halloumi exports from Northern Cyprus (million US dollars). Source: the Cyprus Turkish Chamber of Industry.

The major hellim/halloumi export countries are Kuwait, Saudi Arabia, and Turkey (Figure 3), where hellim/halloumi is known, and the exports significantly increased since the Cyprus Turkish Chamber of Industry obtained the geographic indication of hellim/halloumi, allowing for production only to take place in Cyprus.

Although the majority of hellim/halloumi is exported to the Middle East, there is also a significant amount of hellim/halloumi exported to the United Kingdom, Australia, and the United States, where a majority of Cypriots now live after war in 1974. In addition, a significant part of hellim/halloumi

Sustainability **2018**, *10*, 3183

is consumed in the domestic market. According to the Cyprus Turkish Chamber of Industry [29], in Northern Cyprus, people consumer around 12 kg of hellim/halloumi per capita in a year, which is more than any type of imported cheese, showing the importance of this traditional product in consumers' choices.

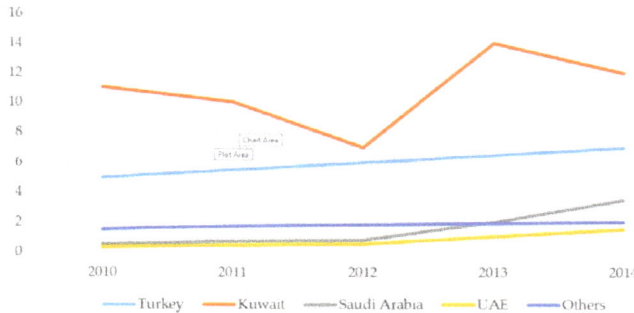

Figure 3. Hellim/halloumi export countries (million US dollars). Source: the Cyprus Turkish Chamber of Industry.

4. Methodology

The authors aimed to describe and clarify human experiences as it appears in people's lives, using a qualitative approach [51], as the field, rather than the laboratory, is the natural context for CCT research [41]. In order to achieve the aims of this study, the authors conducted semi-structured interviews, which were audio-recorded and transcribed. Purposive sampling was used to select the participants for the semi-structured interviews, as suggested by Teddlie and Yu [52], and Tongco [53]. More specifically, a homogeneous sampling technique was applied [54] in order to answer the research question that is aimed at understanding and explaining the consumption practices among the post-millennial generation. To define the sample size, the authors followed the suggestion of Guest, Bunce, and Johnson [55], who propose that saturation occurs in around 12 participants in homogeneous groups. A similar sample size is used in various qualitative researches [56–58]. After the first ten interviews, the authors recognized that the information given starts to repeat, and there are no new insights brought. Finally, a sample of 12 participants was included, and the main criteria was that the participants belonged to the post-millennials generation (also referred as Gen Z), defined as individuals born after 1997 [39], and had reached the age of 18 at the time of interview. Each of the interviews lasted up to an hour. The profiles of the respondents are represented in Table 1.

Table 1. Profile of interviewees.

Interviewee	Gender	Birth Year	Occupation
1	Male	1999	Working in a family business
2	Male	2000	Student
3	Female	1997	Working
4	Male	1999	Student
5	Female	1998	Student/part time working
6	Female	1999	Establishing business
7	Male	1999	Student
8	Female	1997	Student
9	Female	1999	Student
10	Male	1998	Student/part time working
11	Male	2000	Student
12	Female	1999	Student/working in family business

Firstly, the participants were asked to describe the attributes of ideal hellim/halloumi cheese, based on given criteria (price, taste, freshness, milk used in production, odor, traditional recipe used, origin, and production method). Then, questions were asked in order to understand the building blocks of the cultural and emotional attachment towards hellim/halloumi cheese and the consumption habits that build loyalty towards hellim/halloumi cheese. The detailed set of questions that were asked in order to evaluate the consumption practices of traditional hellim/halloumi cheese amongst post-millennials can be found in Appendix A.

Data triangulation has been applied in order to facilitate the validation and reliability of the collected data, as well as to facilitate a comprehensive understanding of the phenomena [59]. Apart from focused observations and semi-structured interviews, the author used theoretical data gathered from relevant sources, as well as newspapers, magazines, and documentation regarding hellim/halloumi cheese. Content analysis, defined as "a subjective interpretation of text data through systematic classification process of coding and identifying themes or patterns" by Hsieh and Shannon [60], was then applied to analyze the outcomes of the research.

5. Results

5.1. Memories That Surround the Food

Memories have an important role in assessing food and its taste [61], and in affecting the present eating practices [62]. Post-millennials are in their early 20's, but their hellim/halloumi consumption is hugely affected by memories of their family, especially of their ancestors, friends, and others. First of all, almost all of the interviewees had some memories of hellim/halloumi being produced in villages by their grandparents.

"I have never made it myself, but I have seen my grandma doing it. The process looks so complicated" (Female, born in 1999).

"My grandma always made hellim. She prepared it in spring time, as it is the season of the best goat and sheep milk. Then she stored in in big bottles, and served to everybody" (Male, born in 2000).

"I think those times there was not much choices in the supermarkets. Everybody made hellim at home" (Male, born in 1998).

As seen in the previous quotes, young Cypriots have an understanding of the hellim/halloumi production process and the milk used, because they have witnessed it from their ancestors. Moreover, two of the respondents living in villages have been involved in hellim/halloumi production, which allows young individuals to experience the cultural act themselves.

"I was helping my grandma to make hellim when I was a kid. I can still make. Actually, in my family we are still doing it sometimes" (Male, born in 1999).

"We have always been keeping our own animals for milk and meat purposes. First it was my grandma making hellim. She taught it to my mom. And my mom taught it to me and my sister. Every season, around April, we produce our hellim. It the best time because the milk is fatty. Hellim becomes hard, heavy and strong taste" (Male, born in 1999).

The post-millennials that have been involved in hellim/halloumi production have a deep understanding of all of the factors that affect the taste of hellim/halloumi, as it is shared knowledge amongst multiple generations. Also, the post-millennials, although consuming modern varieties of hellim/halloumi that is purchased in stores, have distinct memories regarding the taste of mature (referred to as 'old') hellim/halloumi, as described in the quotes below.

"In general, there are two main types of hellim- fresh and mature. Nowadays, we are mostly eating the fresh hellim that is purchased from supermarkets. It tastes mild, salty and rubbery. But I remember in my childhood our grandma always served mature hellim to us. It is very hard, salty cheese that has a goat milk aroma. I used to eat it when I was a kid, but now the taste is just too strong for me" (Female, born in 1997).

"You can use old hellim with macaroni dishes, but really the taste is too much. I prefer fresh, packaged hellim for my everyday use" (Male, born in 1999).

Understandably, the younger generations are not very familiar with the 'old' taste of this traditional product, as they have mostly consumed modernized variations of hellim/halloumi, therefore, the taste of mature hellim/halloumi, although familiar, is too strong for post-millennials, who prefer milder, softer varieties of hellim/halloumi.

5.2. Food That Is a Ritual

Rituals are actions that are frequently repeated [63], and make distinction between primitive understandings of food as a fuel for the body to more symbolic meanings of consumption [64]. Preparing and eating a food is a social act, shared with family, friends, and other people around, which allows one to feel sense of belonging an inclusion [65]. Post-millennials have developed their own rituals that involve hellim/halloumi as a distinctive and integral breakfast food, as well as a central ingredient in several other traditional foods.

"Since I was small, I saw from my dad. He was always buying hellim and we were using it in sandwiches. Every morning before school he was making us sandwich or toast with hellim and tomatoes. I still use it the same way" (Female, born in 1998).

"When you light your barbeque and people are coming over for lunch or dinner, it is impossible not to serve grilled hellim in pita bread. It is an essential part of a barbeque party" (Male, born in 2000).

"Since we are small kids, my mom always cooks macaroni with chicken, and puts graded hellim on the top. The smell of that food fills the house and it makes you feel so good. If one cook it at home, the smell fills up all the village, and always I had all my friends over for lunch that day" (Female, born in 1999).

Hellim/halloumi, as mentioned by the interviewees, is essential part of breakfast, lunch, and dinner that is shared amongst friends, family, and other people. Currently, the majority of post-millennials are still living with the families, and do not have their own households. Therefore, as admitted by many, purchasing hellim/halloumi is done by their parents. When asked if they would continue to purchase hellim/halloumi even when having their own household, the answers were positive.

"Of course, I would continue to buy it. I don't want my kids to grow up and not know the taste of hellim sandwich or fried hellim" (Female, born 1999).

Unanimously, the post-millennials agreed that hellim/halloumi is a food that they would like to continue using in their own houses, and they would teach consuming hellim to their children, therefore the rituals surrounding hellim/halloumi seem to be sustainable.

5.3. Food That Builds Identity

Identity is "that part of an individual's self-concept which derives from his knowledge of his membership of a social group (or groups) together with the value and emotional significance attached to that membership" [66]. People have a tendency to attach symbolic meaning to the foods that they consume, especially in the case of traditional foods [67]. Moreover, according to Fischler, [68] "food is central to our sense of identity". Identity building is acknowledged to be a dynamic process [69] accumulating and changing over time. The hellim/halloumi consumption process seems to be an identity builder for the Turkish Cypriot post-millennials, who admit the following:

"Hellim is very important. I mean, I cannot imagine my life without hellim" (Female, born in 1999).

"Without hellim it is not possible" (Male, born in 2000).

"Hellim is not just a food. Hellim is Cypriot" (Male, born in 1998).

These quotes describe how young Cypriots recognize that what they eat actually is a construct of what they are. Young people, especially the ones belonging to the post-millennial generation, are often seen as technologically inclined, and belonging to the most global generation, while this is not the true when it comes to consuming food. Post-millennials love their traditional food and, interestingly, pay a lot of attention to what they consume and how they consume, developing a value of the food, that allows them to identify themselves with it.

Another aspect of identity building, which is reflected in the below mentioned quotes, allows us to have a deeper understanding on how hellim/halloumi allows people to identify themselves as a Cypriots:

"When my aunty comes from London for her summer holidays in Cyprus, first thing we always serve is hellim with watermelon. And when she goes back, she always takes a lot of hellim with her" (Male, born in 2000).

"If I travel somewhere, I need my hellim with me" (Female, born in 1997).

As seen in the previous quotes, hellim/halloumi is carried to different locations and has an almost symbolic meaning that allows one to feel and sense Cyprus even when abroad. Nevertheless, hellim/halloumi consumption is seen as an act that can differentiate the Turkish Cypriots from other nationalities that have a huge political and economic influence in the island of Cyprus, as follows:

"Hellim is the only thing differentiating us from Turkish people. What else do we have? Nothing" (Female, born in 1999).

"We speak Turkish. We eat Turkish food. We have Turkish people in this island more than we have Cypriots. We watch Turkish TV. We listen to Turkish music. I don't know, we are always, you know, always seen as being Turkish. When we pass to Greek side, they call us Turks. I mean, hellim, hellim is something else. When you go to Turkey and say you are from Cyprus, they immediately recall hellim, because it is something to do with Cyprus, not Turkey" (Male, born in 1998).

In the above quotations, a strong willingness of young people to be identified with their ethnic group, Turkish Cypriots, rather than Turkish, has been expressed. With Turkey being a 'mother country' [70] and the political, economic, and social burdens that Turkish Cypriots face, their food has become their identity that distinguish them from harsh influences.

6. Conclusions and Discussion

Loyalty towards traditional foods is not a well explored field in academic literature, especially from a cultural perspective. From the insights obtained from the interviews amongst young consumers (referred to as post-millennials or Gen Z), the current research allows us to supplement and to progress the understanding of loyalty towards traditional food products from a consumer culture theory perspective.

According to the responses of the interviewees, it is possible to conclude that hellim consumption is the part of the culture that they belong to, an inevitable piece of their rituals, and it is transmitted from old generations to new ones. Looking from a consumer perspective, hellim as a product, has a significant place in the minds of customers and it is an unchanging complement of specific consumption behaviors.

There is no doubt that young consumers are driven by the sociocultural and symbolical motivators for the hellim consumption as a result of their 'learnt' experiences from their ancestors. Post-millennials had attributed meaning to their hellim consumption, each unique to their own, after their experiences from their childhood. Their interpersonal relationships with their family and their experiences with hellim consumption are the great drivers for deciding how to spend their money (on hellim). These experiences can be explained as "personally unique events that are associated with significant emotional experiences" [47].

As specified in the existing literature, consumption behaviors are a way of distinct identity construction of consumers/consumer groups. Food as a trade product is related to the economy and to consumption practices, whereas food as a sign of culture, is related to tradition, identity, and uniqueness, and it is possible to say that it can be used in the construction of personal identities [71]. Presently, in the situation of the Turkish Cypriot community, the way food is produced, served, and consumed can be a sign of a distinct identity and the culture of an individual, community, or nation [72]. While food and drink can be indicators of group culture and identity, because it says something meaningful about people, it can be considered as an often open-ended process of social identification [73]. However, on the other hand, while food can gather people together with the similar tastes and preferences, it can also divide people based on their dietary and religious preferences, which also contributes to identity creation [72]. This can be an example of how the Turkish Cypriot community is differentiating themselves from the Turkish community, by consuming the traditional hellim/halloumi cheese.

Undeniably, traditional foods play an important role in the economic, social, and environmental sustainability. Therefore, loyalty towards traditional foods is the key driving factor to ensuring all aspects of sustainability, especially amongst young consumers, as they are the ones imposing significant changes in the market place. Firstly, as observed in the relevant literature [38,47], and in the current empirical investigation, loyalty towards traditional food products is build based on (1) the memories that surround the food, (2) the rituals that preparing and eating a food involve, and (3) the identity that it builds, allowing people feel sense of belonging to their ethnic group. Our findings confirm the previous outcomes of Collin-Lachaud and Kjeldgaard's [37] research, which confirmed loyalty as being a social and emotional process, rather than behavioral.

It is possible to explain this situation by a CCT perspective, whereby culture is not defined as only "homogenous system of collectively shared meanings, ways of life, and unifying values", but the "heterogeneous allocation of meanings and the reasons of overlapping cultural groupings" [41].

Understanding the mentality behind hellim consumption is important for companies in order to engage buyers to consume their own product brands. An example can be given from answers of the interviewees, where the post-millennials like to consume the new kind of hellim, however they find the traditional old hellim as an inevitable complementary for some meals. While some like the taste of the 'old' hellim and find the 'new' hellim to be 'rubbery', the others are more used to the taste of the new hellim.

As discussed before, post-millennials do not select the most traditional type of hellim/halloumi cheese, which is 'old' hellim/halloumi in brine, giving preference to the modern varieties of the packaged and less salted hellim/halloumi, because of the milder and lighter taste and odor. It shows us that in the very near future, the market demand will drive producers to produce more of the modern varieties of hellim/halloumi, and less of the 'old' varieties. Firstly, there is the issue of production planning, as producers have to refocus from producing traditional varieties of hellim/halloumi, to producing modern varieties of the specialty cheese. Secondly, and most importantly, although innovations in traditional food products are a market driven process [8], it poses risks to traditional food products in sustaining its traditional taste, recipe, odor, and even the production method. Therefore, the Cyprus Turkish Chamber of Industry and other certification bodies and involved stakeholders have to pay additional attention to ensure the sustainability and originality of the hellim/halloumi cheese production methods and taste.

7. Limitations and Suggestions for Future Research

This particular study focuses on post-millennials as an emerging market segment, which is not a well explored sector in the literature currently. Not much is known about post-millennials, and we have built this study to understand the fundamentals of the consumption habits of young consumers, focusing only on the traditional product in particular geographical locations. Obviously, there is more to explore regarding loyalty amongst post-millennials, especially because they have just entered their adulthood and are becoming independent customers slowly.

The results of the current study should be evaluated with caution, as the current study is limited to only one particular region and traditional food product in this region. However, the results generated from the study provide a general overview of post-millennials and their consumption habits of traditional food from cultural perspective. Further studies extending to different product ranges apart from traditional products and different geographical locations are welcome.

For future research, we also suggest exploring post-millennials and their consumptions habits more in-depth. Firstly, we suggest exploring the building blocks that lead to brand equity and loyalty, like satisfaction, experience, service quality, and even emotional attachment. Studies dealing with how young consumers make purchase decisions would be highly appreciated.

Author Contributions: Introduction, Z.G. Theoretical background, A.B. Methodology, Z.G. Results and findings, A.B. and Z.G. Conclusions and discussion, Z.G. and A.B.

Funding: This research received no external funding.

Conflicts of Interest: The authors declare no conflict of interest.

Appendix A. Sample Questions for Semi-Structured Interviews

1. Background information, namley: age, family status, occupation, and current place of living.
2. How could you define hellim/halloumi? What does it mean for you? How important is hellim/halloumi for you, your family, and your country?
3. Is hellim/halloumi 'Cypriot'? Is it part of the Cypriot culture? Is it important for Cypriots? What other foods you can name that are also of a Cypriot origin?
4. How would you describe the taste of perfect hellim/halloumi cheese? How would you describe the odor of perfect hellim/halloumi cheese? How would you describe the texture of perfect hellim/halloumi cheese?
5. How do you normally consume hellim/halloumi? What are the foods you use it for? What is your favorite way to consume it? What foods do you prepare yourself, and what foods do others prepare for you?

6.   How do you use different types of hellim/halloumi in your household? Who likes mature hellim/halloumi? Who consumes fresh hellim/halloumi? Are there any differences between the hellim/halloumi preferences in your family? Which foods do you prefer with mature hellim/halloumi? Which foods do you prefer with fresh hellim/halloumi cheese?

7.   How important is hellim/halloumi in your family? Is it part of some important family events/food/traditions? Do you use it at your family gatherings? Do you use it as a gift for someone?

8.   Who purchases hellim/halloumi in your household? Do you ever buy it? When you buy it, which type of hellim/halloumi do you prefer?

9.   When you think of your childhood, what memories do you have regarding hellim/halloumi cheese? Has it always been part of your diet? Do you remember eating it when you were small? Do you remember the foods you were eating with hellim/halloumi? Do you remember who was preparing them?

10.   Do you know how hellim/halloumi is prepared? Do you know the recipe of hellim/halloumi? Have you ever witnessed the hellim/halloumi production process? Have you ever been involved in the hellim/halloumi production process?

11.   When you have your own family, will you continue to consume hellim/halloumi? Would you teach your kids to eat and prepare foods with hellim/halloumi?

References

1.   Guerrero, L.; Guàrdia, M.D.; Xicola, J.; Verbeke, W.; Vanhonacker, F.; Zakowska-Biemans, S.; Sajdakowska, M.; Sulmont-Rosse, C.; Issanchou, S.; Contel, M.; et al. Consumer-driven definition of traditional food products and innovation in traditional foods. A qualitative cross-cultural study. *Appetite* **2009**, *52*, 345–354. [CrossRef] [PubMed]

2.   Yi, Y.M.; Gim, T.H. What Makes an Old Market Sustainable? An Empirical Analysison the Economic and Leisure Performances of Traditional Retail Markets in Seoul. *Sustainability* **2018**, *10*, 1779. [CrossRef]

3.   Bruntland, G.H. *Our Common Future: The World Commission on Environment and Development*; Oxford University Press: Oxford, UK, 1987.

4.   Finkbeiner, M.; Schau, E.M.; Lehmann, A.; Traverso, M. Towards life cycle sustainability assessment. *Sustainability* **2010**, *2*, 3309–3322. [CrossRef]

5.   Patapiou, N. Leonardo Donà in Cyprus. A Future Doge in the Karpass Peninsula (1557). *Cyprus Today Q. Cult. Rev. Minist. Educ. Cult.* **2006**, *44*, 3–18.

6.   Osam, N.; Kasapoglu, U.M. Hallumi: The Origin Analysis of a Cultural Entity. *Milli Folklor* **2010**, *87*, 170–180.

7.   Gellynck, X.; Kühne, B. Innovation and collaboration in traditional food chain networks. *J. Chain Netw. Sci.* **2008**, *8*, 121–129. [CrossRef]

8.   Jordana, J. Traditional foods: Challenges facing the European food industry. *Food Res. Int.* **2000**, *33*, 147–152. [CrossRef]

9.   Kühne, B.; Vanhonacker, F.; Gellynck, X.; Verbeke, W. Innovation in traditional food products in Europe: Do sector innovation activities match consumers' acceptance? *Food Qual. Preference* **2010**, *21*, 629–638. [CrossRef]

10.   Żakowska-Biemans, S.; Sajdakowska, M.; Issanchou, S. Impact of innovation on consumers liking and willingness to pay for traditional sausages. *Pol. J. Food Nutr. Sci.* **2016**, *66*, 119–128. [CrossRef]

11.   Di Monaco, R.; Cavella, S. Differences in liking of traditional salami: The effect of local consumer familiarity and relation with the manufacturing process. *Br. Food J.* **2015**, *117*, 2039–2056. [CrossRef]

12.   European Commission. Available online: http://europa.eu/rapid/press-release_IP-15-5448_en.htm (accessed on 2 July 2018).

13.   Welz, G. Contested Origins: Food Heritage and the European Union's Quality Label Program. *Food Cult. Soc.* **2013**, *16*, 265–279. [CrossRef]

14.   Garanti, Z. Marketing Hellim/Halloumi Cheese: A Comparative Study of Northern and Southern Cyprus. *Econ. Sci. Rural Dev. Conf. Proc.* **2016**, *43*, 65–72.

15.   Koop Milk. Available online: http://www.koopsut.com (accessed on 15 August 2018).

16.   Bowen, S.; Zapata, A.V. Geographical indications, terroir, and socioeconomic and ecological sustainability: The case of tequila. *J. Rural Stud.* **2009**, *25*, 108–119. [CrossRef]

17. Bowen, S. Embedding local places in global spaces: Geographical indications as a territorial development strategy. *Rural Sociol.* **2010**, *75*, 209–243. [CrossRef]
18. Dokuzlu, S. Geographical indications, implementation and traceability: Gemlik table olives. *Br. Food J.* **2016**, *118*, 2074–2085. [CrossRef]
19. Stasi, A.; Nardone, G.; Viscecchia, R.; Seccia, A. Italian wine demand and differentiation effect of geographical indications. *Int. J. Wine Bus. Res.* **2011**, *23*, 49–61. [CrossRef]
20. Blackwell, M. The relationship of geographical indications with real property valuation and management. *Prop. Manag.* **2007**, *25*, 193–203. [CrossRef]
21. Adinolfi, F.; De Rosa, M.; Trabalzi, F. Dedicated andgeneric marketing strategies: The disconnection between geographical indications and consumer behavior in Italy. *Br. Food J.* **2011**, *113*, 419–435. [CrossRef]
22. Belletti, G.; Marescotti, A.; Vakoufaris, H. Linking protection of geographical indications to the environment: Evidence from the European Union olive-oil sector. *Land Use Policy* **2015**, *48*, 94–106. [CrossRef]
23. Josling, T. The war on terroir: Geographical indications as a transatlantic trade conflict. *J. Agric. Econ.* **2006**, *57*, 337–363. [CrossRef]
24. Zhao, X.; Kneafsey, M.; Finlay, D. Food safety and Chinese geographical indications. *Br. Food Journal.* **2016**, *118*, 217–230. [CrossRef]
25. Aggarwal, R.; Singh, H.; Prashar, S. Branding of geographical indications in India: A paradigm to sustain its premium value. *Int. J. Law Manag.* **2014**, *56*, 431–442. [CrossRef]
26. Donner, M.; Fort, F.; Vellema, S. From Geographical Indications to Collective Place Branding in France and Morocco. In *The Importance of Place: Geographical Indications as a Tool for Local and Regional Development*; Springer: Cham, Switzerland, 2017; pp. 173–196.
27. Belletti, G.; Marescotti, A.; Brazzini, A. Old World Case Study: The Role of Protected Geographical Indications to Foster Rural Development Dynamics: The Case of Sorana Bean PGI. In *The Importance of Place: Geographical Indications as a Tool for Local and Regional Development*; Springer: Cham, Switzerland, 2017; pp. 253–276.
28. Biénabe, E.; Marie-Vivien, D. Institutionalizing geographical indications in southern countries: Lessons learned from Basmati and Rooibos. *World Dev.* **2017**, *98*, 58–67. [CrossRef]
29. Cyprus Turkish Chamber of Industry. Available online: http://www.kibso.org/index.php/tr/ (accessed on 20 June 2018).
30. Preedy, V.R.; Watson, R.R.; Patel, V.B. *Handbook of Cheese in Health: Production, Nutrition and Medical Sciences*; Wageningen Academic Publishers: Wageningen, The Netherlands, 2013.
31. Osorio, M.T.; Koidis, A.; Papademas, P. Major and trace elements in milk and Halloumi cheese as markers for authentication of goat feeding regimes and geographical origin. *Int J. Dairy Technol.* **2015**, *68*, 573–581. [CrossRef]
32. Sutek. Available online: https://www.sutek.org (accessed on 12 August 2018).
33. EU Prices of Raw Milk. Available online: https://ec.europa.eu/agriculture/sites/agriculture/files/market-observatory/milk/pdf/eu-raw-milk-prices_en.pdf (accessed on 15 August 2018).
34. Vanhonacker, F.; Lengard, V.; Hersleth, M.; Verbeke, W. Profiling European traditional food consumers. *Br. Food J.* **2010**, *112*, 871–886. [CrossRef]
35. Veloutsou, C.; McAlonan, A. Loyalty and or disloyalty to a search engine: The case of young Millennials. *J. Consum. Mark.* **2012**, *29*, 125–135. [CrossRef]
36. Nowak, L.; Thach, L.; Olsen, J.E. Wowing the millennials: Creating brand equity in the wine industry. *J. Prod. Brand Manag.* **2006**, *15*, 316–323. [CrossRef]
37. Collin-Lachaud, I.; Kjeldgaard, D. Loyalty in a Cultural Perspective: Insights from French Music Festivals. In *Consumer Culture Theory*; Emerald Group Publishing Limited: Bingley, UK, 2013; pp. 285–295.
38. Schouten, J.W.; McAlexander, J.H. Subcultures of consumption: An ethnography of the new bikers. *J. Consum. Res.* **1995**, *22*, 43–61. [CrossRef]
39. Thompson, M.L. Smartphones: Addiction, or Way of Life? *J. Ideol.* **2017**, *38*, 3–17.
40. Moisander, J.; Valtonen, A.; Hirsto, H. Personal interviews in cultural consumer research–post-structuralist challenges. *Consum. Mark. Cult.* **2009**, *12*, 329–348. [CrossRef]
41. Arnould, E.J.; Thompson, C.J. Consumer culture theory (CCT): Twenty years of research. *J. Consum. Res.* **2005**, *31*, 868–882. [CrossRef]
42. Catulli, M.; Cook, M.; Potter, S. Consuming use orientated product service systems: A consumer culture theory perspective. *J. Clean. Prod.* **2017**, *141*, 1186–1193. [CrossRef]

43. Askegaard, S.; Linnet, J.T. Towards an Epistemology of Consumer Culture Theory: Phenomenology and the Context of Context. *Mark. Theory* **2011**, *11*, 381–404. [CrossRef]

44. Fitchett, J.A.; Patsiaouras, G.; Davies, A. Myth and ideology in consumer culture theory. *Mark. Theory* **2014**, *14*, 495–506. [CrossRef]

45. Bajde, D. Consumer Culture Theory: Ideology, Mythology and Meaning in Technology Consumption. *Int. J. Actor Netw. Theory Technol. Innov.* **2014**, *6*, 10–25. [CrossRef]

46. Patterson, M.; Schroeder, J. Borderlines: Skin, tattoos and consumer culture theory. *Mark. Theory* **2010**, *10*, 253–267. [CrossRef]

47. Ahola, E.K. How is the concept of experience defined in consumer culture theory? Discussing different frames of analysis. *Kuluttajatutkimus* **2005**, *1*, 91–108.

48. World Cheese Market. Available online: http://www.pmfood.dk/upl/9735/WCMINFORMATION.pdf (accessed on 27 June 2018).

49. Top 20 Cheese Exporting Countries. Available online: https://www.worldatlas.com/articles/top-20-cheese-exporting-countries.html (accessed on 27 June 2018).

50. Countries Who Consume the Most Cheese. Available online: https://www.worldatlas.com/articles/countries-who-consume-the-most-cheese.html (accessed on 27 June 2018).

51. Polkinghorne, D.E. Language and meaning: Data collection in qualitative research. *J. Couns. Psychol.* **2005**, *52*, 129–137. [CrossRef]

52. Teddlie, C.; Yu, F. Mixed methods sampling: A typology with examples. *J. Mix. Methods Res.* **2007**, *1*, 77–100. [CrossRef]

53. Tongco, M.D. Purposive sampling as a tool for informant selection. *Ethnobot. Res. Appl.* **2007**, *31*, 147–158. [CrossRef]

54. Palinkas, L.A.; Horwitz, S.M.; Green, C.A.; Wisdom, J.P.; Duan, N.; Hoagwood, K. Purposeful sampling for qualitative data collection and analysis in mixed method implementation research. *Adm. Policy Ment. Health Ment. Health Serv. Res.* **2015**, *42*, 533–544. [CrossRef] [PubMed]

55. Guest, G.; Bunce, A.; Johnson, L. How many interviews are enough? An experiment with data saturation and variability. *Field Methods* **2006**, *18*, 59–82. [CrossRef]

56. Pfeiffer, C.; Speck, M.; Strassner, C. What Leads to Lunch—How Social Practices Impact (Non-) Sustainable Food Consumption/Eating Habits. *Sustainability* **2017**, *9*, 1437. [CrossRef]

57. Torquati, B.; Tempesta, T.; Vecchiato, D.; Venanzi, S. Tasty or Sustainable? The Effect of Product Sensory Experience on a Sustainable New Food Product: An Application of Discrete Choice Experiments on Chianina Tinned Beef. *Sustainability* **2018**, *10*, 2795. [CrossRef]

58. Górska-Warsewicz, H.; Żakowska-Biemans, S.; Czeczotko, M.; Świątkowska, M.; Stangierska, D.; Świstak, E.; Bobola, A.; Szlachciuk, J.; Krajewski, K. Organic Private Labels as Sources of Competitive Advantage—The Case of International Retailers Operating on the Polish Market. *Sustainability* **2018**, *10*, 2338. [CrossRef]

59. Patton, M.Q. Enhancing the quality and credibility of qualitative analysis. *Health Serv. Res.* **1999**, *34*, 1189–1208. [PubMed]

60. Hsieh, H.F.; Shannon, S.E. Three approaches to qualitative content analysis. *Qual. Health Res.* **2005**, *15*, 1277–1288. [CrossRef] [PubMed]

61. Orlando, G. From the risk society to risk practice: Organic food, embodiment and modernity in Sicily. *Food Cult. Soc.* **2018**, *21*, 144–163. [CrossRef]

62. Chan, S.C. Food, memories, and identities in Hong Kong. *Identities Glob. Stud. Cult. Power* **2010**, *17*, 204–227. [CrossRef]

63. Marshall, D. Food as ritual, routine or convention. *Consum. Mark. Cult.* **2005**, *8*, 69–85. [CrossRef]

64. Levy, S.J. Revisiting the marketing domain. *Eur. J. Mark.* **2002**, *36*, 299–304. [CrossRef]

65. Isherwood, B.; Douglas, M. *The World of Goods*; Allen Lane: London, UK, 1979.

66. Tajfel, H. *Human Groups and Social Categories: Studies in Social Psychology*; CUP Archive: Cambridge, UK, 1981.

67. Pillen, H.; Tsourtos, G.; Coveney, J.; Thodis, A.; Itsiopoulos, C.; Kouris-Blazos, A. Retaining Traditional Dietary Practices among Greek Immigrants to Australia: The Role of Ethnic Identity. *Ecol. Food Nutr.* **2017**, *56*, 312–328. [CrossRef] [PubMed]

68. Fischler, C. Food, self and identity. *Inf. Int. Soc. Sci. Counc.* **1988**, *27*, 275–292. [CrossRef]

69. Phinney, J.S. Ethnic identity in adolescents and adults: Review of research. *Psychol. Bull.* **1990**, *108*, 499–514. [CrossRef] [PubMed]

Sustainability **2018**, *10*, 3183

70. Vural, Y.; Rustemli, A. Identity fluctuations in the Turkish Cypriot community. *Mediterr. Politics* **2006**, *11*, 329–348. [CrossRef]

71. Stajcic, N. Understanding culture: Food as a means of communication. Hemispheres. *Stud. Cult. Soc.* **2013**, *28*, 77–87.

72. Perry, M.S. Feasting on Culture and Identity: Food Functions in a Multicultural and Transcultural Malaysia. *3L Southeast Asian J. Engl. Lang. Stud.* **2017**, *23*, 184–199. [CrossRef]

73. Wilson, T.M. *Food, Drink and Identity in Europe*; Rodopi: Amsterdam, The Netherlands, 2006.

sustainability

MDPI

Review

Mainstreaming Underutilized Indigenous and Traditional Crops into Food Systems: A South African Perspective

Tafadzwanashe Mabhaudhi [1,*], Tendai Polite Chibarabada [1,2], Vimbayi Grace Petrova Chimonyo [1,3], Vongai Gillian Murugani [1], Laura Maureen Pereira [4,5], Nafiisa Sobratee [4], Laurencia Govender [1], Rob Slotow [4,6] and Albert Thembinkosi Modi [1]

[1] Centre for Transformative Agricultural and Food Systems, School of Agricultural, Earth and Environmental Sciences, University of KwaZulu-Natal, P. Bag X01, Scottsville 3209, Pietermaritzburg, South Africa; tendaipolite@gmail.com (T.P.C.); vimbayic@gmail.com (V.G.P.C.); vgmurugani@gmail.com (V.G.M.); GovenderL3@ukzn.ac.za (L.G.); modiat@ukzn.ac.za (A.T.M.)

[2] Soil, Crop and Climate Sciences, University of the Free State P.O Box 339, Bloemfontein 9300, South Africa

[3] Plant Soil and Microbial Sciences Department, Michigan State University, 1066 Bogue St A286, East Lansing, MI 48824, USA

[4] School of Life Sciences, University of KwaZulu-Natal, P. Bag X01, Scottsville 3209, Pietermaritzburg, South Africa; pereira.laura18@gmail.com (L.M.P.); SobrateeN@ukzn.ac.za (N.S.); Slotow@ukzn.ac.za (R.S.)

[5] Centre for Food Policy, City University of London, Northampton Square, London EC1V 0HB, UK

[6] Department of Genetics, Evolution & Environment, University College, London WC1E 6BT, UK

* Correspondence: mabhaudhi@ukzn.ac.za; Tel.: +27-(0)-33-26-05-442

Received: 30 September 2018; Accepted: 24 December 2018; Published: 31 December 2018

Abstract: Business as usual or transformative change? While the global agro-industrial food system is credited with increasing food production, availability and accessibility, it is also credited with giving birth to 'new' challenges such as malnutrition, biodiversity loss, and environmental degradation. We reviewed the potential of underutilized indigenous and traditional crops to bring about a transformative change to South Africa's food system. South Africa has a dichotomous food system, characterized by a distinct, dominant agro-industrial, and, alternative, informal food system. This dichotomous food system has inadvertently undermined the development of smallholder producers. While the dominant agro-industrial food system has led to improvements in food supply, it has also resulted in significant trade-offs with agro-biodiversity, dietary diversity, environmental sustainability, and socio-economic stability, especially amongst the rural poor. This challenges South Africa's ability to deliver on sustainable and healthy food systems under environmental change. The review proposes a transdisciplinary approach to mainstreaming underutilized indigenous and traditional crops into the food system, which offers real opportunities for developing a sustainable and healthy food system, while, at the same time, achieving societal goals such as employment creation, wellbeing, and environmental sustainability. This process can be initiated by researchers translating existing evidence for informing policy-makers. Similarly, policy-makers need to acknowledge the divergence in the existing policies, and bring about policy convergence in pursuit of a food system which includes smallholder famers, and where underutilized indigenous and traditional crops are mainstreamed into the South African food system.

Keywords: agro-ecology biodiversity; climate resilience; health

1. Introduction

Agriculture became the backbone of food systems more than 10,000 years ago, as humans shifted from hunting and gathering to growing and cultivating food [1]. It has since been instrumental to rising civilizations and their growing populations, and, over time, has evolved as humans have acquired more knowledge and innovation. The Industrial Revolution, which started in Europe in the 1800s, was an important event characterized by an improvement of farming practices and the invention of new tools [1]. Following this, there was an occupational shift as people left farming to work in the factories, cities mushroomed, a strong low to middle class emerged, and economic and income growth occurred. This necessitated a shift in the food system to allow greater production to supply the labor force in cities [2].

By the mid-1900s, world population was ≈ 2.5 billion. Whereas the industrialized countries were largely food secure, countries in Latin America, Asia and Africa continued to suffer from chronic hunger [3]. The Green Revolution led to development of high-yielding varieties of major cereal crops [maize (*Zea mays*), wheat (*Triticum aestivum*), rice (*Oryza sativa*)] that were responsive to additional inputs such as fertilizers and water, resulting in the birth of an agro-industrial food regime [3,4]. To an extent, Asian and Latin American countries benefited from the Green Revolution, saving them from famine [3,5]; however, this impacted smallholder agriculture with a shift towards greater dependency on the agro-industrial food system [6,7]. Inevitably, the successes of the Green Revolution and subsequent emergence of the agro-industrial food system gave birth to 'new' challenges, such as environmental pollution and degradation [3,5], loss of biodiversity [5,8], and malnutrition [9]. What is noticeable is that sub-Saharan Africa (SSA) was left behind in the first Green Revolution, and there is now a controversial, but concerted effort to enable a 'new' green revolution for Africa [10,11].

In spite of the technological advances, the global food system is failing to meet the basic food needs of the world's citizens equitably. These challenges currently dominate the global Sustainable Development Agenda. Additional technological improvements towards the end of the 20th century, combined with a consolidation of capital in the food system to a few multi-national corporations, further entrenched the agro-industrial food system into what McMichael [12] refers to as the 'corporate food regime'. The current global food system remains a diverse mixture of localized and industrialized systems of interconnected food chains [2,13–15]; however, the majority of these systems are centered on a handful of crop choices [16]. As a result, modern food systems are more vulnerable to economic and climatic shocks, as they may not always have the requisite diversity and redundancy to be able to buffer these risks—i.e., they are not resilient [17]. This is especially true in developing economies that are currently experiencing climate variability and change [18].

A lesser documented outcome of the global agro-industrial food system has been the post-colonial replacement, and subsequent relegation, of underutilized indigenous and traditional crops through the introduction of exotic and, now considered "major" crops, that were often higher yielding, but also more input intensive [19,20]. This led to neglect of underutilized indigenous and traditional crop species that had previously formed the basis of local food systems, especially in the global South. The displacement of indigenous and traditional crop species by a few major crops has inevitably contributed, in part, to the limited successes of the global food systems, especially in underdeveloped regions of the world [21]. This is also evident by the minimal influence exerted by smallholder rural farmers, also referred to as family farmers, who are custodians of underutilized indigenous and traditional crops. While the global agro-industrial food system has recognized the role played by smallholder rural farming systems [22], these groups of farmers marginally influence the system, and are at great risk from economic and climatic shocks [23]. This is because they have limited access to the modernized inputs, techniques, and markets necessary to participate in the production of major crops [24,25]. The erosion of agro-biodiversity, combined with an emphasis on input-intensive cropping systems has, arguably, lowered the resilience of food systems in the global South [26]. In this paper we have adopted the South African definition for underutilized indigenous and traditional crops as described by Modi and Mabhaudhi [27]. They defined underutilized indigenous and traditional crops as "crops that have either originated in South Africa or those that have become "indigenized" over many years (>10 decades) of cultivation as well as natural and farmer selection within South Africa." The term 'indigenous' has also often been used to refer to crops that may have originated elsewhere but have undergone extensive domestication locally, thus giving rise to local variations, i.e., 'naturalized/indigenized crops' [28]. Indigenized crops are sometimes referred to as traditional crops [28]. For examples of underutilized indigenous and traditional crops, the reader is referred to Chivenge et al. [29]. Underutilized indigenous and traditional crops are often characterized by limited development relative to their potential. Consequently, they have poorly developed and understood value chains; however, this varies across geographic and socio-economic settings.

South Africa is one of the few African countries that has been embedded in the global agro-industrial food system for decades [30]. Despite this consolidation, South African food prices remain too high for the majority of her people, who, consequently, cannot afford to purchase adequate food, leaving 21.3% of the population with poor access to food [31]. Furthermore, the concerns regarding environmental degradation, loss of biodiversity, and vulnerability to climate change, have prompted a call to rethink the current configuration of the South African food system [32]. A focus on reinvigorating underutilized indigenous and traditional crops, and bringing these to the market, has been suggested as an entry point for improving diets and making them more sustainable [32–34].

In this regard, we reviewed the status of South Africa's food system with the aim to identify opportunities for mainstreaming underutilized indigenous and traditional crops into the food system. The specific objectives were to review the current status of the food system and its limitations, and to identify opportunities for mainstreaming indigenous crops for environmental sustainability and improved health outcomes, resilient agricultural systems, and agro-ecological biodiversity. Within the context of this review, the term "mainstreaming" is used to refer to the integration/inclusion of underutilized indigenous and traditional crops into the dominant food system. However, such integration or inclusion should occur in a way that allows them to retain the attributes that make them attractive and transformative while benefiting from the support mechanisms that exist within the dominant food system.

2. Methodology

2.1. Literature Review

The review followed the PRISMA guideline (www.prisma-statement.org) for a structured review. A mixed-method review approach, which included combining quantitative and qualitative research, was used to compile the review. Scientific journal articles and book chapters were obtained from databases such as JStor®, Scopus, ScienceDirect®, and Cab Direct®, while technical reports and other forms of literature were obtained from Google™ and South African Government Gazettes. The main search key terms included "Food systems in South Africa", "Indigenous crops and Food Systems", "Traditional crops and Food Systems", "Agro-ecology biodiversity and Food Systems", "Food Systems and Health", "Food system and Environment". Search terms were set to be in the title, keywords, or abstract. Results from all the databases totaled 13,145 papers (Figure S1), and were exported to Mendeley desktop (Elsevier, USA), where duplicates were immediately removed. Further to this, false hits, publications in languages other than English, and articles that had only abstracts available were also removed. Following this, there were 276 articles remaining (Figure S1). The literature was then subjected to review by relevant experts who further filtered the papers, and also added peer reviewed literature relevant for the review that was not obtained in the literature search (Figure S1). Although the review has a particular focus on South Africa, literature and case studies from outside South Africa, especially developing countries, were obtained, and also used as examples in the review. Eventually 127 articles consisting of scientific research, working papers, Government Gazettes, and popular articles, were used to compile the review. The review was then structured into three sections; the current status of food systems in South Africa (Section 3), opportunities for underutilized indigenous and traditional crops into a new food systems paradigm (Section 4), and, lastly, the study recommendations (Section 5).

2.2. A Systemic Analysis of the Opportunities for Underutilized Indigenous and Traditional Crops in the Current SA Food System

As a means to gain a 'real world' perspective on the ways in which the food system can be leveraged to create an enabling environment for the empowerment of historically underprivileged farmer communities and the inclusion of underutilized indigenous and traditional crops in smallholder farming, qualitative system dynamics models were used to show the inter-linkages in the system. The Vensim PLE x32 software (Ventana Systems Inc., Harvard MA, USA) was used to construct the system maps, which combines causal loop diagramming (CLD) and stock accumulation in a qualitative portrayal of the dynamics of the food system. In the CLD, arrows show the influence of one variable on another—a change in the cause leads to a change in the effect. The polarity of the arrows indicates the factual relationship between any two nodes, which illustrates the causal link. Simple stock and flow networks are also used to depict accumulation, and the corresponding rate of change over time. In trying to understand a particular 'system of interest', the interplay of balancing and reinforcing loops gives rise to a realistic multi-loop system that explains behavior through time [35]. In the present paper, the system of interest refers to the mainstreaming of underutilized indigenous and traditional crops into the food system in post-apartheid South Africa.

3. The Current State of Food Systems and Underutilized Indigenous and Traditional Crops in South Africa

Overall, the current status of the food system in South Africa remains a reflection of the legacy of apartheid policies. Land ownership and, therefore, crop production for commercial consumption was largely the preserve of white commercial farmers, while smallholder farmers' produce was largely for own-consumption and sale in informal markets [36]. The result is a dichotomous food system where the commercial sector supplies the dominant food system while the smallholder producers channel their produce to household consumption and alternative food systems that are poorly developed. Consequently, smallholder producers and low-income households also have to rely on the dominant food system [37,38]. Despite reference in agricultural transformation policies to improve smallholder farming that is still practiced by previously disadvantaged South African smallholder farmers, the food system remains largely untransformed. The objective of this section was to assess the current status of South Africa's food system and its limitations, and to identify opportunities for mainstreaming underutilized indigenous and traditional crops with regards to food and nutrition security, agriculture and economic exclusion, environment and policy.

3.1. Food and Nutrition Security

Food and nutrition security is the ability to obtain safe, nutritious foods to meet the basic dietary requirements of an individual, in order to perform daily duties [39], while malnutrition refers to deficiencies (undernutrition), excesses (overnutrition), or imbalances in a person's intake of nutrients [40]. Although the current food system has the capacity to feed the South African population [41], food remains inaccessible to ≈ 26% of the population [31]. Approximately 16 million people are food insecure, and the trends show that there has been no evidence of a decline in these numbers, and the majority of the food insecure population in South Africa is in rural areas [42]. Recent statistics on malnutrition show that 43% of children under five are malnourished [stunting (27%), wasting (3%) and overweight (13%)]. In addition to this, 68% of women in South Africa are overweight [43]. This highlights the flaws in the food systems and the country's nutrition agenda. Over-nutrition has been linked to several non-communicable diseases such as diabetes, hypertension and cardiovascular disease [9,44].

While there is evidence that food production is increasing, this is not enough to feed the growing population and the food insecure (Figure 1). To feed the growing population, and to close the food insecurity gap, food production has to increase by at least 30% of the current production (Figure 1). The gap between achieved food production and the desired food production seems to be increasing compared to previous decades (Figure 1). While agriculture is the main source of livelihood for poor rural households [31,45], it is under marginal conditions, and often cannot sustain subsistence [29,46]. In addition, these households have limited buying power and cannot cope with high food prices [8]. Current agricultural policies have been structured to promote production of cash crops such as maize, wheat, and sugarcane (*Saccharum officinarum*) to enable them to sell the surplus after own consumption [37,38]. These crops are energy dense, and, if consumed alone, do not provide adequate nutrition and lack micronutrients [47]. This is all evident in the high prevalence of all forms of malnutrition seen in young children and women in South Africa [42]. In this regard, underutilized indigenous and traditional crops could be an alternative to bridge the food and nutrition security gap, especially in the rural areas. Several underutilized indigenous and traditional crops have been reported to be nutrient-dense with good adaptability to marginal conditions; hence, they are more likely to be a sustainable and nutritious source of food [29].

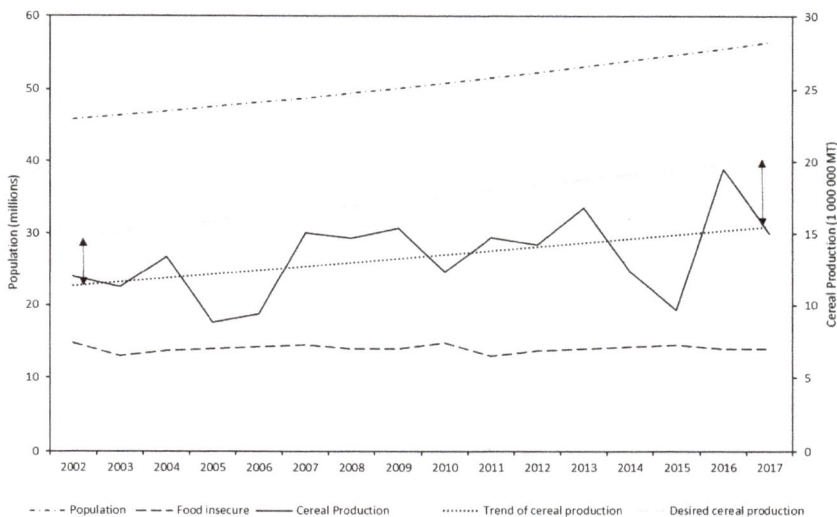

Figure 1. Population in South Africa, population of food insecure and the current and desired trend in food production (Figure developed by authors using population and food insecurity data from the General Household Survey [31] and cereal production data from www.grainsa.co.za.

The Agricultural Policy in South Africa recognizes the role of smallholder farmers; however, it is silent on the issues of underutilized indigenous and traditional crops. This is consistent in other strategy documents issued by the Department of Agriculture [48]. Instead, these documents focus on developing an emerging class of smallholder farmers with commercial aspirations. Though not mentioned explicitly, it can be surmised that this would be through growing major crops, which fit into the dominant food system. These same priorities are reflected in the National Development Plan which prioritizes the commercialization of crops which are in line with the dominant food system [49]. The National Food and Nutrition Security Policy has similar priorities, as it speaks about efficient agricultural production, and is largely silent about the cultivation of underutilized indigenous and traditional crops and development of alternative food systems that include these crops [50]. These policies do, however, advocate for the consumption of nutrient-dense underutilized indigenous and traditional crops [50].

Although current policies reflect a developmental agenda and aim to include and develop small family farmers, they do not fully consider their limitations and opportunities for inclusion into the dominant food systems. Such inconsistencies may be a symptom of the divergent views held by policy-makers on how to address problems of equity, food and nutrition security, and poverty reduction. These different views may be influenced by the success past government policy played in supporting commercial farmers, so that South Africa would be food secure [51]. An unintended consequence has been the resulting skewness of household food insecurity, as income poor rural households who should access food through self-production, purchase most of their food from markets [34]. This is consistent with views of policy-makers worldwide, who perceive the inclusion of small farmers in food systems as a means to improve rural livelihoods, but who are torn between including them in the dominant food systems or creating alternative food systems for them [52]. There is space for underutilized indigenous and traditional crops in this discourse, however, they are often excluded from policy and strategic documents, as shown in their exclusion from The Plant Improvement Act [53].

What has emerged from our systems analysis (Figure 2), is that current policy frameworks have inevitably resulted in dual outcomes of increased "hidden hunger" and reduced poor household food and nutrition security, leading to increased malnutrition and all of its associated consequences (Figure 2). Key factors leading to this emergent outcome are Apartheid policies, policy framework, dominant intensive food system, and the pervasive limitations faced by poor family farmers to be food secure (Figure 2). A counterintuitive effect of the post-apartheid policy implementation has been the democratization of agriculture that enabled underprivileged smallholder producers to embark on sugarcane (*Saccharum officinarum*) monocropping. However, this shift decreased food crop production and dietary diversity. All of these realities would need to be addressed to counter the vicious cycle to malnutrition, as depicted at the top right of Figure 2. The recognition of the importance of Indigenous Knowledge Systems, for instance, through a formalization of the indigenous knowledge base, could be used as an entry point to leverage the incorporation of underutilized indigenous and traditional crops within policy implementation processes, in order to improve access to dietary diversity among the previously disadvantaged. This group is currently caught in a vicious circle of malnutrition, which is exacerbated by poor socio-economic outcomes for household food and nutrition security, and an unsustainable reliance on big food producers and the informal peri-urban food market system. The various consequences of malnutrition for human wellbeing are exacerbated by their inherent vulnerability in the socio-economic system.

3.2. Agriculture and Economic Exclusion

Current agricultural activities recognize the need for increased crop productivity to fight poverty, unemployment and food and nutrition insecurity [48]. However, many of them remain modelled on green revolution ideology, that emphasizes efficiency and productivity over resilience [54]. In South Africa, current policies, funding opportunities, and research interests are still trying to push yield potential of a few major crops, and are not geared towards the development of an indigenous food crop sector. This has made the promotion of underutilized indigenous and traditional crops within current crop and food systems challenging. It also explains the apparent low biodiversity in the country's agro-industrial system. There is growing evidence that shows linkages between low agro-biodiversity and the failures of current food systems to deliver adequate quantities of healthy, nutritionally balanced food, especially to underprivileged people [34,44,55,56]. The global focus on a few crops has resulted in reduced nutrition, which has become more conspicuous in rural areas [57]. Diversity of diet, founded on diverse farming systems, delivers better nutrition and greater health, with additional benefits for human productivity, livelihoods and wellbeing [58,59]. Agricultural biodiversity will be essential to cope with malnutrition, and to establish more sustainable food systems. The inclusion of underutilized indigenous and traditional crops and associated alternative food systems into policy frameworks can result in improvements of agro-biodiversity.

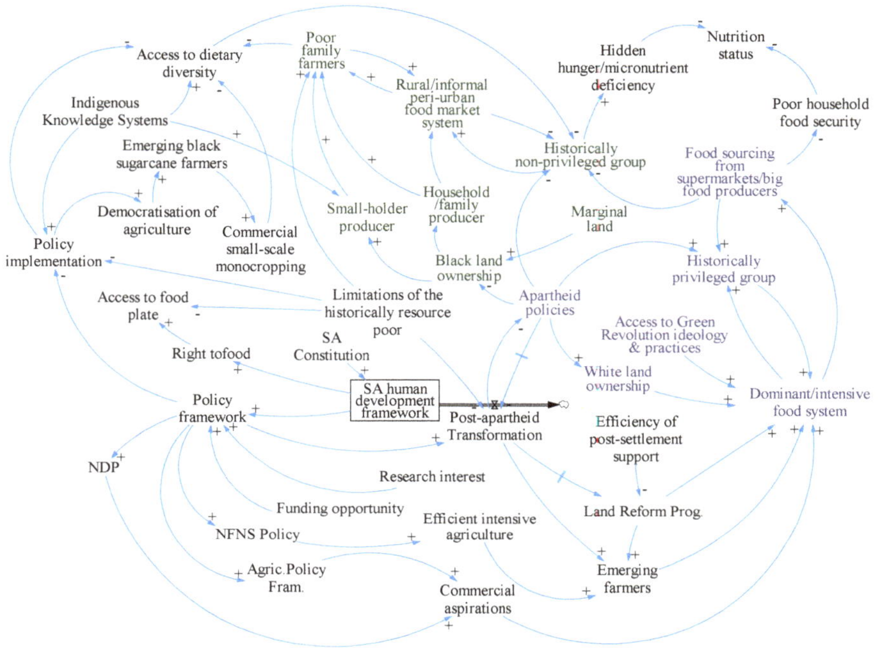

Figure 2. System map depicting the current dichotomous nature of the food system in South Africa. Nodes in blue and green color indicate the dominant and the smallholder producer food systems, respectively. The black nodes represent the opportunities and limitations of the current policy landscape in enabling the smallholder communities to create and sustain a viable food system. An 'equal' sign on an arrow denote systemic delay in the causal relationship. Agric. Policy Fram.: Agricultural Policy Framework; Land Reform Prog.: Land Reform Programme; NDP: National Development Plan; NFNS: National Food and Nutrition Security.

The importance of smallholder farmers to food systems, and their participation in local food systems, must be emphasized. The large-scale commercial farming sector dominates production of agricultural commodities, both for the 'formal' and 'informal' segments of the agro-food system in South Africa [60]. Given the dominance of large-scale and corporate farming business, the mainstream agricultural supply chains are failing to distribute enough food for the country on their own [34]. A paradigm shift is required for the future of farming and food systems in South Africa if government is to deliver on its promises of food and nutrition security for all. A dynamic approach is needed, moving away from a rigid industrial approach; putting smallholder producers at the center of food systems. Extensive evidence suggests smallholder farming systems, also referred to as family farming systems, can help feed a growing South African population [61]. Their participation requires unblocking ideological barriers biased in favor of industrial agriculture; understanding the ways to facilitate and augment agro-ecological practices, local, and traditional knowledge systems; and, re-orientating and prioritizing local agro-food systems [62]. Supporting local food chains for underutilized indigenous and traditional crops is vital to improve local demand, and improve opportunities for smallholder farmers to increase participation in national and regional food chains. Ultimately, supporting local agricultural systems will improve the interest of farmers in cultivating underutilized indigenous and traditional crops, which have the potential to be profitable cash crops [63,64].

The inequality reflected in the production system is mirrored in consumption practices. Socio-economic factors contribute significantly to what one purchases from the food system for household consumption [65].

Yet, what is offered by the current food system in many countries is often not representative of a nation's cross-section of cultural and religious dietary requirements [66]. This affects both producers and consumers. The exclusion of underutilized indigenous and traditional crops from the dominant food system, likely due to poorly developed value chains, disadvantages smallholder farmers, particularly female farmers who are responsible for the conservation, production, and processing of underutilized indigenous and traditional crops [67]. The establishment of such value chains would be beneficial to those households who reside in urban and peri-urban areas, where vegetables and other fresh produce are scarce and therefore expensive [68]. Consumers belonging to minority and marginalized groups, who have little buying power, often find that the dominant food system does not offer what they require [69]. These consumers often turn to informal food systems to purchase their food choices, which includes underutilized indigenous and traditional crops that are often not available from the dominant food system [68].

3.3. Environment

The ongoing intensification of agricultural production in South Africa has had particularly notable effects on the environment through release of greenhouse gases, pollution, loss in species biodiversity and erosion. In South Africa, it is estimated that agriculture contributes $\approx$12% of global anthropogenic greenhouse gas (GHG) emissions, releasing 21,714 GgCO$_2$eq in 2010 [70]. In general, agricultural production, including indirect emissions associated with land use changes and direct emissions from land clearing, contributes 80%–86% of total food system emissions while processing, transporting, storing, cooking and disposing of food contributes the remaining 14%–20% [58,71]. There were no statistics to separate GHG emissions of smallholder and commercial agriculture. However, it is hypothesized that commercial agriculture could possibly contribute to more GHG emissions due to its reliance on external inputs and energy to drive machinery [70]. Animal production contributes a bigger carbon and water footprint compared to plant production. This higher carbon footprint is associated with their feeding, processing and the release of methane gas by ruminants such as cows (*Bos Taurus*) [58,72]. Per ton of product, animal sourced foods have up to a 20 times larger water footprint than crop products [73]. In South Africa, there is an increase in the consumption of animal sourced foods as socio-economic status improves and the replacement of traditional diets by more Westernized diets occurs [74]; hence an increase in health and environmental burdens. The environmental benefits of reducing the fraction of animal sourced foods in food systems are known [75]. Transitioning toward more plant-based food systems could reduce food-related greenhouse gas emissions by 29%–70% [75]. In support of this notion and to further diversify current plant-based food systems, we advocate for the inclusion of nutrient-dense underutilized indigenous and traditional crops [50].

Other concerns are the contribution of agriculture to fine particulate matter in the form of ammonia emissions from animal production and manure processing activities, and, to a lesser extent, fertilizer use [76]. Ammonia emissions have an atmospheric lifespan which ranges from days to weeks, and pollute whole regions in the process, affecting both ecosystems and human health [77–79]. In addition, the application of agrochemicals and fertilizer to increase yield in the dominant food system is associated with possible contamination of soil and water through the wrong application or over-usage of these chemicals [80,81]. Uncontrolled application of pesticides can kill other non-target and beneficial organisms such as bacteria, fungi, and earthworms. Microbial biomass is a labile component of soil organic matter and has an important role in the soil nutrient element cycle [82]. Agrochemicals can move from agricultural fields into nearby streams, rivers and lakes where their toxicity could pose a risk to aquatic ecosystems. These agrochemicals can vary significantly in their toxicity towards aquatic organisms as well as their mobility in the environment—properties which are influenced by their chemical make-up and other climatic, geographic and land management factors. Nitrogen and phosphorus release from agricultural fields also results in eutrophication of aquatic ecosystems, leading to the loss of biodiversity, imbalance of species distribution, shifts in the structure of food chains, and impairment of fisheries [83]. Within avian populations found in the country, 11 species are listed as critically endangered and 43 species as vulnerable [75]. The sensitivity of avian species to agrochemical pollutants has also been widely reported [75]. There is need to moderate and regulate the use of agrochemicals in agriculture to reduce the impacts of food systems to the environment. Underutilized indigenous and traditional crops are less susceptible to pests and diseases and require less fertilizer interventions; hence, they can mitigate the negative environmental impacts of agrochemicals.

While underutilized indigenous and traditional crops may offer some reprieve to environmental issues, there is a need to complement efforts to mainstream them into the dominant food system with sound agricultural practices for the system as a whole. This would ensure current impacts of agriculture on environmental degradation are minimized and food systems become more sustainable. Additional considerations in this regard include, but are not limited to, the use of sustainable and climate smart agriculture techniques that speak to adaptation, mitigation and sustainable intensification of production systems.

3.4. Policy

The policies governing the food system reflect a favorable environment for big businesses, and have made it conducive for a few players to dominate the food system [84]. Many policy-makers favor the commercial agenda because its actors have shown that it is productive; improving national food security, reducing unemployment, and contributing to the national gross domestic product (GDP) [68,85]. However, the profit-making aspects of this food system, which make it attractive, also expose its cost in the form of environmental harm, and inequitable distribution, which results in household food and nutrition insecurity [52,86]. The inability of the dominant food system to distribute produced food adequately to ensure household food security is glaring [68,87]. Thus, two types of policy gaps are revealed, on the one hand policies with divergent goals, and, secondly, policy gaps at critical points of the food system, which have largely been left to big business to address.

It is evident that the status quo is unsustainable. The food system is under pressure to achieve equitable distribution of food produced in the food system, and to feed the growing population using the resources already dedicated to agriculture [85]. Arguments have been made for the inclusion of underutilized indigenous and traditional crops in the existing food system. The first is that there is potential to increase crop diversity and thereby to increase dietary diversity, thus achieving food and nutrition security outcomes [5]. Secondly, advocates for an inclusive food system propose the strengthening of local food systems [88]. This move is seen as setting communities and nations on the path to achieving food sovereignty, which is seen as a key ingredient to achieving both national and household food and nutrition security [85]. These arguments show that there is room for underutilized indigenous and traditional crops in the current food system, however, their inclusion cannot happen effectively in the current policy environment.

It can be argued that policies which seek to enhance smallholder farmer participation in the dominant food system may also expand the inclusion of underutilized indigenous and traditional crops in the food system. However, this is not so, as often when smallholder farmers participate in activities in the dominant food system, they do so by adopting its few crops [67]. These include either staple crops like hybrid maize, or cash crops like sugarcane (*Saccharum officinarum*) or exotic vegetables such as peppers (*Capsicum spp.*) that are more water intensive than alternative traditional crops. The extension system exacerbates these trends, as it is not designed to promote the kind of knowledge that is required to invigorate traditional farming practices that are conducive to cultivating underutilized indigenous and traditional crops.

The South African government amended the National Health Act in 2003, as part of its mandate to ensure basic health services were available to everyone. As part of its mandate, it aims to educate people on healthy food, monitor dietary patterns, nutrient intakes, and nutrition status indicators to promote human health and to prevent diseases [89]. However, because the South African food system is largely driven by market forces, these are not always aligned with nutrition goals. Inorganically processed food has been a major component of diets in South Africa, and has been associated with overweight in both children and adults, as well as chronic diseases such as cancer [90]. This gave rise to a growing market for organic food, which is often very expensive and is beyond the reach of the majority of the population. Ironically, rural farming systems closely resemble organic farming as a result of minimal use of synthetic inputs; however, while their production systems may resemble organic farming, for farming systems to be considered organic, they need certification, which this group of farmers cannot afford [91]. In addition, the extension system is designed to modernize their production systems and increase the use of inorganic inputs like pesticides and fertilizers [92,93].

There is a clear need for a shift in the policy environment, one that recognizes that the 'business as usual' attitude, and piece-meal policy inclusions, will not result in the improved presence of underutilized indigenous and traditional crops in the food system; unless they are intentionally included in the food system [52]. This demands that policy-makers recognize that transformative approaches are required, and that this will require follow-through and not mere rhetoric. This process can be initiated by policy-makers acknowledging the divergence in the existing policies and taking it as an opportunity to bring about policy convergence in pursuit of an inclusive food system. Getting underutilized indigenous and traditional crops onto the policy-makers' agenda could lead to their inclusion in future policies; however, advocates for such progressive ideas often lack the power to do so [84]. There may be need for an alliance with actors with more power to influence agenda setting in the policy arena. Once such items are on the policy agenda, policy-makers can also direct the research agenda to fill the research gap on underutilized indigenous and traditional crops' breeding, processing etc. [94,95]. However, merely getting underutilized indigenous and traditional crops onto the agenda is not enough, the resulting policies should be implemented to bring about inclusivity and equity in the food system, and to strengthen its contribution to the local economy and the GDP. South Africa has had a draft Indigenous Food Crops Strategy in the works since April 2014, but as yet, nothing has come of it.

In addition, government needs to develop policies to address areas in the food system which they may have previously perceived as outside their purview, e.g., the distribution and marketing of food. The policy-makers need to adopt a transformative approach, which embraces informal traders and transporters, who, while already active in the food systems, operate illegally [68]. These informal traders and transporters are an important conduit for including underutilized indigenous and traditional crops in the current system, and, thus, their activities must not be stifled, but, rather, must be incorporated into the dominant food system. Embracing these informal traders and creating a policy environment which makes it conducive for them to distribute food, including underutilized indigenous and traditional crops, to poor communities, will prevent the proliferation of food shortages in underserviced communities.

4. Way Forward for Diverse and Indigenous Food Systems

Given the evidence presented above, in what ways can the paradigm shift be steered to support vulnerable populations, and improve their wellbeing? What type of evidence should be created to support policy-makers in creating the opportunities for supporting smallholder farming? Local/family farming is a subset of the smallholder system, which itself is embedded in the wider socio-economic and agro-ecological systems (Figure 3). The quality of the food plate of vulnerable populations can be improved through the introduction, or re-introduction, of the underutilized indigenous and traditional crops into the food systems. In order to create an enabling environment for the change to happen at the local level, a multi-disciplinary approach, combining research and practitioner-led interventions, is proposed to inform policy makers of the landscape that needs to be created (Figure 3).

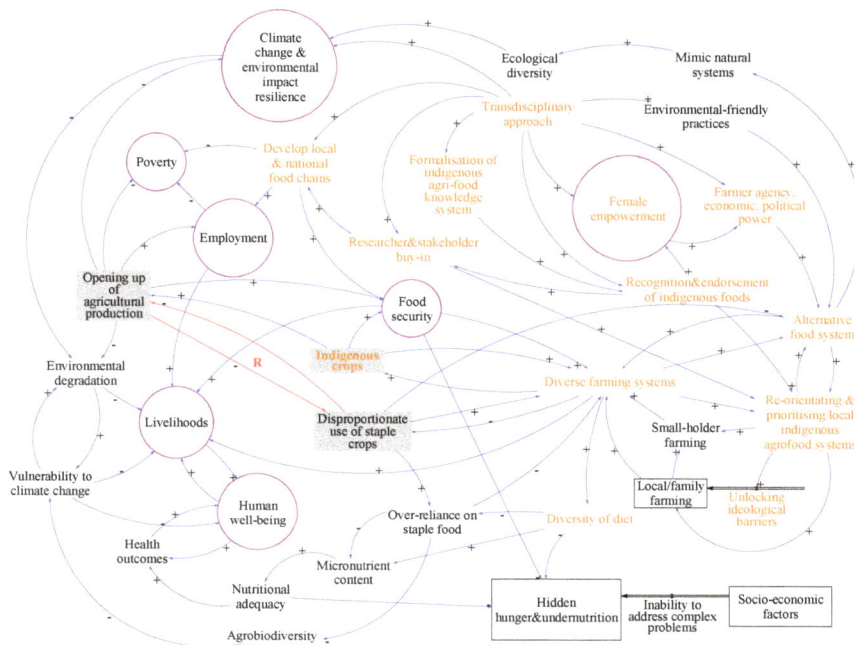

Figure 3. Leveraging the current SA food system for the resource poor. The orange nodes denote the interventions that expand beyond the parochial, to change the system towards sustainability. Variables with a gray background reveal the important nexus points of the system. Encircled nodes in purple color demonstrate how the some of the subcomponents of the system converge towards achieving specific SDG relevance. By opening up diverse agricultural production, the current state which is characterized by disproportionate use of staple crops could be counteracted within a virtuous cycle (denoted by the red arrows) that promote diversified crop production.

4.1. Addressing Health and Nutrition through Diversity

The current food system in South Africa has been shown to lack dietary diversity and has exposed ≈26% of the population to food and nutrition insecurity (c.f. Section 3.2). Young children and women of child-bearing age in rural areas are often most vulnerable, as they lack access to a diversified diet which leads to malnutrition [42]. On a global scale, there are approximately 7000 known and documented edible species of plants [96]; however, due to globalization, there is a decline in the consumption of underutilized indigenous and traditional crops [97]. With climate change, and the fluctuation in food prices [96], it becomes very important to support traditional crops and farming systems [97]. Currently, traditional farming systems are poorly developed and not well-marketed [97]. Including underutilized indigenous and traditional crops into a new food system will increase dietary diversity. Several underutilized indigenous and traditional crops, especially vegetables, have high nutritional value, and could improve the nutritional status of many impoverished individuals [95].

African leafy vegetables such as amaranth (*Amaranthus* spp.) Chinese cabbage (*Brassica rapa*), black nightshade (*Solanum nigrum*), Jew's mallow (*Corchorus olitorius*), cowpea leaves (*Vigna unguiculata*), pumpkin leaves (*Cucurbita spp.*), tsamma melon (*Citrullus lanatus*), and spider flower (*Cleome gynandra*) have been reported to be good sources of vitamins, fiber and iron [98]. These vegetables have also been reported to be used in other parts of South Africa but have been facing marginalization due to low economic value [99,100]. These crops could contribute to alleviating malnutrition in South Africa [98]. Cowpea and bambara groundnut (*Vigna subterranea*) are other examples of underutilized indigenous and traditional crops that are gaining popularity [101]. Although crops such as bambara groundnut are gaining popularity, not many consumers are familiar with the production and preparation of this crop, which may pose a challenge to uptake [101]. By incorporating underutilized indigenous and traditional crops into the food system there will be an increase in dietary diversity, improvement in nutritional status, and a reduction in household food and nutrition insecurity [102]. However, basic knowledge on the production and preparation of underutilized indigenous and traditional crops should be provided to farmers and impoverished individuals, to improve acceptance. When underutilized indigenous and traditional crops are incorporated into a new food system, there should be a transdisciplinary approach, so that new technologies can be used to add value to the products [103].

4.2. Addressing Socio-Economic and Environmental Concerns through Agro-Biodiversity

The consumption of underutilized indigenous and traditional crops in many communities in South Africa and worldwide was adversely affected by their perception as 'poverty foods' or food for the elderly [100,104]. Colonization and globalization contributed to the introduction of foods which were considered 'modern' and, therefore, more attractive than traditional foods [100]. However, some of these 'modern' foods were not as nutritious as underutilized indigenous and traditional crops, and, subsequently, narrowed dietary diversity [88]. In addition, the consumption of underutilized indigenous and traditional crops declined as they were not included in the dominant food systems and could only be purchased in alternative and informal food systems [68]. These alternative food systems, however, were also slowly displaced with the proliferation of regional supermarket chains in both urban and rural South Africa [68].

The agro-industrial food system has inadvertently disempowered farmers, particularly smallholder farmers, many of whom are female [71]. Thus, any attempts to include underutilized indigenous and traditional crops in the food system should, ideally, begin with imparting skills which will increase farmer agency, and economic and political power [71,84]. Such initiatives would not only equip farmers to demand services and opportunities, but would position them as equal partners in the exchange of knowledge between themselves and researchers and other stakeholders [84]. This is important because elderly farmers, who are custodians of underutilized indigenous and traditional crops' conservation and knowledge in most communities [100], are needed to document the different species of underutilized indigenous and traditional crops and their uses. Publishing and recording the information on locally available underutilized indigenous and traditional crops and their uses has been associated with an increase in the use of underutilized indigenous and traditional crops in local diets [88], which created demand for underutilized indigenous and traditional crops in the beneficiary communities [88,105]. The endorsing of underutilized indigenous and traditional crops by influential persons and bodies can also improve their demand in the food system, thus creating opportunities for inclusion in the value chain [88]. This has been observed in some southern African countries, which are battling with both malnutrition and the high incidence of non-communicable diseases [100]. Medical endorsements of underutilized indigenous and traditional crops, which are positioned as healthy and nutrient-rich foods [100], have led to increased consumption, particularly by wealthier members of society [95]. Mainstreaming into the diet of wealthier people would, presumably, also make these crops more attractive and aspirational to poor households, thereby overcoming any previous stigma (see above). Such a surge in consumption of underutilized indigenous and traditional crops can potentially reduce malnutrition and the prevalence of non-communicable diseases, while simultaneously stimulating market opportunities for farmers [85,95].

Cultivating underutilized indigenous and traditional crops in an inclusive food system could contribute significantly to addressing environmental concerns. Agro-ecological practices and other farming systems which mimic nature would be instrumental in reducing impacts [85]. By their nature, such farming systems promote the growth of a multiplicity of edible and medicinal plants which are indigenous to a region [67,84,85]. Such crops, which had been relegated to alternative food systems, and are therefore not widely available, can once again be introduced and integrated into the dominant food system. Increasing their prominence in the food system would also serve as a means of fostering their conservation, as these species would otherwise be lost through underutilization and land use changes [88,106]. Furthermore, a system which is based on species diversity is resilient and can withstand different threats and shocks, climatic or otherwise [16,107].

This sustainability extends to agricultural livelihoods in the event of an environmental shock, as some crops would survive these events, and, thus, reduce farmer vulnerability [16]. Such benefits would be appealing to farmers who are increasingly at risk of being affected by climate change events, but who have no access to insurance [108]. The added benefit of growing underutilized indigenous and traditional crops is that they provide dietary diversity, which has a lower carbon footprint than animal sources [58]. Indigenous species are agroecologically adapted to their local environment, and often grow spontaneously with few added inputs, which significantly reduces the pollution of soil and water from the introduction of agricultural chemicals [88]. Furthermore, when the cultivation of underutilized indigenous and traditional crops is coupled with little land disturbance, this would then reduce the extent to which farmers disturb the ecosystem, reducing environmental degradation.

The environmental benefits of including underutilized indigenous and traditional crops in the food system notwithstanding, a significant constraint in the adoption of these strategies is the diminishing knowledge surrounding their uses, and perceptions of low productivity [84,109]. Documenting and popularizing the benefits of planting underutilized indigenous and traditional crops, for instance legumes as part of crop rotations or intercrop systems, could appeal to both environmentally conscious and resource poor farmers [110]. For those farmers who have concerns about the productivity of farming systems which mimic nature, research is showing that, when well-managed, they perform better than monoculture farming systems, which are common in the dominant food system [85]. The resulting crop diversity and high yields may be an adequate incentive for those who wish to adopt sustainable agricultural practices on a large scale, but had previously found them not to be financially rewarding [107]. Thus, incorporating underutilized indigenous and traditional crops into the production system will result in improved sustainability and resilience in the food system [102].

Figure 3 elicits and captures the transdisciplinary approach to re-orient and prioritize local indigenous agro-food systems, namely: (i) the creation of an alternative and diverse food system that empowers local smallholder famers; (ii) that caters to the nutritional health and livelihood of the resource poor; and, (iii) that addresses climate and environmental change impacts and builds resilience. The influence of including a transdisciplinary approach, recognizing/endorsing underutilized indigenous and traditional crops, developing local food chains, resulting in diverse farming systems and alternative food systems, would create a virtuous cycle of increased household food and nutrition security, improved livelihoods, and reduced malnutrition (Figure 3), thereby reducing vulnerability, inequalities, and improving human wellbeing.

4.3. Agriculture-Environment-Health Nexus

Based on the review on the current status of the South African food system (c.f. Section 3) there was scant evidence of an appreciation of the linkages between agriculture, environment, and health in the current agro-industrial food system. The three sectors need to work together in order to address common issues such as improved agricultural productivity, food and nutritional security, reduced environmental degradation, improved human health outcomes, and improved human wellbeing in general. It is clear that agriculture is responsible for increasing food production and influencing healthy diets. Consumers are important players in the food system and their demands can influence production, as shown by the increased demand for animal sourced foods in countries experiencing rapid economic growth [111]. It is possible that improving consumer education on the link between certain foods and their carbon footprint can reduce the demand for those foods. There is evidence to show that reducing the consumption of beef results in lower carbon and ammonia emissions, despite increasing food demand [112].

However, agriculture is not just about growing food for consumption, there are also aspects of the environment that are key for agriculture [113,114]. Key inputs in agriculture include water and land, thus, any consideration of increasing food production will need to consider water and land use [113,114]. Given the challenges of water scarcity and associated challenges in expansive agriculture, there is a need for an agriculture-environment-health nexus approach to address the sustainable provision of enough nutritious food for supporting healthy diets (Figure 4).

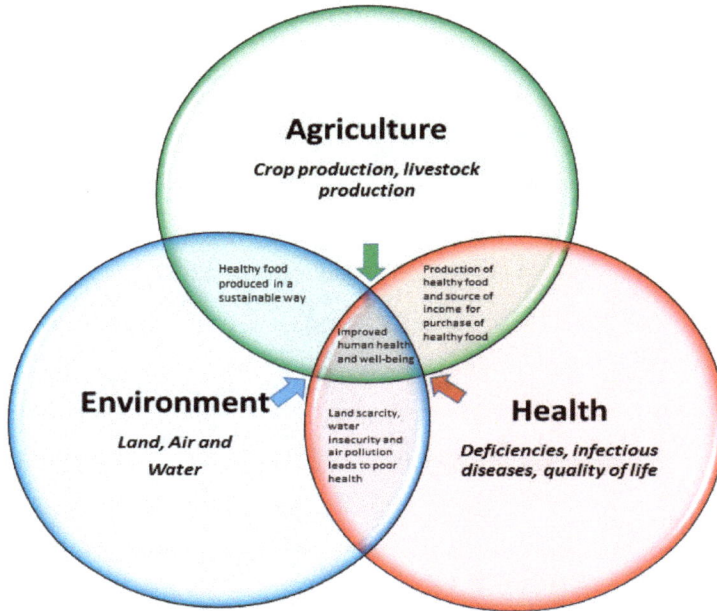

Figure 4. Agriculture-Environment-Health Nexus showing the interconnectedness and linkages between agriculture, environment and health sectors. The nexus outcome of improved human health and well-being is an outcome of a sustainable and healthy food system inclusive of underutilized indigenous and traditional crops.

The contribution of agriculture to food and nutrition security and, ultimately, human health, is not only through the direct provision of food. Benefits of agriculture can also be indirect, for example, agriculture is already a source of income for ≈65% of South African households [115]. In rural parts of South Africa, smallholder farming systems have been associated with women as their source of livelihood and employment [29]. Such income can be used by women to buy food that can improve food and nutrition security, and health. According to the Organization for Economic Co-operation and Development (OECD) agriculture plays an important role for poverty reduction and economic development in developing countries [116,117].

Conceptually, analyses of the agriculture-environment-health nexus require a food systems approach that recognizes the interdependence and interconnectivity of these sectors. Furthermore, this approach emphasizes that the food system comprises a set of activities ranging from production to consumption, and that each of these activities, such as the processing of food, impacts all of these areas. Underutilized indigenous and traditional crops engender several of the key nexus points within the agriculture-environment-health nexus (Figure 4). Firstly, with regards to the agriculture-environment linkages, underutilized indigenous and traditional crops are often produced within niche agro-ecologies, and hence require less landscape modification [103]. This also supports the notion that they support environmentally sustainable and resilient agriculture [103]. Similarly, with regards to the agriculture-health linkages, the inclusion of underutilized indigenous and traditional crops offers prospects for broadening food and nutrition security (dietary diversity) [45] as well as income security for smallholder farmers [103] (c.f. Section 3.2). With respect to the environment-health linkages, as already established, the adaptation of most underutilized indigenous and traditional crops to harsh agro-ecologies and marginal production areas, means that they exert less pressure on already scarce land and water resources [29]. Lastly, owing to these attributes, underutilized indigenous and traditional crops have been reported to fit within the water-food-nutrition-health nexus as they can be used as part of efforts to improve human health and nutrition is water scarce environments [103].

5. Conclusions and Recommendations

A sustainable and healthy food system delivers food and nutrition security for all, in a way that is economically, socially, and environmentally sound, so as not to compromise food and nutrition security for future generations [118]. Underutilized indigenous and traditional crops can support and strengthen the existing food system, as they in particular are considered as economically, socially, and environmentally sound [28]. Several underutilized indigenous and traditional crops are nutrient dense and adapted to marginal conditions, suggesting that they could be used to champion sustainable and resilient agriculture and food systems for smallholder farmers residing in these environments [28].

The mainstreaming of underutilized indigenous and traditional crops should not seek to transform them into "new major crops" but should recognize their attributes that make them desirable as well as the role played by their current custodians in conserving them. In this regard, the fact that value chains for underutilized indigenous and traditional crops are currently poorly developed creates opportunities to diversify the current food systems, thus creating new employment, market, and distribution opportunities, and promoting autonomous pathways out of poverty. While the agro-industrial food system has inadvertently excluded smallholder farmers and women in rural areas, underutilized indigenous and traditional crops offer the opportunity for such groups of farmers to re-enter the system. Importantly, the significant role played by women in the production and conservation of underutilized indigenous and traditional crops [119] offers opportunities for women empowerment through their inclusion in the food system. Promoting gender equality and women's empowerment is inextricably linked to the strengthening of sustainable food systems to fight hunger and malnutrition, and improving the lives and livelihoods of rural populations. The mainstreaming of underutilized indigenous and traditional crops into the food system would support women to diversify their landscapes in a sustainable way, feed their own households, and to provide nutritious food at local markets. In this regard, promotion and inclusion (i.e., mainstreaming) of underutilized indigenous and traditional crops could contribute towards addressing Sustainable Development Goals related to social, economic, and environmental issues; specifically, SDGs 1, 2, 3, 8, and 15 [120].

It is acknowledged that smallholder farmers have low literacy and numeracy skills, and that this may hamper efficient production and the adoption of new technologies. Initiatives taken to integrate smallholder farmers into the agro-industrial food system require a strong training component to address underdeveloped skills [121]. There is a positive correlation between farmer training and technology adoption, and this has often resulted in positive economic and environmental outcomes [112,121]. This calls for all the role players, from researchers to policy makers, to view these sectors in a nexus approach, and not to view them as separate, siloed, sectors.

Our systems analysis revealed the importance of inequalities and power imbalances, especially as a legacy of Apartheid discriminative policies, resulting in reduction in diversity of crops, reduced access to the food system, and reduced diversity of diet; these have outcomes of reduced household food security and increased "hidden hunger", both of which feed into malnutrition and all its consequences (Figure 3) in a vicious cycle. When we include underutilized indigenous and traditional crops into the food system, the increased diversity in crops, improved local value chain, and diversified food system, result in increased household food security, improved livelihoods, and reduced hidden hunger (Figure 3), a virtuous cycle.

Policy-makers need to transform policy-making processes to represent the interests of different food system actors, promote underutilized indigenous and traditional crops, and support systems at all stages of the food system. This is an important point about the policy-making process. As we advocate for transdisciplinary research, which involves the various actors' knowledge co-creation and co-development, this approach should also be applied for policy making. It should be truly participatory, and provide for influence by beneficiaries as required by transformation, and by societies' values, as per global human rights and the SA constitution. The current policy environment makes little mention of underutilized indigenous and traditional crops, and is thus unfavorable to actors who grow, process, and distribute underutilized indigenous and traditional crops in informal food systems. In addition, the current policy environment leaves critical aspects of the food system, such as marketing and distribution, to the private sector, which may have a different agenda. Furthermore, it leaves the holders of the indigenous and local knowledge about these diverse crops vulnerable to corporate patents that do not take into account access and benefit sharing legislation [122–125]. Transformative change should be underpinned by robust scientific evidence. Thus, the current body of research that has been generated on underutilized indigenous and traditional crops, as highlighted in this review, needs to be translated for uptake by policy-makers. This highlights a gap in terms of the role of researchers—that there is not effective translation of work to make it accessible and relevant to policy makers, or to policy processes. With the point above, this means that two key actors are missing in the policy process, the scientists (at the start of the policy value chain), and the beneficiaries (at the end of the value chain). Another element that is often ignored or marginalized is indigenous knowledge systems, which provide an alternative information/evidence source to western-driven science; indigenous knowledge also needs to be made accessible and translated for decision-makers [126].

The different power relations in the food system must be recognized, as these may influence the success of any interventions. Smallholder farmers who currently produce underutilized indigenous and traditional crops are disempowered compared to other food system players, because of their poverty, and gendered access to resources and opportunities, as they are predominantly female. There is a need to rebrand and reposition underutilized indigenous and traditional crops as an important component of a sustainable and healthy food system.

Our review highlights the unintended consequence of a commercialized food system, based on a reduced number of crops, as reinforcing inequality and imbalances. While increasing national food security and stimulating national GDP, rather than making food cheaper and more accessible to all members of society, such a food system creates imbalances, reduces household food and nutrition security, and exacerbates existing inequalities. Therefore, instead of improving the wellbeing of all, as envisaged in the Sustainable Development Goals, such a food system is disempowering, increases vulnerability of the most vulnerable, and perpetuates and creates legacy imbalances in outcomes such as health, wealth, and education, which will have long-term effects on national development and nation building. Such insights, explored in the South African context, have similar implications for other developing countries faced with competing policy agendas of increasing agricultural production for commercial growth and development, versus ensuring affordable and household food and nutrition security for the most vulnerable citizenry, thereby decreasing social cohesion, and increasing the threat of social instability and conflict. Future studies should focus on conducting an econometric assessment to determine a cost-benefit analysis for mainstreaming underutilized indigenous and traditional crops into the South African food system. This would go some way in informing policy of the value of such action.

Supplementary Materials: The following are available online at http://www.mdpi.com/2071-1050/11/1/172/s1, Figure S1: title, Table S1: title, Video S1: PRISMA 2009 Checklist.

Author Contributions: Conceptualization: T.M. and T.P.C.; Methodology: T.M., T.P.C. and N.S.; Investigation and writing-original draft preparation: T.M., T.P.C., V.G.P.C., V.M., L.P., N.S. and L.G.; Critical Review and Redrafting: T.M., T.P.C., A.T.M. and R.S.

Funding: This research was funded by the Wellcome Trust through the Sustainable and Healthy Food Systems (SHEFS) Project (Grant number-205200/Z/16/Z), the Water Research Commission of South Africa (WRC Project Number K5/2493//4), and the Adaptation Fund through the uMngeni Resilience Project. The APC was funded by the Wellcome Trust through the Sustainable and Healthy Food Systems (SHEFS) Project Grant number-205200/Z/16/Z).

Acknowledgments: We would like to acknowledge Prof Alan Dangour from the London School of Hygiene and Tropical Medicine for providing valuable input and direction during the critical review and redrafting.

Conflicts of Interest: The authors declare no conflict of interest.

References

1. Hueston, W.; McLeod, A. Overview of the global food system: Changes over time/space and lessons for future food safety. In *Improving Food Safety through a One Health Approach: Workshop Summary*; National Academies Press: Washington, DC, USA, 2012; p. 189.

2. Sustainable Development Solutions Network. *Solutions for Sustainable Agriculture and Food Systems*; Sustainable Development Solutions Network: New York, NY, USA, 2013.

3. Pingali, P.L. Green Revolution: Impacts, limits, and the path ahead. *Proc. Natl. Acad. Sci. USA* **2012**, *109*, 12302–12308. [CrossRef] [PubMed]

4. Evenson, R.E.; Gollin, D. Assessing the impact of the Green Revolution, 1960 to 2000. *Science* **2003**, *300*, 758–762. [CrossRef]

5. Kerr, R.B. Lessons from the old Green Revolution for the new: Social, environmental and nutritional issues for agricultural change in Africa. *Prog. Dev. Stud.* **2012**, *12*, 213–229. [CrossRef]

6. Sanz-Cañada, J.; Muchnik, J. Geographies of origin and proximity: Approaches to local agro-food systems. *Cult. Hist. Digit. J.* **2016**, *5*. [CrossRef]

7. Greenberg, S. Agrarian reform and South Africa's agro-food system. *J. Peasant Stud.* **2015**, *42*, 957–979. [CrossRef]

8. Dudley, N.; Alexander, S. Agriculture and biodiversity: A review. *Biodiversity* **2017**, *18*, 45–49. [CrossRef]

9. Gómez, M.I.; Barrett, C.B.; Raney, T.; Pinstrup-Andersen, P.; Meerman, J.; Croppenstedt, A.; Carisma, B.; Thompson, B. Post-green revolution food systems and the triple burden of malnutrition. *Food Policy* **2013**, *42*, 129–138. [CrossRef]

10. Dawson, N.; Martin, A.; Sikor, T. Green revolution in sub-Saharan Africa: Implications of imposed innovation for the wellbeing of rural smallholders. *World Dev.* **2016**, *78*, 204–218. [CrossRef]

11. Blaustein, R.J. The green revolution arrives in Africa. *AIBS Bull.* **2008**, *58*, 8–14. [CrossRef]

12. McMichael, P. A food regime analysis of the 'world food crisis'. *Agric. Hum. Values* **2009**, *26*, 281–295. [CrossRef]

13. Pimbert, M. *Fair and Sustainable Food Systems*; International Institute for Environment and Development: London, UK, 2012.

14. Ackerman, K.; Conard, M.; Culligan, P.; Plunz, R.; Sutto, M.-P.; Whittinghill, L. Sustainable food systems for future cities: The potential of urban agriculture. *Econ. Soc. Rev.* **2014**, *45*, 189–206.

15. Zalewski, R.I.; Skawińska, E. Towards sustainable food system. *Acta Sci. Pol. Oeconomia* **2016**, *15*, 187–198.

16. MacFall, J.; Lelekacs, J.M.; LeVasseur, T.; Moore, S.; Walker, J. Toward resilient food systems through increased agricultural diversity and local sourcing in the Carolinas. *J. Environ. Stud. Sci.* **2015**, *5*, 608–622. [CrossRef]

17. Gordon, L.J.; Bignet, V.; Crona, B.; Henriksson, P.J.G.; Van Holt, T. Rewiring food systems to enhance human health and biosphere stewardship. *Environ. Res. Lett.* **2017**, *12*. [CrossRef]

18. Vermeulen, S.J.; Campbell, B.M.; Ingram, J.S.I. Climate change and food systems. *Annu. Rev. Environ. Resour.* **2012**, *37*, 195–222. [CrossRef]

19. Seburanga, J.L. Decline of indigenous crop diversity in colonial and postcolonial Rwanda. *Int. J. Biodivers.* **2013**, *2013*, 401938. [CrossRef]

20. Kerr, R.B. Lost and Found Crops: Agrobiodiversity, Indigenous Knowledge, and a Feminist Political Ecology of Sorghum and Finger Millet in Northern Malawi. *Ann. Assoc. Am. Geogr.* **2014**, *104*, 577–593. [CrossRef]

21. Shelef, O.; Weisberg, P.J.; Provenza, F.D. The Value of Native Plants and Local Production in an Era of Global Agriculture. *Front. Plant Sci.* **2017**, *8*, 2069. [CrossRef]

22. Shiferaw, B.; Prasanna, B.M.; Hellin, J.; Bänziger, M. Crops that feed the world 6. Past successes and future challenges to the role played by maize in global food security. *Food Secur.* **2011**, *3*, 307–327. [CrossRef]

23. Nkonya, E.; Place, F.; Kato, E.; Mwanjololo, M. Climate Risk Management Through Sustainable Land Management in Sub-Saharan Africa. In *Sustainable Intensification to Advance Food Security and Enhance Climate Resilience in Africa*; Springer International Publishing: Cham, Switzerland, 2015; pp. 75–111.

24. Schmitz, C.; Biewald, A.; Lotze-Campen, H.; Popp, A.; Dietrich, J.P.; Bodirsky, B.; Krause, M.; Weindl, I. Trading more food: Implications for land use, greenhouse gas emissions, and the food system. *Glob. Environ. Chang.* **2012**, *22*, 189–209. [CrossRef]

25. Fischer, A.R.H.; Beers, P.J.; Latesteijn, H.V.; Andeweg, K.; Jacobsen, E.; Mommaas, H.; Van Trijp, H.C.M.; Veldkamp, A. Transforum system innovation towards sustainable food. A review. *Agron. Sustain. Dev.* **2012**, *32*, 595–608. [CrossRef]

26. Coq-Huelva, D.; Sanz-Cañada, J.; Sánchez-Escobar, F. Values, conventions, innovation and sociopolitical struggles in a local food system: Conflict between organic and conventional farmers in Sierra de Segura. *J. Rural Stud.* **2017**, *55*, 112–121. [CrossRef]

27. Modi, A.T.; Mabhaudhi, T. *Water Use and Drought Tolerance of Selected Traditional and Indigenous Crops*; Water Research Commision of South Africa: Pretoria, South Africa, 2013; ISBN 978-1-4312-0434-2.

28. Mabhaudhi, T.; Chimonyo, V.G.P.; Modi, A.T. Status of Underutilised Crops in South Africa: Opportunities for Developing Research Capacity. *Sustainability* **2017**, *9*, 1569. [CrossRef]

29. Chivenge, P.; Mabhaudhi, T.; Modi, A.T.; Mafongoya, P. The potential role of neglected and underutilised crop species as future crops under water scarce conditions in Sub-Saharan Africa. *Int. J. Environ. Res. Public Health* **2015**, *12*, 5685–5711. [CrossRef]

30. Greenberg, S. Corporate power in the agro-food system and the consumer food environment in South Africa. *J. Peasant Stud.* **2017**, *44*, 467–496. [CrossRef]

31. Statistics South Africa. *General Household Survey 2017*; Statistics South Africa: Pretoria, South Africa, 2018.

32. Drimie, S.; Pereira, L. Advances in Food Security and Sustainability in South Africa. In *Advances in Food Security and Sustainability*; Barling, D., Ed.; Academic Press: Burlington, NJ, USA, 2016; pp. 1–31. ISBN 9780128098639.

33. Mbhenyane, X.G. *The contribution of 'Indigenous Foods' to the elimination of hidden hunger and food insecurity: An illusion or innovation? 2016*; Stellenbosch University: Stellenbosch, South Africa, 2016.

34. Pereira, L.M. *The Future of South Africa's Food System: What is Research Telling Us?* Southern Africa Food Lab: Stellenbosch, South Africa, 2014.

35. Morecroft, J.D.W. Metaphorical models for limits to growth and industrialization. *Syst. Res. Behav. Sci.* **2012**, *29*, 645–666. [CrossRef]

36. Pereira, L.; Drimie, S. Governance Arrangements for the Future Food System: Addressing Complexity in South Africa. *Environ. Sci. Policy Sustain. Dev.* **2016**, *58*, 18–31. [CrossRef]

37. Baiphethi, M.; Jacobs, P. *Agrekon—The Contribution of Subsistence Farming to Food Security in South Africa*; Routledge: Abingdon, UK, 2009.

38. Drimie, S. Understanding South African Food and Agricultural Policy Implications for Agri-Food Value Chains, Regulation and Formal and Informal Livelihoods. 2016. Available online: http://www.plaas.org.za/plaas-publication/wp39-foodagripolicy-drimie (accessed on 16 June 2018).

39. Food and Agriculture Organization (FAO). The Rome declaration on world food security. *Popul. Dev. Rev.* **1996**, *22*, 14–17. [CrossRef]

40. International Food Policy Research Institute. *Global Nutrition Report 2016: From Promise to Impact: Ending Malnutrition by 2030*; International Food Policy Research Institute: Washington, DC, USA, 2016.

41. Altman, M.; Hart, T.G.; Jacobs, P.T. Household food security status in South Africa. *Agrekon* **2009**, *48*, 345–361. [CrossRef]

42. Shisana, O.; Labadarios, D.; Simbayi, L.; Zuma, K.; Dhansay, A.; Reddy, P.; Parker, W.; Hoosain, E.; Naidoo, P.; Hongoro, C.; et al. *South African National Health and Nutrition Examination Survey (SANHANES-1)*; HSRC Press: Cape Town, South Africa, 2014; ISBN 9780796924766.

43. National Department of Health. *ICF South Africa Demographic and Health Survey 2016: Key Indicators*; National Department of Health: Pretoria, South Africa, 2016.

44. Stuart, G.; van den Bold, M. Agriculture, Food Systems, and Nutrition: Meeting the Challenge. *Glob. Chall.* **2017**, *1*, 1600002. [CrossRef]

45. Govender, L.; Pillay, K.; Siwela, M.; Modi, A.; Mabhaudhi, T. Food and Nutrition Insecurity in Selected Rural Communities of KwaZulu-Natal, South Africa—Linking Human Nutrition and Agriculture. *Int. J. Environ. Res. Public Health* **2016**, *14*, 17. [CrossRef] [PubMed]

46. Hadebe, S.T.; Modi, A.T.; Mabhaudhi, T. Drought Tolerance and Water Use of Cereal Crops: A Focus on Sorghum as a Food Security Crop in Sub-Saharan Africa. *J. Agron. Crop Sci.* **2016**, *203*, 177–199. [CrossRef]

47. Alders, R.; Kock, R. What's food and nutrition security got to do with wildlife conservation? *Aust. Zool.* **2018**, *39*, 120–126. [CrossRef]

48. Department of Agriculture. *Strategic Plan for South African Agriculture 2001*; Department of Agriculture: Pretoria, South Africa, 2002.

49. National Planning Commission. *National Development Plan 2030: Our future—Make It Work*; National Planning Commission: Pretoria, South Africa, 2012.

50. Department of Agriculture Forestry and Fisheries (DAFF). *National Policy on Food and Nutrition Security*. Available online: https://www.nda.agric.za/docs/media/NATIONAL%20POLICYon%20food%20and%20nutririon%20security.pdf (accessed on 15 June 2018).

51. Cousins, B.; Scoones, I. Contested paradigms of 'viability' in redistributive land reform: Perspectives from Southern Africa. *J. Peasant Stud.* **2010**, *37*, 31–66. [CrossRef]

52. Pitt, H.; Jones, M. Scaling up and out as a pathway for food system transitions. *Sustainability* **2016**, *8*, 1025. [CrossRef]

53. National Department of Agriculture. *The Plant Improvement Act, 1976 (Act No. 53 of 1976)*; National Department of Agriculture: Pretoria, South Africa, 2017.

54. Musvoto, C.; Nortje, K.; De Wet, B.; Mahumani, B.K.; Nahman, A. Imperatives for an agricultural green economy in South Africa. *S. Afr. J. Sci.* **2015**, *111*, 1–8. [CrossRef]

55. Rosin, C.; Stock, P.; Campbell, H. *Food Systems Failure: The Global Food Crisis and the Future of Agriculture*; Routledge: Abingdon, UK, 2013.

56. Goodman, D.; Watts, M. Agrarian questions: Global appetite, local metabolism: Nature, culture, and industry in fin-de-siecle agro-food systems. In *Globalising Food*; Taylor & Francis: Abingdon, UK, 2013.

57. Grubben, G.; Klaver, W. Vegetables to combat the hidden hunger in Africa. *Chronica Hortic.* **2014**, *54*, 9.

58. Meybeck, A.; Gitz, V. What diets for sustainable food systems? *Cah. Nutr. Diet.* **2016**, *51*, 304–314. [CrossRef]

59. Meybeck, A.; Gitz, V. Sustainable diets within sustainable food systems. *Proc. Nutr. Soc. USA* **2017**, *76*, 1–11. [CrossRef]

60. Pereira, L.M.; Wynberg, R.; Reis, Y. Agroecology: The Future of Sustainable Farming? *Environ. Sci. Policy Sustain. Dev.* **2018**, *60*, 4–17. [CrossRef]

61. Affholder, F.; Poeydebat, C.; Corbeels, M.; Scopel, E.; Tittonell, P. The yield gap of major food crops in family agriculture in the tropics: Assessment and analysis through field surveys and modelling. *Field Crops Res.* **2013**, *143*. [CrossRef]

62. Machum, S. Shifting practices and shifting discourses: Policy and small-scale agriculture in sustainable food systems past and present. In Proceedings of the 10th European IFSA Symposium, Aarhus, Denmark, 1–4 July 2012.

63. Baena, M.; Galluzzi, G.; Padulosi, S. Improving Community Livelihoods by Recovering and Developing Their Traditional Crops. Available online: https://www.bioversityinternational.org/e-library/publications/detail/improving-community-livelihoods-by-recovering-and-developing-their-traditional-crops/ (accessed on 17 July 2018).

64. Padulosi, S.; Thompson, J.; Rudebjer, P. *Fighting Poverty, Hunger and Malnutrition with Neglected and Underutilized Species: Needs, Challenges and the Way Forward: Neglected and Underutilized Species*; Bioversity International: Rome, Italy, 2013.

65. Crush, J.; McCordic, C. The Hungry Cities Food Purchases Matrix: Household Food Sourcing and Food System Interaction. *Urban Forum* **2017**, *28*, 421–433. [CrossRef]

66. Tieman, M.; Hassan, F.H. Convergence of food systems: Kosher, Christian and Halal. *Br. Food J.* **2015**. [CrossRef]

67. Adhikari, L.; Hussain, A.; Rasul, G. Tapping the potential of neglected and underutilized food crops for sustainable nutrition security in the mountains of Pakistan and Nepal. *Sustainability* **2017**, *9*, 291. [CrossRef]

68. Smit, W. Urban governance and urban food systems in Africa: Examining the linkages. *Cities* **2016**, *58*, 80–86. [CrossRef]

69. Raja, S.; Morgan, K.; Hall, E. Planning for equitable Urban and regional food systems. *Built Environ.* **2017**, *43*, 309–314. [CrossRef]

70. Department of Environmental Affairs. *GHG National Inventory Report South Africa 2000–2010*; Department of Environmental Affairs: Pretoria, South Africa, 2014.

71. O'Kane, G.; Wijaya, S.Y. Contribution of farmers' markets to more socially sustainable food systems: A pilot study of a farmers' market in the Australian Capital Territory (ACT), Australia. *Agroecol. Sustain. Food Syst.* **2015**, *39*, 1124–1153. [CrossRef]

72. Alsaffar, A.A. Sustainable diets: The interaction between food industry, nutrition, health and the environment. *Food Sci. Technol. Int.* **2016**, *22*, 102–111. [CrossRef]

73. Gerbens-Leenes, P.W.; Mekonnen, M.M. The water footprint of poultry, pork and beef: A comparative study in different countries and production systems. *Water Resour. Ind.* **2013**, *1–2*, 25–36. [CrossRef]

74. Ronquest-Ross, L.-C.; Vink, N.; Sigge, G.O. Food consumption changes in South Africa since 1994. *S. Afr. J. Sci.* **2015**, *111*, 1–12. [CrossRef]

75. Quinn, L.P.B.; De, J.; Fernandes-Whaley, M.; Roos, C.; Bouwman, H.; Kylin, H.; Pieters, R.; van den Berg, J. Pesticide Use in South Africa: One of the Largest Importers of Pesticides in Africa. In *Pesticides in the Modern World—Pesticides Use and Management*; Stoytcheva, M., Ed.; InTech: Rijeka, Croatia, 2011.

76. Bouwman, A.F.; Boumans, L.J.M.; Batjes, N.H. Estimation of global NH3 volatilization loss from synthetic fertilizers and animal manure applied to arable lands and grasslands. *Glob. Biogeochem. Cycles* **2002**, *16*, 1–8. [CrossRef]

77. Chowdhury, M.A.; de Neergaard, A.; Jensen, L.S. Potential of aeration flow rate and bio-char addition to reduce greenhouse gas and ammonia emissions during manure composting. *Chemosphere* **2014**, *97*, 16–25. [CrossRef] [PubMed]

78. Oonincx, D.G.A.B.; van Itterbeeck, J.; Heetkamp, M.J.W.; van den Brand, H.; van Loon, J.J.A.; van Huis, A. An Exploration on Greenhouse Gas and Ammonia Production by Insect Species Suitable for Animal or Human Consumption. *PLoS ONE* **2010**, *5*, e14445. [CrossRef] [PubMed]

79. Gonzalez-Garza, D.; Rivera-Tinoco, R.; Bouallou, C. Comparison of ammonia, monoethanolamine, diethanolamine and methyldiethanolamine solvents to reduce CO2 greenhouse gas emissions. *Chem. Eng. Trans.* **2009**, *18*, 279–284. [CrossRef]

80. Mittal, S.; Kaur, G.; Vishwakarma, G.S. Effects of environmental pesticides on the health of rural communities in the Malwa Region of Punjab, India: A review. *Hum. Ecol. Risk Assess. Int. J.* **2014**, *20*, 366–387. [CrossRef]

81. Carvalho, F.P. Pesticides, environment, and food safety. *Food Energy Secur.* **2017**, *6*, 48–60. [CrossRef]

82. Snapp, S.S.S.; Grabowski, P.; Chikowo, R.; Smith, A.; Anders, E.; Sirrine, D.; Chimonyo, V.; Bekunda, M. Maize yield and profitability tradeoffs with social, human and environmental performance: Is sustainable intensification feasible? *Agric. Syst.* **2018**, *162*, 77–88. [CrossRef]

83. Tilman, D. Global environmental impacts of agricultural expansion: The need for sustainable and efficient practices. *Proc. Natl. Acad. Sci. USA* **1999**, *96*, 5995–6000. [CrossRef]

84. Fernandez, M.; Goodall, K.; Olson, M.; Méndez, V.E. Agroecology and alternative agri-food movements in the United States: Toward a sustainable agri-food system. *Agroecol. Sustain. Food Syst.* **2013**, *37*, 115–126. [CrossRef]

85. Tudge, C. Agroecology and biodiversity: The kind of farming we need to secure our food supply is also wildlife friendly. *Biodiversity* **2014**, *15*, 290–293. [CrossRef]

86. Prosperi, P.; Allen, T.; Cogill, B.; Padilla, M.; Peri, I. Towards metrics of sustainable food systems: A review of the resilience and vulnerability literature. *Environ. Syst. Decis.* **2016**, *36*, 3–19. [CrossRef]

87. Ofori, D.A.; Gyau, A.; Dawson, I.K.; Asaah, E.; Tchoundjeu, Z.; Jamnadass, R. Developing more productive African agroforestry systems and improving food and nutritional security through tree domestication. *Curr. Opin. Environ. Sustain.* **2014**, *6*, 123–127. [CrossRef]

88. Kuhnlein, H.V. Food system sustainability for health and well-being of Indigenous Peoples. *Public Health Nutr.* **2015**, *18*, 2415–2424. [CrossRef] [PubMed]

89. Department of Health. *Republic of South Africa: National Health Bill*; Department of Health: Pretoria, South Africa, 2003.

90. Holmes, M.D.; Dalal, S.; Sewram, V.; Diamond, M.B.; Adebamowo, S.N.; Ajayi, I.O.; Adebamowo, C.; Chiwanga, F.S.; Njelekela, M.; Laurence, C. Consumption of processed food dietary patterns in four African populations. *Public Health Nutr.* **2018**, *21*, 1529–1537. [CrossRef] [PubMed]

91. Tung, O.J.L. Organic food certification in South Africa: A private sector mechanism in need of state regulation? *Potchefstroom Electron. Law J.* **2016**, *19*. [CrossRef]

92. Vignola, R.; Harvey, C.A.; Bautista-Solis, P.; Avelino, J.; Rapidel, B.; Donatti, C.; Martinez, R. Ecosystem-based adaptation for smallholder farmers: Definitions, opportunities and constraints. *Agric. Ecosyst. Environ.* **2015**, *211*, 126–132. [CrossRef]

93. Erenstein, O. Smallholder conservation farming in the tropics and sub-tropics: A guide to the development and dissemination of mulching with crop residues and cover crops. *Agric. Ecosyst. Environ.* **2003**, *100*, 17–37. [CrossRef]

94. Blay-Palmer, A.; Knezevic, I. Research priorities for future food systems: A sustainable food systems perspective from Ontario, Canada. *Urban Agric. Mag.* **2015**, 35–37. [CrossRef]

95. Ebert, A.W. Potential of underutilized traditional vegetables and legume crops to contribute to food and nutritional security, income and more sustainable production systems. *Sustainability* **2014**, *6*, 319–335. [CrossRef]

96. Johns, T.; Powell, B.; Maundu, P.; Eyzaguirre, P.B. Agricultural biodiversity as a link between traditional food systems and contemporary development, social integrity and ecological health. *J. Sci. Food Agric.* **2013**, *93*, 3433–3442. [CrossRef]

97. Bisht, I.S.; Mehta, P.S.; Negi, K.S.; Verma, S.K.; Tyagi, R.K.; Garkoti, S.C. Farmers' rights, local food systems, and sustainable household dietary diversification: A case of Uttarakhand Himalaya in north-western India. *Agroecol. Sustain. Food Syst.* **2018**, *42*, 77–113. [CrossRef]

98. Van Jaarsveld, P.; Faber, M.; van Heerden, I.; Wenhold, F.; Jansen van Rensburg, W.; van Averbeke, W. Nutrient content of eight African leafy vegetables and their potential contribution to dietary reference intakes. *J. Food Compos. Anal.* **2014**, *33*, 77–84. [CrossRef]

99. Van Rensburg, W.S.J.; Van Averbeke, W.; Slabbert, R.; Faber, M.; Van Jaarsveld, P.; Van Heerden, I.; Wenhold, F.; Oelofse, A. African leafy vegetables in South Africa. *Water SA* **2007**, *33*, 317–326. [CrossRef]

100. Cloete, P.C.; Idsardi, E.F. Consumption of indigenous and traditional food crops: Perceptions and realities from South Africa. *Agroecol. Sustain. Food Syst.* **2013**, *37*, 902–914. [CrossRef]

101. Chibarabada, T.; Modi, A.; Mabhaudhi, T. Expounding the Value of Grain Legumes in the Semi- and Arid Tropics. *Sustainability* **2017**, *9*, 60. [CrossRef]

102. Adhikari, P.; Araya, H.; Aruna, G.; Balamatti, A.; Banerjee, S.; Baskaran, P.; Barah, B.C.; Behera, D.; Berhe, T.; Boruah, P.; et al. System of crop intensification for more productive, resource-conserving, climate-resilient, and sustainable agriculture: Experience with diverse crops in varying agroecologies. *Int. J. Agric. Sustain.* **2018**, *16*, 1–28. [CrossRef]

103. Mabhaudhi, T.; Chibarabada, T.; Modi, A. Water-Food-Nutrition-Health Nexus: Linking Water to Improving Food, Nutrition and Health in Sub-Saharan Africa. *Int. J. Environ. Res. Public Health* **2016**, *13*, 107. [CrossRef]

104. Venter, S.L.; van Rensburg, W.S.J.; van den Heever, E.; Vorster, H.J.; Allemann, J. Indigenous crops with potential but under-utilized in South Africa. In *Breeding of Neglected and Under-Utilized Crops, Spices and Herbs*; Science Publishers, Inc.: Enfield, UK, 2007; pp. 391–427. ISBN 9781578085095.

105. Fernandes, M.; Galloway, R.; Gelli, A.; Mumuni, D.; Hamdani, S.; Kiamba, J.; Quarshie, K.; Bhatia, R.; Aurino, E.; Peel, F.; et al. Enhancing linkages between healthy diets, local agriculture, and sustainable food systems: the school meals planner package in Ghana. *Food Nutr. Bull.* **2016**, *37*, 571–584. [CrossRef] [PubMed]

106. Allen, T.; Prosperi, P. Modeling sustainable food systems. *Environ. Manag.* **2016**, *57*, 956–975. [CrossRef]

107. Rotz, S.; Fraser, E.D.G. Resilience and the industrial food system: analyzing the impacts of agricultural industrialization on food system vulnerability. *J. Environ. Stud. Sci.* **2015**, *5*, 459–473. [CrossRef]

108. Poulton, P.L.; Dalgliesh, N.P.; Vang, S.; Roth, C.H. Resilience of Cambodian lowland rice farming systems to future climate uncertainty. *Field Crops Res.* **2016**, *198*, 160–170. [CrossRef]

109. Magrini, M.-B.; Anton, M.; Cholez, C.; Corre-Hellou, G.; Duc, G.; Jeuffroy, M.-H.; Meynard, J.-M.; Pelzer, E.; Voisin, A.-S.; Walrand, S. Why are grain-legumes rarely present in cropping systems despite their environmental and nutritional benefits? Analyzing lock-in in the French agrifood system. *Ecol. Econ.* **2016**, *126*, 152–162. [CrossRef]

110. Stagnari, F.; Maggio, A.; Galieni, A.; Pisante, M. Multiple benefits of legumes for agriculture sustainability: An overview. *Chem. Biol. Technol. Agric.* **2017**, *4*, 2. [CrossRef]

111. Masuda, T.; Goldsmith, P.D. China's meat and egg production and soybean meal demand for feed: An elasticity analysis and long-term projections. *Int. Food Agribus. Manag. Rev.* **2012**, *15*, 33–54.

112. Sheppard, S.C.; Bittman, S. Linkage of food consumption and export to ammonia emissions in Canada and the overriding implications for mitigation. *Atmos. Environ.* **2015**, *103*, 43–52. [CrossRef]

113. Latruffe, L. *Competitiveness, Productivity and Efficiency in the Agricultural and Agri-Food Sectors*; Organisation for Economic Co-operation and Development: Paris, France, 2010.

114. Lankoski, J. *Alternative Payment Approaches for Biodiversity Conservation in Agriculture*; Organisation for Economic Co-operation and Development: Paris, France, 2016.

115. FAO; IFAD; UNICEF; WFP; WHO. *The State of Food Security and Nutrition in the World 2017*; Food and Agriculture Organization: Rome, Italy, 2017.

116. Cervantes-Godoy, D.; Dewbre, J. *Economic Importance of Agriculture for Poverty Reduction*; Organisation for Economic Co-operation and Development: Paris, France, 2010.

117. Dewbre, J.; Cervantes-Godoy, D.; Sorescu, S. *Agricultural Progress and Poverty Reduction: Synthesis Report*; Organisation for Economic Co-operation and Development: Paris, France, 2011.

118. High Level Panel of Experts. *Sustainable Fisheries and Aquaculture for Food Security and Nutrition*; Food and Agriculture Organization: Rome, Italy, 2014.

119. Padulosi, S.; Ravi, S.B.; Rojas, W.; Valdivia, R.; Jager, M.; Polar, V.; Gotor, E.; Mal, B. Experiences and lessons learned in the framework of a global UN effort in support of neglected and underutilized species. *Acta Hortic.* **2013**, 517–531. [CrossRef]

120. Mabhaudhi, T.; O'Reilly, P.; Walker, S.; Mwale, S. Opportunities for Underutilised Crops in Southern Africa's Post–2015 Development Agenda. *Sustainability* **2016**, *8*, 302. [CrossRef]

121. Giannakis, E.; Bruggeman, A. Exploring the labour productivity of agricultural systems across European regions: A multilevel approach. *Land Use Policy* **2018**, *77*, 94–106. [CrossRef]

122. Wynberg, R.; Laird, S.A. Fast Science and Sluggish Policy: The Herculean Task of Regulating Biodiscovery. *Trends Biotechnol.* **2018**, *36*, 1–3. [CrossRef] [PubMed]

123. Wynberg, R. One step forward, two steps back: Implementing access and benefit-sharing legislation in South Africa. In *Routledge Handbook of Biodiversity and the Law*; McManis, C.R., Ong, B., Eds.; Routledge: Oxford, UK, 2017; p. 421. ISBN 131553083X.

124. Wynberg, R. Biotechnology and the commercialisation of biodiversity in Africa. In *International Environmental Law and Policy in Africa*; Chaytor, B., Gray, K.E., Eds.; Environmental Law Centre: Bonn, Germany, 2003; pp. 83–102.

125. Wynberg, R. Rhetoric, Realism and Benefit-Sharing. *J. World Intellect. Prop.* **2005**, *7*, 851–876. [CrossRef]

126. Dei, G.J.S. Indigenous knowledge and economic production: The food crop cultivation, preservation and storage methods of a west african community. *Ecol. Food Nutr.* **1990**, *24*, 1–20. [CrossRef]

sustainability

MDPI

Review

Contribution of *Schinziophyton rautanenii* to Sustainable Diets, Livelihood Needs and Environmental Sustainability in Southern Africa

Alfred Maroyi

Medicinal Plants and Economic Development (MPED) Research Centre, Department of Botany,
University of Fort Hare, Private Bag X1314, Alice 5700, South Africa; amaroyi@ufh.ac.za; Tel.: +27-71-9600326

Received: 16 January 2018; Accepted: 4 February 2018; Published: 26 February 2018

Abstract: *Schinziophyton rautanenii* is a multipurpose plant species in Southern Africa which provides numerous ecosystem goods and services. This review evaluated the contribution of the species to sustainable diets, livelihood needs and environmental sustainability throughout the geographical range of the species. The literature relevant to the study was obtained from scientific databases such as ScienceDirect, SciFinder, Pubmed, Google Scholar, Medline and SCOPUS. Literature was also obtained from the University of Fort Hare library, dissertation search engines like ProQuest, Open-thesis, OATD, and EThOS. *S. rautanenii* is an essential source of food, herbal medicines, income, oil, timber and wood. The species provides substantial health, economic and ecological benefits to local communities that depend on the species as a source of livelihood needs. This study represents a holistic view on multiple ecosystem goods and services that are derived from *S. rautanenii* forming an essential component of the 2030 Agenda for sustainable development goals (SDGs) adopted by the United Nations General Assembly. Use, cultivation and management of *S. rautanenii* in Southern Africa offers enormous potential for contributing to the fulfillment of the SDGs, resulting in improved food security, household nutrition and health, income, livelihoods, ecological balance, sustainable diets and food systems.

Keywords: ecosystem goods and services; environmental sustainability; livelihood needs; *Schinziophyton rautanenii*; Southern Africa; sustainable development goals; sustainable diets

1. Introduction

Sustainable use of plant biodiversity for food and medicines, as a source of other goods and ecosystem services play a crucial role in the fight against poverty, hunger and poor health and ensuring environmental sustainability while providing basic food and primary healthcare needs to local communities. Sustainable plant biodiversity management is central to the fulfillment of the 2030 Agenda for Sustainable Development Goals (SDGs), particularly the following [1]:

Goal 1: end poverty in all its forms everywhere,

Goal 2: end hunger, achieve food security and improved nutrition and promote sustainable agriculture,

Goal 3: ensure healthy lives and promote well-being for all at all ages,

Goal 12: ensure sustainable consumption and production patterns, and

Goal 15: protect, restore and promote sustainable use of terrestrial ecosystems, sustainably manage forest, combat desertification, and halt and reverse land degradation and halt biodiversity loss.

Use, cultivation and management of wild edible plant species in Southern Africa offers enormous potential for contributing to the fulfillment of the above-mentioned SDGs, resulting in improved food security, household nutrition and health, income, livelihoods, ecological balance, sustainable diets and food systems.

According to the Food and Agriculture Organization [2], "sustainable diets" are those diets with low environmental impacts which contribute to food and nutrition security and to healthy life for present and future generations. Sustainable diets are protective and respectful of biodiversity and ecosystems, culturally acceptable, accessible, economically fair and affordable; nutritionally adequate, safe and healthy; while optimizing natural and human resources (Figure 1). The key dimensions of sustainable diet emphasize the interdependencies of food procurement, production, consumption, requirements and nutrient recommendations, needs, accessibility and availability in surrounding ecosystems. Therefore, sustainable diets can advocate the utilization of food resources with lower water and carbon footprints, and encourage utilization of agricultural biodiversity, particularly local and indigenous food resources [2]. Agricultural biodiversity is known to perform many closely interrelated environmental and socio-economic functions; among these are livelihood needs, food security, supporting resilient rural economies, productive and environmental sustainability [3]. Agricultural biodiversity is a multi-dimensional subject with utilization and management of the environment and its components resulting in producing economic and social benefits from various components of agricultural biodiversity. Agricultural biodiversity is a fundamental component of many farming systems dotted throughout the world and encompasses several types of biological entities tied to agricultural systems, including the following [4,5]:

- i. genetic resources, the essential component of all living organisms,
- ii. edible plants and crops,
- iii. freshwater fish and livestock,
- iv. soil organisms which are important to soil structure, fertility, quality and health,
- v. variety of bacteria, fungi and insects which are crucial in controlling diseases and pests of animals and plants,
- vi. agroecosystem features and types required various ecosystem productivity processes, ecosystem stability and nutrient cycling, and
- vii. landscapes and undomesticated resources that are characterized by numerous ecosystem goods and services.

Two kinds of agricultural biodiversity are recognized: domesticated (managed) and non-domesticated (unmanaged) agricultural biodiversity. The cultivated agricultural biodiversity combined with their wild relatives are important as they supply local people with different types of genetic resources required for food and essential plant and animal products. People throughout the world consume about 7000 plant species, with 150 species regarded as commercially valuable, while 103 species provide 90 percent of the world's major food crops [6]. Research by FAO [7] reveals that more than 50 percent of the carbohydrates and proteins needed by the world's human population is provided by three food crops, namely rice, maize and wheat. But in Southern Africa and other marginal environments of developing agricultural economies, many less well-known plant species are consumed, domesticated and managed in home gardens contributing to the livelihood needs of the poor and to the development of agricultural biodiversity. Such plant species include *S. rautanenii* (Schinz) Radcl.-Sm., a multipurpose food, medicinal and timber plant species characterized by promising socio-economic potential for food security of local communities, poverty alleviation and meeting important livelihood needs in rural, peri-urban and marginalized areas in Southern Africa. *S. rautanenii* is a component of non-domesticated (unmanaged) agricultural biodiversity, with its fruits, seeds and other important plant parts collected mainly from the wild although in some countries in Southern Africa the species is domesticated and managed in home gardens. This review will add valuable literature on plant species that are important for local livelihood needs and income generation by the rural poor and those people living in marginalized areas in Southern Africa. Scientific evidence for the livelihood and health benefits of useful plant species such as *S. rautanenii* is important as this species will contribute towards environmental management, carbon sequestration by sustainable use and conservation of such species. This review evaluates the contribution of a useful plant species, *S. rautanenii* to sustainable diets, including nutritional, social, religious, provisioning and environmental issues. The literature reviewed in this study highlight the important issues required to strengthen the contribution of such useful plant species to sustainable diets, livelihood needs and environmental sustainability in rural, peri-urban and marginalized areas.

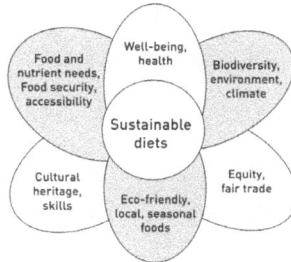

Figure 1. Venn diagram showing the key components of a sustainable diet [2].

2. Materials and Methods

Literature search for information relevant to the contribution of *S. rautanenii* to sustainable diets, livelihood needs and environmental sustainability in Southern Africa was carried out from June to December 2017. Emphasis was placed on literature from the geographical range of the species that included Namibia, Malawi, South Africa, Mozambique, the Democratic Republic of Congo, Tanzania, Botswana, Zambia, Angola and Zimbabwe. The information was obtained from the main online scientific sites including Science Direct, SciFinder, Pubmed, Google Scholar, Medline, and SCOPUS. Searches were also undertaken in the library, University of Fort Hare and the dissertation search engines like ProQuest, Open-thesis, OATD and EThOS. The keywords used in the search included "*S. rautanenii*", the synonym of the species "*Ricinodendron rautanenii* Schinz", English common names "false balsa", "featherweight tree", "manketti tree", "manketti nut tree" and "mongongo nut". Additional search was also carried out using the keywords "ethnobotany + *S. rautanenii*", "medicinal uses + *S. rautanenii*", "traditional uses + *S. rautanenii*", "indigenous knowledge + *S. rautanenii*", "local knowledge + *S. rautanenii*", "traditional ecological knowledge + *S. rautanenii*", "environmental sustainability + *S. rautanenii*", "livelihood needs + *S. rautanenii*" and "sustainable diets + *S. rautanenii*". A total of 41 articles matched the inclusion criteria and were included in the review (Figure 2).

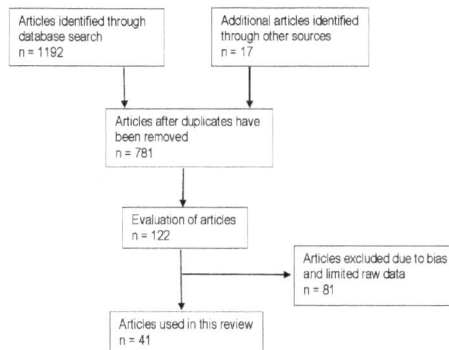

Figure 2. Flow diagram with the number of selected articles.

3. *S. rautanenii* Constitute an Essential Food Crop

According to Robbins and Campbell [8] the nuts of *S. rautanenii* have been used as a staple food source in the Kalahari desert in Botswana for at least 7000 years. The nuts form the mainstay of the vegetable diet of the !Kung bushmen of Botswana who are known to consume between 100 and 300 of these nuts per day [9–11]. The pulp of the drupe-like fruits, fresh or naturally dried is available all year round providing an estimated 40% of the total diet of local people in north western Botswana [12,13]. Similar findings were obtained by Hailwa [14] who evaluated the importance of non-timber forest products (NTFPs) in Namibia and showed that *S. rautanenii* is a valuable source of food to several people in the San community in the country, particularly those who are not involved in farming activities. Research by Saxon and Chidiamassamba [15] showed that the consumption of the nuts increases in difficult times such as droughts and during civil wars. Therefore, the fruits of *S. rautanenii* constitute an important food source in Southern Africa, regarded as an essential component of the diet of some people in the region, especially the poor and those living in marginalized areas. Fruits and nuts of *S. rautanenii* are also widely used in Botswana, Malawi, Mozambique, Namibia, Zambia and Zimbabwe [12,15–20] as a strategy by the rural people to reduce food insecurity and as an important source of supplementary food particularly during drought.

The nut or seed is the most popular part of *S. rautanenii* fruit. It is eaten raw or dried and most pleasant when roasted [11]. A meal or paste is sometimes made by pounding the nuts or porridge or fermented alcoholic drink is made from the pulp (mesocarp) [10,11,20]. *S. rautanenii* fruits are famous for producing liquor in Namibia, where water is added to peeled or unpeeled fruits and leaving the mixture to ferment for at least three days up to two weeks in a plastic bucket. The fermented juice is then heated and distilled to produce fermented alcoholic drink [20]. The kernel is also used as a food thickener agent for stews and soups that accompany meat, fish and vegetables [10,11,15]. The oil extracted from the nuts is used for cooking [21,22]. The fruit pulp and nut of *S. rautanenii* is highly nutritious by virtue of high amounts of carbohydrates, protein and vitamins (Table 1), which are comparable to other well-known indigenous fruits with edible kernels such as *Sclerocarya birrea* (A. Rich.) Hochst. [23,24] and a commercial fruit crop, cashew nuts, *Anacardium occidentale* L. [25]. In terms of mineral content, the mesocarp and seed tissues of *S. rautanenii* are excellent sources of essential minerals such as phosphorus, magnesium, potassium, iron, sodium, copper, zinc and calcium [10]. Several amino acids and fatty acids (Table 2) have been identified from the mesocarp and seed tissues of *S. rautanenii* [10,17,23–28]. Compared to conventional nut and oil crops such as *A. occidentale*, *S. rautanenii* is a rich source of macro- and micronutrients (Tables 1 and 2). Reference is also made to the recommended dietary allowance (RDA) representing the average daily intake of essential nutrients that are sufficient to meet the nutrient requirements of a health person (Tables 1 and 2). When compared with the World Health Organization (WHO) protein RDA values, *S. rautanenii* demonstrated low proportions of several essential amino acids, a trend also demonstrated by *A. occidentale* and *S. birrea* [23–25]. The carbohydrates, protein, vitamins, different minerals, amino acids and fatty acids identified from the fruit pulp and seed of *S. rautanenii* imply that the species is essential in meeting the nutritional needs of local communities at risk of food insecurity, macro- and micronutrient malnutrition. According to the United Nations International Children's Emergency Fund (UNICEF) [29], malnutrition is a major challenge in Eastern and Southern Africa where 40% of minor children, particularly those under the age of five years are suffering from it. Research by Chivandi et al. [30] showed that edible fruits obtained from indigenous fruit trees like *S. rautanenii* are important sources of macro- and micronutrients and health enhancing chemicals that have chemical, pharmacological and biological properties that mitigate some of the medical and biological effects of malnutrition. Unfortunately, indigenous edible fruits like *S. rautanenii* are currently underutilized, neglected by researchers, agricultural extension officers and policy makers. Their promotion could assist in a protracted fight against malnutrition, domestication and environmental sustainability in Southern Africa. These are some of the initial steps that can be taken in trying to address the SDGs including goals 1, 2, 3, 12 and 15.

Table 1. Nutritional composition of the mesocarp and seed tissues of *S. rautanenii* compared with nutritional values of kernels of *S. birrea* and *A. occidentale* and the recommended dietary allowance (RDA).

	S. rautanenii		S. birrea	A. occidentale	Recommended Dietary
Chemical Composition	Mesocarp	Seed			Allowance (RDA)
Ash % of dry weigh	5.6	4.1	5.05 ± 0.61	2.5	-
Calcium g/100 g	104	193	51.73 ± 6.0	41.0	1000–1300
Copper g/100 g	1.6	2.82	1.07 ± 0.10	-	1–2
Energy value (kJ per 100 g)	1424	2692	1545.36	2525	-
Fat % of dry weigh	1.2	57.3	1.30 ± 0.15	-	300
Fibre % of dry weigh	2.5	2.5	-	3.6	25–38
Fructose g/100 g	0.5	-	-	-	130
Glucose g/100 g	0.2	-	0.21 ± 0.01	-	130
Iron g/100 g	4.3	3.7	8.83 ± 0.15	5.7	8–15
Magnesium g/100 g	266	527	24.53 ± 2.06	248.8	310–320
Moisture % of dry weight	8.4	4.2	-	3.8	-
Nicotinic acid mg/100 g	4.79	0.31	-	1.31	16–35
Phosphorus mg/100 g	62.0	845	0.18 ± 0.02	502.5	1250
Potassium mg/100 g	2666	673	44.54 ± 0.41	622.5	4700
Protein % of dry weigh	9.4	26.0	3.31 ± 0.10	21.3	34
Riboflavin mg/100 g	0.21	0.21	-	0.028	0.3–1.6
Sodium g/100 g	1.86	3.1	14.88 ± 6.0	1.44	2300
Sucrose g/100 g	29.8	-	0.76 ± 0.21	6.3	130
Thiamine mg/100 g	0.49	0.31	-	0.477	1–2
α-tocopherol (Vitamin E)	-	29	-	0.45	20
γ-tocopherol mg/100 g	-	536	-	5.07	20
Total carbohydrate g/100 g	72.9	5.9	90.35 ± 0.77	20.5	130
Vitamin C mg/100 g	14.7	-	0.49 ± 0.3	0.13	46
Zinc g/100 g	1.79	4.09	2.96 ± 1.0	5.3	8–11

Source: Wehmeyer [10]; Hassan et al. [23]; Mariod and Abdelwahab [24]; Rico et al. [25].

Table 2. Amino acids and fatty acid composition of seeds and mesocarp tissues of *S. rautanenii* compared with nutritional values of kernels of *S. birrea* and *A. occidentale* and the RDA.

	S. rautanenii		S. birrea	A. occidentale	Recommended Dietary
Chemical Composition	Mesocarp	Seed			Allowance (RDA)
Amino acids					
Alanine g/100 g	0.4	1.0	1.81	0.82	-
Arginine g/100 g	0.7	3.5	6.12	2.22	-
Aspartic acid g/100 g	0.4	2.4	4.87	1.89	-
Cysteine g/100 g	-	0.1	4.1	-	28–43
Glutamic acid g/100 g	0.7	4.2	1.42	4.60	-
Glycine g/100 g	0.2	1.2	2.75	0.89	-
Histidine g/100 g	0.1	0.7	2.68	0.47	21–32
Isoleucine g/100 g	0.2	0.7	3.3	0.80	28–43
Leucine g/100 g	0.2	1.4	4.8	1.47	63–93
Lysine g/100 g	0.2	0.7	1.18	0.97	58–89
Methionine g/100 g	0.8	0.4	4.1	0.37	28–43
Phenylalanine g/100 g	0.2	1.3	2.5	0.93	54–84
Proline g/100 g	0.2	1.2	2.06	0.75	-
Serine g/100 g	0.3	1.3	2.43	1.11	-
Threonine g/100 g	0.2	1.0	1.31	0.74	32–49
Tyrosine g/100 g	0.1	0.5	1.68	0.63	37–58
Valine g/100 g	0.2	1.8	3.06	1.12	10
Fatty acid					
Arachidic acid (C20:0) %	-	0.2–0.6	-	0.63	0.1
α eleostearic acid (C18:3) %	-	21.7	-	-	-
β eleostearic acid (C18:3) %	-	21.7	-	-	-
Erucic (C22:1n9) %	-	21.5 ± 0.84	0.38	-	-
Gondoic acid (C20:1) %	-	0.3	0.14–0.70	-	-
Linoleic acid (C18:2) %	-	37.8–51.9	4.3–5.93	17.77	6.7
Linolenic acid (C18:3) %	-	26.6	0.12	0.13	1.4
Margaric acid (C17:0) %	-	0.08	-	-	-
Myristic acid (C14:0) %	-	0.03 ± 0.01	0.1–2.12	-	-

Table 2. *Cont.*

Chemical Composition	S. rautanenii Mesocarp	S. rautanenii Seed	S. birrea	A. occidentale	Recommended Dietary Allowance (RDA)
Myristoleoic acid (C14:1n7) %	-	0.01 ± 0.0	-	-	-
Oleic acid (C18:1n9) %	-	15.2 ± 1.53	4.13–67.3	-	-
Oleic acid (C18:1) %	-	17.7–24.4	-	60.7	-
Palmitic acid (C16:0) %	-	8.8–11.95	14.2–22.6	10.02	-
Palmitoleic acid (C16:1) %	-	0.06	-	-	-
Stearic acid (C18:0) %	-	3.0–11.77	8.84–50.8	8.93	-

Source: Wehmeyer [10]; Chivandi et al. [17]; Hassan et al. [23]; Mariod and Abdelwahab [24]; Rico et al. [25]; Mitei et al. [26]; Gwatidzo et al. [27,28].

4. Medicinal Uses and Ethnopharmacology of *S. rautanenii*

Local people in Southern Africa use different parts of *S. rautanenii* as herbal medicine for back pain, cancer, fever, infertility, measles, skin diseases, skin cleanser, skin moisturizer, sleepless nights, sores and stomachache (Table 3). In Mozambique and Zimbabwe, the hard outer nut shells of *S. rautanenii* fruits are popular as divining "bones" [15,31]. Different plant parts of *S. rautanenii* widely used as herbal concoctions have medicinal value probably due to their inherent constituents such as alkaloids, anthraquinones, coumarins, flavonoids, phenols, saponins and triterpenes that have been identified from the bark and root extracts of the species [32]. The author found total phenolic compound content of the root and bark of *S. rautanenii* to be 48.6 ± 1.2 and 41.4 ± 1.2 gallic acid equivalent (GAE) µg/mL, respectively. Alkaloid content of the bark and roots of *S. rautanenii* was 2.1% and 9.5%, respectively. Saponin content of the bark and roots of *S. rautanenii* was 13.1% and 2.0%, respectively [32]. Dushimemaria [32] evaluated antioxidant activities of *S. rautanenii* extracts using 2,2′-azino-bis(3-ethylbenzothiazoline-6-sulphonic acid) (ABTS) assay. Antioxidant activities of the bark and root extracts of *S. rautanenii* were 200 and 900 ascorbic acid equivalent (AAE) µg/mL, respectively [32]. The author also evaluated anti-protease activities of *S. rautanenii* extracts using the GIBEX screens-to-nature assay. This study reports high anti-protease activity of all methanolic extracts derived from *S. rautanenii*. Dushimemaria [32] evaluated anti-cancer activities of *S. rautanenii* extracts against renal, breast and melanoma cancer cell models. The melanoma cell line UACC-62 displayed significant sensitivity towards the aqueous root and bark extracts of the species with low half maximal inhibitory concentration (IC$_{50}$) values of 116.7 µg/mL and 128.7 µg/mL, respectively. The authors also reported that organic root extract of the species demonstrated significant activities towards UACC-62 melanoma cell line with IC$_{50}$ value of 102.6 µg/mL which was not significantly different from the IC$_{50}$ value of 102.4 µg/mL demonstrated by the MCF-7 breast cancer cells. Dushimemaria [32] evaluated cytotoxicity activities of *S. rautanenii* extracts against a human fetal lung fibroblast cell lines. The aqueous and organic root extracts of the species demonstrated some activities with the IC$_{50}$ values of 315.5 µg/mL and 444.8 µg/mL against the human fetal lung fibroblast cell, respectively. Similarly, Dushimemaria and Mumbengegwi [33] evaluated antiproliferative properties of *S. rautanenii* extracts using the brown planaria (*Dugesia dorotocephala*). Treatment of *D. dorotocephala* using the bark extracts of the species revealed growth inhibiting properties at high extract concentration when compared with the negative control, while growth promoting properties were exhibited at lower extract concentration when compared with the negative control. Dushimemaria [32] evaluated toxicity activities of *S. rautanenii* extracts in vivo in fresh water flatworm planaria (*D. dorotocephala*). This depicted that *S. rautanenii* organic bark extract negatively affected planaria regeneration, indicating that as extract concentration increased, the cytotoxicity effects of *S. rautanenii* bark extracts were more visibly seen as affecting planaria regeneration in comparison to low extract concentration. *S. rautanenii* root extract affected the planaria growth in a concentration dependent manner. These anti-cancer and antiproliferative activities demonstrated by *S. rautanenii* extracts [32,33] support the traditional use of the species as herbal medicine against cancer in Namibia [32].

The oil extracted from *S. rautanenii* is also a highly valued emollient in Southern Africa that nourishes and protects the skin and therefore is widely used as a cream, lotion and ointment [34]. In Southern Africa, the oil extracted from *S. rautanenii* is mainly applied on the skin because of its hydrating properties, regenerating and restructuring attributes that contribute to ultraviolet (UV) protection for the skin and hair [35]. Zimba et al. [36] argued that the presence of vitamin E, and fatty acids such as linoleic and eleostearic acids [10], makes the oil useful for hydrating and protecting the skin which may play a vital role in the treatment of atopic diseases, eczema and other diseases where the oil results in the reduction of inflammation and promotion of tissue regeneration and cellular repair, leading to reduced itching, redness and scarring, as well as prevention of keloids. The *S. rautanenii* has naturally high zinc content and, for years, the San communities of the Kalahari region in Namibia have used it to safeguard themselves from the desert sun; it also moisturizes and conditions the skin offering additional protection from the sun [36]. Skin supplementation with anti-oxidants usually result in the reduced photo damage and photo aging due to free-radical oxidative stress [37]. Zimba et al. [36] argued that the oil extracted from *S. rautanenii* is important for aromatherapy and cosmetic applications mainly because they result in long shelf life and oxidative stability of the skin. It has been shown that products rich in fatty acids such as linoleic acid boost antimicrobial properties of the skin, especially in people suffering from atopic dermatitis. In acneic skin, particularly in follicular hyperkeratinisation, topical application of *S. rautanenii* seed oil result in reduced production of oleic acid and squalene leading to normal production of linoleic acid levels [36]. However, detailed phytochemical studies of *S. rautanenii* and its pharmacological properties aimed at illustrating the correlation between its ethnomedicinal uses and pharmacological properties are required. Such detailed chemical, nutritional and toxicological evaluations are required before *S. rautanenii* can be used as an alternative dietary source, herbal medicine or source of aromatherapy products. There is need for extensive in vitro and in vivo experiments to validate the existing pharmacological activities of the species. Toxicological evaluations of the species are required aimed at establishing any potential toxic components and side effects associated with utilization of the species as food, traditional medicine and/or emollient.

Table 3. Ethnomedicinal uses of *S. rautanenii*.

Medicinal Applications	Plant Parts Used	Country	References
Back pain	Bark	Namibia	Elago and Tjaveondja [38]
Cancer	Not specified	Namibia	Dushimemaria [32]
Fever	Bark	Namibia	Elago and Tjaveondja [38]
Infertility	Not specified	Namibia	Msangi [39]
Measles	Leaves	Namibia	Cheikhyoussef and Embashu [19]
Skin diseases	Aerial parts	South Africa	Juliani et al. [40]; Vermaak et al. [41]; Lall and Kishore [42]
Skin cleanser	Aerial parts	South Africa	Juliani et al. [40]; Vermaak et al. [41]; Lall and Kishore [42]
Skin moisturizer	Aerial parts	South Africa	Juliani et al. [40]; Vermaak et al. [41]; Lall and Kishore [42]
Sleepless nights	Bark	Namibia	Elago and Tjaveondja [38]
Sores	Not specified	Namibia	Dushimemaria et al. [43]
Stomachache	Bark	Namibia	Elago and Tjaveondja [38]

5. Integration of *S. rautanenii* into Formal Agricultural Production Systems

S. rautanenii has multiple uses, including oil production, source of edible fruits and kernels, source of timber and wood as well as different plant parts used as herbal medicines. These multiple uses of the species and its social, religious and ecological functions in the ecosystem resulted in some households sparing the species during land clearing for farming and maintaining the species in arable agricultural fields in many local communities of Southern Africa. Research by Van den Eynden [44] revealed that plant management in agricultural arable fields like home gardens can be divided into two major groups: cultivated (planted) or tolerated (not planted). Cultivated species are cared for and tendered by home garden owners during their entire life cycle [44]. They are usually sown as seeds and require regular watering and weeding to promote quick growth. Tolerated species are those plant species that are deliberately spared by home garden owners and their households during digging, weeding and land clearing activities in agricultural fields or home gardens because of the benefits or usefulness they provide to households and the environment [45]. Tolerated species like *S. rautanenii* are tendered and conserved in home gardens and agricultural fields, and they also grow naturally in the wild. *S. rautanenii* also makes an excellent live fence and grow vigorously from truncheons, or coppice shoots, cutting considerable time to fruit bearing, compared to time it would take if they are grown from seeds [16,46]. In Zambia, *S. rautanenii* is widely cultivated and used as a live fence around arable agricultural fields and animal enclosures [36]. The National Council for Scientific Research (NCSR) in Zambia has a Tree Improvement Research Centre which has several projects throughout the country focusing on the domestication of *S. rautanenii* using plant tissue culture techniques [47]. In Namibia, Du Plessis [48] advocated for the domestication and marketing of *S. rautanenii* in the country. Research by Keegan and van Staden [11] showed that the seeds of *S. rautanenii* are dormant and the removal of the endocarp is a prerequisite for germination. Therefore, applications of ethrel (2-chloroethane phosphonic acid) and gaseous ethylene to seeds of *S. rautanenii* resulted in rapid germination. The authors found that gibberellic acid was effective only once the testa had been scarified. These results suggest that the embryos are unable to resume growth owing to a hormonal imbalance in which ethylene plays a primary role in seed germination of *S. rautanenii* [11].

Domestication of *S. rautanenii* is necessary for the diversification of subsistence agriculture in southern Africa, which could play a big role in attaining and achieving the SDGs, particularly goals 1, 2, 3, 12 and 15. At the same time, the species can also be used to halt and reverse the increasing degradation of ecosystems in Southern Africa and contribute towards carbon sequestration while providing socio-economic and livelihood needs to local communities. For plant species such as *S. rautanenii* that are in high demand and characterized by increasingly limited supplies, sustainable use of such species can be an effective conservation alternative. *S. rautanenii* is listed by Mark et al. [49] as one of the timber species that is at risk of disappearing because it is actively harvested for commercial trade. Similarly, the tree was designated a protected species in Namibia in the early 1990s in terms of the existing forest legislation since 1952 [50], probably because of its socio-economic importance as food, timber and herbal medicine. Over-exploitation, indiscriminant collection, uncontrolled deforestation and habitat destruction are all responsible for negatively affecting the population size of *S. rautanenii*. Therefore, to reach an effective compromise between conservation of the species on one hand and sustainable utilization on the other, serious consideration should be given to establishing commercially viable stands of large numbers of the tree in Malawi, Zambia, Botswana, Mozambique, Zimbabwe and Namibia where the species is widely used by local communities. The overall trend of useful plant loss observed in the past and the need to develop effective conservation strategies for such species has led to the emergence of several paradigms and principles of conservation. One of them is the principle of "conservation through use or through trade", an important conservation strategy aimed at providing monetary incentives or other forms of benefits to local communities for all species and habitats they are conserving and managing in a sustainable manner [13,51]. Such a conservation strategy can be applied to multipurpose species such as *S. rautanenii* throughout its distributional range. Research by Dovie et al. [52] revealed that when such a conservation approach is adopted, it is more successful as a livelihood strategy when social and community beliefs and rights are understood and addressed in the conservation programmes.

6. Commercial Potential of *S. rautanenii*

Research by Bennett [53] showed that about 200,000 people are currently employed in gathering *S. rautanenii* fruits and trading in this species, and its products have the potential of generating close to US$20 million in Southern Africa. *S. rautanenii* and its products are traded in informal markets in Malawi, Mozambique, Zambia and Zimbabwe [54] while oil extracted from the species is exploited commercially in Botswana and Namibia [17]. In Zambia, the wood and timber obtained from the species is used in producing curios and other crafts that are marketed in tourist resorts [55]. Kivevele and Huan [56] evaluated the fuel properties of biodiesel from *S. rautanenii* seeds and the authors found that, most of the determined fuel attributes of the species fulfilled the minimum chemical requirements of global biodiesel standards. The biodiesel produced from *S. rautanenii* seed oil may be used as a substitute of the common mineral diesel. Seed oil extracted from *S. rautanenii* has over the years been used in producing lubricants, soaps and personal cosmetic care products, as well as health products used in the topical treatment of various ailments and conditions such as hair dandruff, muscle spasms, varicose veins and wounds [17,36]. Therefore, commercialization of the oil derived from the species may result in several economic activities that positively impact local communities and may contribute to household economy and their livelihoods. Nemarundwe et al. [57] argued that promotion of natural resources utilization should not only emphasize the need to sustainably use these resources to meet the livelihood needs of local people, but also encourage communities to effectively manage their natural resources and protect the environment. Creating markets for *S. rautanenii* will ultimately lead to "conservation through use or through trade" resulting in increased awareness and economic importance associated with the species. It is evident that commercialization, formal trade and socio-economic activities will result in a positive impact on local communities, conservation, and the natural products industry as a whole [57].

The land where *S. rautanenii* trees are indigenous and growing is known not to be suitable for agricultural activities like cultivation of crops [56]. The development of additional uses and external markets for *S. rautanenii* will benefit local communities by providing several local, regional and international markets for the products derived from the species. Shackleton et al. [58] argued that the growing interest in the use and commercialization of NTFPs is a result of studies that have shown natural products as being crucial alternatives to agricultural products which may be very low due to poor and erratic rainfall. Therefore, commercialization of productions derived from NTFPs such as *S. rautanenii* will result in creating economic opportunities to people in rural areas and marginalized areas, thus improving household income and employment opportunities. The majority of poor people in rural, marginal and peri-urban areas have no access to industries and other economic activities such that during off-cropping season they have no other sources of income other than harvesting and selling *S. rautanenii* fruits. Nemarundwe et al. [57] argued that trade in natural products such as *S. rautanenii* should be promoted since this commodity is more accessible to the poor, there are minimal barriers to entry and there are few and in some cases no harvesting costs other than labor and time. For harvesting and processing of *S. rautanenii* and its products, local people already possess the required skills. Therefore, trade in *S. rautanenii* builds on rich traditional ecological knowledge, skills and technologies which may contribute towards the promotion of new externally facilitated products and markets.

Small and medium-sized enterprises (SMEs) and local communities that are marketing *S. rautanenii* at a commercial scale still rely on the traditional methods to crack the species' nuts to obtain the oil-rich kernels. This rudimentary system for cracking the nut is a major barrier to commercializing of *S. rautanenii* products. According to Nemarundwe et al. [57] traditional cracking of *S. rautanenii* nuts involve hitting the nut between two stones, a bigger stone below and a smaller one used as a hammer. Kernel extraction is a slow, tedious, difficult, laborious and a delicate operation that requires skill, experience and patience as use of excessive force may result in shattering the kernel leading to low output [57]. The same nut-cracking method is used to extract the oil-rich kernels of *S. birrea* commonly known as marula, a species of significant commercial prominence in Southern Africa. One of the *S. birrea* products that is widely traded at local, regional and international markets is marula cream liqueur. Marula cream liqueur is exported to nearly 150 countries and is the world's second most popular cream liqueur to Baileys Irish cream [59]. Therefore, the current extraction methods of *S. birrea* and *S. rautanenii* kernels should be improved, ideally with semi-manual equipment that will retain manual labor and also benefit the local economy. For *S. rautanenii* to realize its full local, regional and international market potential, there is need for development of necessary infrastructure, processing technology, product packaging and preservation, government support in terms of extension services, policies and guidelines related to horticultural development and marketing.

Sustainability **2018**, *10*, 581

7. Conclusions

This article explored the potential contribution of *S. rautanenii* as a source of multiple ecosystem goods and services that are essential components of the 2030 SDGs approved by the United Nations and sets out for the period 2016–2030. Sustainable diets and food systems are being explored throughout the world as a strategy to orient collective action towards the eradication of hunger and malnutrition in communities and the fulfilment of other SDGs [60]. Several socio-economic and ecological activities are supported by wise utilization of *S. rautanenii* in Southern Africa, including food production, fuel, source of timber and wood as well as different plant parts used as traditional medicines and proving several ecosystem services. Results of this study support previous observations made by Bharucha and Pretty [61] that wild foods make an enormous contribution towards reducing the existing supply and demand of food and nutritional security gap as multipurpose species like *S. rautanenii* provide health, ecosystem services and socio-economic benefits to communities that depend on them. Therefore, there is need to diversify crops in the arable agricultural systems to include the lesser known and underutilized species such as *S. rautanenii* in order to strengthen the food and ecological networks of local communities. National governments and local communities in Southern Africa need to consider the significance of *S. rautanenii* to food security, macro- and micronutrient contribution of the species and its ecological and provisioning role.

Acknowledgments: The author would like to express his gratitude to the National Research Foundation, South Africa (NRF grant number T398) and Govan Mbeki Research and Development Centre (GMRDC, grant number C169), University of Fort Hare for financial support to conduct this study.

Conflicts of Interest: No conflict of interest is associated with this work.

References

1. United Nations (UN) General Assembly. *Resolution Adopted by the General Assembly on 25 September 2015*; United Nations (UN): New York, NY, USA, 2015. Available online: http://www.un.org/en/development/desa/population/migration/generalassembly/docs/globalcompact/A_RES_70_1_E.pdf (accessed on 10 October 2017).

2. Food and Agriculture Organization (FAO). Sustainable Diets and Biodiversity. Directions and Solutions for Policy, Research and Action. In Proceedings of the International Scientific Symposium, Biodiversity and Sustainable Diets. United Against Hunger, Rome, Italy, 3–5 November 2010; Food and Agriculture Organization (FAO): Rome, Italy, 2010.

3. Pimbert, M. *Sustaining the Multiple Functions of Agricultural Biodiversity*; Gatekeeper Series No. 88; International Institute for Environment and Development (IIED): London, UK, 1999.

4. Thrupp, L.A. Linking agricultural biodiversity and food security: the valuable role of agrobiodiversity for sustainable agriculture. *Int. Aff.* **2000**, *76*, 265–281. [CrossRef] [PubMed]

5. Cromwell, E.; Cooper, D.; Mulvany, P. Agriculture, biodiversity and livelihoods: Issues and entry points for development agencies. In *Living off Biodiversity: Exploring Livelihoods and Biodiversity Issues in Natural Resources Management*; Koziell, I., Saunders, J., Eds.; International Institute for Environment and Development (IIED): London, UK, 2001; pp. 75–112, ISBN 978-1-899825-67-7.

6. Zhou, M. Promote conservation and use of underutilized crops. In *Plant Genetic Resources Conservation and Use in China, Proceedings of the National Workshop on Conservation and Utilization on Plant Genetic Resources, Beijing, China, 25–27 October 1999*; Gao, W., Rao, V.R., Zhou, M., Eds.; Institute of Crop Germplasm Resources, Chines Academy of Agricultural Sciences (CAAS): Beijing, China; International Plant Genetic Research Institute (IPGRI) Office for East Asia: Beijing, China, 2001; ISBN 92-9043-463-5.

7. Food and Agriculture Organization (FAO). *Report on the State of the World's Plant Genetic Resources for Food and Agriculture*; Food and Agriculture Organization (FAO): Rome, Italy, 1996.

8. Robbins, L.H.; Campbell, A.C. Prehistory of mongongo nut exploitation in the western Kalahari desert, Botswana. *Botsw. Notes Rec.* **1990**, *22*, 37–42.

9. Palmer, E.; Pitman, P. *Trees for Southern Africa Covering all Known Indigenous Species in Republic of South Africa, South West Africa, Botswana, Lesotho and Swaziland*; A.A. Balkema: Cape Town, South Africa, 1972; ISBN 0869610333.

10. Wehmeyer, A.S. *Ricinodendron rautanenii Schinz, Addendum 1: The Nutrient Composition of Manketti Fruit9 Southern African Plants, No. 4463,000-0010*; Government Printer: Pretoria, South Africa, 1976.

11. Keegan, A.B.; van Staden, J. Dormancy and germination of the manketti nut *Ricinodendron rautanenii* Schinz. *S. Afr. J. Sci.* **1981**, *77*, 262–264.

12. Lee, R.B. Mongongo: The ethnography of a major wild food resource. *Ecol. Food Nutr.* **1973**, *2*, 307–321. [CrossRef]

13. Peters, C.M.; Balick, M.J.; Kahn, F.; Anderson, A.B. Oligarchic forests of economic plants in Amazonia: Utilization and conservation of an important tropical resource. *Conserv. Biol.* **1989**, *3*, 341–349. [CrossRef] [PubMed]

14. Hailwa, J. *Non-Wood Forest Products of Namibia: Data Collected and Analysis for Sustainable Forest Management in ACP Countries-Linking National and International Efforts*; Directorate of Forestry, Ministry of Environment and Tourism: Windhoek, Namibia, 1998.

15. Saxon, G.; Chidiamassamba, C. *Indigenous Knowledge of Edible Tree Products: The Mungomu Tree in Central Mozambique*; Food and Agriculture Organization (FAO): Rome, Italy, 2005.

16. Storrs, A.E.G. *Know Your Trees: Some of the Common Trees in Zambia*; Forestry Department: Ndola, Zambia, 1979; ISBN 9789105324052.

17. Chivandi, E.; Davidson, B.C.; Erlwanger, K.H. A comparison of the lipid and fatty acid profiles from the kernels of the fruit (nuts) of *Ximenia caffra* and *Ricinodendron rautanenii* from Zimbabwe. *Ind. Crop. Prod.* **2008**, *27*, 29–32. [CrossRef]

18. Van Wyk, B.; Gericke, N. *People's Plants: A Guide to Useful Plants of Southern Africa*; Briza Publications: Pretoria, South Africa, 2008; ISBN 9781875093373.

19. Cheikhyoussef, A.; Embashu, W. Ethnobotanical knowledge on indigenous fruits in Ohangwena and Oshikoto regions in Northern Namibia. *J. Ethnobiol. Ethnomed.* **2013**, *9*, 34. [CrossRef] [PubMed]

20. Misihairabgwi, J.; Cheikhyoussef, A. Traditional fermented foods and beverages of Namibia. *J. Ethn. Foods* **2017**, *4*, 145–153. [CrossRef]

21. Engelter, C.; Wehmeyer, A.S. Fatty acid composition of oils of some edible seeds of wild plants. *J. Agric. Food Chem.* **1970**, *18*, 25–26. [CrossRef] [PubMed]

22. Davis, J.B.; Kay, D.E.; Clark, V. *Plants Tolerant of Arid, or Semi-Arid, Conditions with Non-Food Constituents of Potential Use*; London Bookmark: London, UK, 1983.

23. Hassan, L.G.; Dangoggo, S.M.; Hassan, S.W.; Muhammad, S.; Umar, K.J. Nutritional and antinutritional composition of *Sclerocarya birrea* fruit juice. *Niger. J. Basic Appl. Sci.* **2010**, *18*, 222–228. [CrossRef]

24. Mariod, A.A.; Abdelwahab, S.I. *Sclerocarya birrea* (Marula), an African tree of nutritional and medicinal uses: A review. *Food Rev. Int.* **2012**, *28*, 375–388. [CrossRef]

25. Rico, R.; Bulló, M.; Salas-Salvadó, J. Nutritional composition of raw fresh cashew (*Anacardium occidentale* L.) kernels from different origin. *Food Sci. Nutr.* **2016**, *4*, 329–338. [CrossRef] [PubMed]

26. Mitei, Y.C.; Ngila, J.C.; Yeboah, S.O.; Wessjohann, L.; Schimidt, J. NMR, GC-MS and ESI-FTICT-MS profiling of fatty acids and triacylglycerols in some Botswana seed oils. *J. Am. Oil Chem. Soc.* **2008**, *85*, 1021–1032. [CrossRef]

27. Gwatidzo, L.; Botha, B.M.; McCrindle, R.I. Determination of amino acid contents of manketti seeds (*Schinziophyton rautanenii*) by pre-column derivatisation with 6-aminoquinolyl-Nhydroxysuccinimidyl carbamate and RP-HPLC. *Food Chem.* **2013**, *141*, 2163–2169. [CrossRef] [PubMed]

28. Gwatidzo, L.; Botha, B.M.; McCrindle, R.I. Fatty acid profile of manketti (*Schinziophyton rautanenii*) nut oil: Influence of extraction method and experimental evidence on the existence of α-eleostearic acid. *J. Cereals Oilseeds* **2017**, *8*, 33–44. [CrossRef]

29. United Nations Children's Fund (UNICEF). UNICEF Eastern and Southern Africa: Young Child Survival and Development: Nutrition, 2017. Available online: https://www.unicef.org/esaro/5479_nutrition.html (accessed on 21 November 2017).

30. Chivandi, E.; Mukonowenzou, N.; Nyakudya, T.; Erlwanger, K.H. Potential of indigenous fruit-bearing trees to curb malnutrition, improve household food security, income and community health in sub-Saharan Africa: A review. *Food Res. Int.* **2015**, *76*, 980–985. [CrossRef]

31. Gelfand, M.; Mavi, S.; Drummond, R.B.; Ndemera, B. *The Traditional Medical Practitioner in Zimbabwe: His Principles of Practice and Pharmacopoeia;* Mambo Press: Gweru, Zimbabwe, 1985; ISBN 9780869223505.

32. Dushimemaria, F. An Investigation into the Antineoplastic Properties of Schinziophyton rautanenii and Colophospermum mopane. Master's Thesis, University of Namibia, Windhoek, Namibia, 2014.

33. Dushimemaria, F.; Mumbengegwi, D.R. Proposition of a low cost field assay to determine antiproliferate properties of indigenous plants using *Dugesia dorotocephala* (brown planaria). *Sci. Res. Essays* **2015**, *10*, 144–149. [CrossRef]

34. Mohammad, A.; Mahmood, N. Taxonomic perspective of plant species yielding vegetable oils used in cosmetics and skin care products. *Afr. J. Biotechnol.* **2005**, *4*, 36–44. [CrossRef]

35. Gunstone, F.D. *Bailey's Industrial Oil and Fat Products,* 6th ed.; John Wiley & Sons: New York, NY, USA, 2005; ISBN 9780471678496.

36. Zimba, N.; Wren, S.; Stucki, A. Three major tree nut oils of southern central Africa: Their uses and future as commercial base oils. *Int. J. Aromather.* **2005**, *15*, 177–182. [CrossRef]

37. Saral, Y.; Uyar, B.; Ayar, A.; Nazirogly, M. Protective effects of topical alpha tocopherol acetate on UVB irradiation in guinea pigs: Importance of free radicals. *Physiol. Res.* **2002**, *51*, 285–290. [PubMed]

38. Elago, S.N.; Tjaveondja, L.T. A comparative evaluation of the economic contributions and uses of *Strychnos cocculoides* and *Schinziophyton rautanenii* fruit trees to poverty alleviation in mile 20 village of Namibia. *Agric. Food Sci. Res.* **2015**, *2*, 25–31.

39. Msangi, J.P. *Food Security Among Small-Scale Agricultural Producers in Southern Africa;* Springer: New York, NY, USA, 2014; ISBN 978-3-319-09495-3.

40. Juliani, H.R.; Koroch, A.R.; Simon, J.E.; Wamulwange, C. Mungongo cold pressed oil (*Schinziophyton rautanenii*): A new natural product with potential cosmetic applications. *Acta Hortic.* **2007**, *756*, 407–412. [CrossRef]

41. Vermaak, I.; Kamatou, G.P.P.; Komane-Mofokeng, B.; Viljoen, A.M.; Beckett, K. African seed oils of commercial importance: Cosmetic applications. *S. Afr. J. Bot.* **2011**, *77*, 920–933. [CrossRef]

42. Lall, N.; Kishore, N. Are plants used for skin care in South Africa fully explored? *J. Ethnopharmacol.* **2014**, *153*, 61–84. [CrossRef] [PubMed]

43. Dushimemaria, F.; Mumbengegwi, D.R.; Böck, R. Indigenous knowledge of medicinal plants used for the treatment of cancer. In *Indigenous Knowledge of Namibia;* Chinsembu, K.C., Cheikhyoussef, A., Mumbengegwi, D.R., Kandawa-Schulz, M., Kasandra, C.D., Kazembe, L., Eds.; University of Namibia Press: Windhoek, Namibia, 2015; pp. 63–88, ISBN 9789991642055.

44. Van Den Eynden, V. Use and Management of Edible Non-Crop Plants in Southern Ecuador. Ph.D. Thesis, Ghent University, Ghent, Belgium, 2004.

45. Thomas, E.; Van Damme, P. Plant use and management in homegardens and swiddens: Evidence from the Bolivian Amazon. *Agrofor. Syst.* **2010**, *80*, 131–152. [CrossRef]

46. Leakey, R.R.B.; Mesén, J.F.; Tchoundjeu, Z.; Longman, K.A.; Dick, J.M.; Newton, A.; Matin, A.; Grace, J.; Munro, R.C.; Muthoka, P.N. Low-technology techniques for the vegetative propagation of tropical trees. *Commonw. For. Rev.* **1990**, *69*, 247–257. [CrossRef]

47. Lewanika, M.M. State of the art of biotechnology research in Zambia. *Afr. Crop Sci. J.* **1995**, *3*, 299–301. [CrossRef]

48. Du Plessis, P. *Indigenous Vegetables Development Proposal;* National Agricultural Support Services Programm (NASSP) Report No. 005/2004; Ministry of Agriculture, Water and Rural Development: Windhoek, Namibia, 2004.

49. Mark, J.; Newton, A.C.; Oldfield, S.; Rivers, M. *The International Timber Trade: A Working List of Commercial Timber Tree Species;* Botanic Gardens Conservation International: London, UK, 2014.

50. Erkkila, A.; Siiskonem, H. *Forestry in Namibia, 1850–1990;* Silva Carelica 20; University of Joensuu: Joensuu, Finland, 1992; ISBN 951-708-010-7.

51. Wild, R.G.; Mutebi, J. *Conservation through Community Use of Plant Resources: Establishing Collaborative Management at Bwindi Impenetrable and Mgahinga Gorilla National Parks, Uganda;* United Nations Educational, Scientific and Cultural Organization (UNESCO): Paris, France, 1996.

52. Dovie, D.B.; Witkowski, E.; Shackleton, C.M. Knowledge of plant resource use based on location, gender and generation. *Appl. Geogr.* **2008**, *28*, 311–322. [CrossRef]

53. Bennett, B. Natural Products: The New Engine for African Trade Growth: Consultancy to Further Develop the Trade Component of the Natural Resources Enterprise Programme (NATPRO). Regional Trade Facilitation Programme: Windhoek, Namibia, 2006.

54. Cunningham, A.B. *African Medicinal Plants: Setting Priorities at the Interface between Conservation and Primary Health Care*; People and Plants Working Paper 1; United Nations Educational, Scientific and Cultural Organization (UNESCO): Paris, France, 1993.

55. Chidumayo, E.N. Distribution and abundance of a keystone tree, *Schinziophyton rautanenii*, and factors affecting its structure in Zambia, southern Africa. *Biodivers. Conserv.* **2016**, *25*, 711–724. [CrossRef]

56. Kivevele, T.T.; Huan, Z. An analysis of fuel properties of fatty acid methyl ester from manketti seeds oil. *Int. J. Green Energy* **2015**, *12*, 291–296. [CrossRef]

57. Nemarundwe, N.; Ngorima, G.; Welford, L. Cash from the Commons: Improving Natural Product Value Chains for Poverty Alleviation. In Proceedings of the 12th Biennial Conference of the International Association for the Study of Commons (IASC), Cheltenham, UK, 14–18 July 2008. Available online: http://dlc.dlib.indiana.edu/dlc/bitstream/handle/10535/781/Nemarundwe_219401.pdf?sequence=1 (accessed on 8 October 2017).

58. Shackleton, S.; Shanley, P.; Ndoye, O. Invisible but viable: Recognising local markets for non-timber forest products. *Int. For. Rev.* **2007**, *9*, 697–712. [CrossRef]

59. Fellows, P.J.; Axtell, B. *Opportunities in Food Processing: A Handbook for Setting up and Running a Smallscale Business Producing High-Value Foods*; ACP-EU Technical Centre for Agricultural and Rural Cooperation, CTA: Wganingen, The Netherlands, 2014; ISBN 978-92-9081-556-3.

60. Meybeck, A.; Gitz, V. Sustainable diets within sustainable food systems. *Proc. Nutr. Soc.* **2017**, *76*, 1–11. [CrossRef] [PubMed]

61. Bharucha, Z.; Pretty, J. The roles and values of wild foods in agricultural systems. *Philos. Trans. R. Soc. B Biol. Sci.* **2010**, *365*, 2913–2926. [CrossRef] [PubMed]

MDPI

St. Alban-Anlage 66

4052 Basel

Switzerland

Tel. +41 61 683 77 34

Fax +41 61 302 89 18

www.mdpi.com

Sustainability Editorial Office

E-mail: sustainability@mdpi.com

www.mdpi.com/journal/sustainability

www.ingramcontent.com/pod-product-compliance
Lightning Source LLC
Chambersburg PA
CBHW051845210326
41597CB00033B/5787